BLACK WOMEN SCIENTISTS IN THE UNITED STATES

BLACK WOMEN SCIENTISTS IN THE UNITED STATES

Wini Warren

INDIANA
UNIVERSITY
PRESS

Bloomington | Indianapolis

This book is a publication of

Indiana University Press
601 North Morton Street
Bloomington, IN 47404-3797 USA

http://www.indiana.edu/~iupress

Telephone orders   800-842-6796
Fax orders   812-855-7931
Orders by e-mail   iuporder@indiana.edu

The paper used in this publication meets the
minimum requirements of American National
Standard for Information Sciences—Permanence
of Paper for Printed Library Materials,
ANSI Z39.48-1984.

Manufactured in the United States of America

Library of Congress Cataloging-in-Publication Data

Warren, Wini.
    Black women scientists in the United States / Wini
Warren.
        p.   cm.
    Includes bibliographical references and index.
    ISBN 0-253-33603-1 (cl : alk. paper)
    1. Afro-American women scientists Biography.
I. Title.
Q141.W367   1999
500'.82'0973—dc21   99-40264

1 2 3 4 5 04 03 02 01 00 99

To my husband,
Frank J. Warren,
for everything.

Contents

Acknowledgments

I owe a great debt to many people for their encouragement, support, and interest. First and foremost, I wish to thank my husband, Frank J. Warren, who believed in me when believing seemed a pipe dream. He has supported my quest for an education in so many ways it is impossible to recount them all. Ten years ago, when I began college, he traded places with me and became the primary caregiver in our home—cooking, shopping, attending to children and friends, and seeing to my needs. At age sixty-six, he willingly left home and hearth to accompany me to graduate school at Indiana University where, in addition to his other duties, he drove me to and from class every day, reminding me each time that all I had to worry about was getting good grades—he would do all the rest. He has done it all with love, style, and grace.

I wish to thank Dr. Noretta Koertge, of the Department of History and Philosophy of Science, Indiana University, for her uncommon support and friendship during all my time at graduate school, as well as for her generosity, time, patience, and encouragement as the chair of my dissertation committee. I also wish to thank two other professors at Indiana University, each of whom helped mold me into a scholar: Dr. Frederick B. Churchill, whose support, encouragement, and advice over the years have been nonpareil, and Dr. James Capshew, whose knowledge of the history of American science proved invaluable. In addition, I would be remiss if I did not mention the effect on my intellectual growth of Dr. Edward Grant, scholar extraordinary and Emeritus Professor of the Department of History and Philosophy of Science; and the effect on my writing and self-confidence of Dr. Holly Stocking, of the Ernie Pyle School of Journalism, Indiana University, who has been both teacher and friend.

On behalf of my husband and myself, I also want to acknowledge the friendship and support of former Department of History and Philosophy of Science members Dr. John Winnie and Dr. Linda Wessles. Above all, I thank the entire faculty of the Department of History and Philosophy of Science and the Indiana University Foundation for their willingness to take a chance on a "nontraditional" student by providing generous financial support throughout my graduate school years.

Without the uncommon support, encouragement, and mentoring of some very special professors during my undergraduate years at the State University of New York, College at Old Westbury, however, graduate school would not have been possible. I therefore wish to express my deep gratitude to: Dr. Albert Rabil Jr. and Dr. James Llana for uncommon and continued support, and for recognizing the light hiding under a rough housewife's exterior. I also wish to thank other faculty members at SUNY, College at Old Westbury: Professor Carl Grossman, American Studies; Dr. Mervin Keiser; Dr. Jonathan Collett (retired); Dr. Emilia Doyoga (retired); the late Prof. Paul Welcher, of Comparative Humanities; and

Dr. Samuel Von Winbush (retired), Science Department, who first suggested the topic for my dissertation. I also wish to thank Dr. Llana, former convenor of the Comparative Humanities Department, and Mr. Alonzo McCollum, Director of the Equal Opportunity Program, for providing me with adjunct teaching positions that helped support my dissertation research.

I received a generous research grant from The Thanks Be To Grandmother Winifred Foundation, Wainscott, New York, an organization that provides funds for a variety of projects to women over fifty-five years of age; my thanks to Foundation President Mrs. Deborah Ann Light and the members of the committee — I could not have completed this project without their help.

I also wish to thank all those who helped me with my data gathering, particularly those who afforded me the time for interviews or correspondence. In particular, I wish to acknowledge the late Dr. Margaret Strickland Collins, an uncommon friend; Dr. Joan Murrell Owens, Dr. Gloria Long Anderson, Dr. Prince Rivers, Dr. Shirley Malcolm, Dr. David Telfer, and Dr. Gwendolyn W. Pla for their help and hospitality; Dr. Jewel Plummer Cobb and Dr. Willie Pearson Jr., for early encouragement; Dr. Shirley Vining Brown, Dr. Patricia Clark Kenschaft, Dr. Christine Darden, and Dr. Muriel Poston for supplying material; Ms. Linda Skidmore and Ms. Gaelyn Davidson of the National Academy of Science/National Research Council for the opportunity to meet women who helped my study; and Ms. Joellen El Bashir, Dr. Janet Sims-Wood, Ms. Avril Madison, and Ms. Kathy Jenkins for their expertise with archival materials at the Moorland-Spingarn Research Center, Howard University. I wish also to thank my friends Elizabeth Green, Karen Rader-Powers, Deborah Rush, and Midge Willis for their assistance and encouragement; my friend Alan Hunter for eleventh-hour editorial help; Faruk Curtis for archival assistance; and math teacher Marie Broderick of Northport High School for extraordinary dedication to the teaching profession.

I thank my children — Douglas, Ellen, Jesse, Deborah, and Rebecca; my brother, Jon Bigall; and my sons- and daughter-in-law, Jerry Doyle, Darrel Fitzgerald, David Anderson, and Susan Warren for their love and encouragement. I thank my granddaughter, Miranda Warren, for brightening up every day; and my late uncle, Herb Middlecamp, for kindling the torch.

Wini Warren
Northport, New York
August 1998

Introduction

When I undertook this project, my goal was to identify Black women scientists in the United States during the period between 1900 and 1960.[1] I hoped to provide insights on the extent to which such women, once considered nontraditional participants, had become part of the American scientific enterprise. For a brief while, completing my doctoral dissertation was uppermost in my mind; but as I delved deeper into the lives of these women scientists that changed. As time passed, my research subjects became more than just discrete bits of information to plug into my dissertation; rather they became fascinating "friends." In some cases this was understandable—I met some of the women. In other cases, I came to "know" them through the scattered clues they had left behind. At that point, my research became a passion to rescue as many of these women as I could from historical neglect.

Over the more than six years I have worked on this project, I have added and dropped women from my list any number of times. Yet despite these additions and subtractions, I have kept some women on my list because I just couldn't bear to eliminate them. They were women on whom I had acquired little more than tantalizing bits of information. Most often those bits appeared in magazines directed at a Black readership; such articles—often a photograph with a brief caption—whetted my researcher's appetite. At times, I found myself daydreaming, wondering all sorts of things to try to fill in the gaps: What had happened to this or that young woman, whose face had seemed so full of promise as she faced the camera over her laboratory equipment? How had she subsequently disappeared from notice? What else could she have done with her science training? Had she given up on her professional career? If I could track her down, would she have had experiences that would provide inspiration for others? Sadly, in some cases I found no information; it was almost as if some women had never existed. Still, I held on to my meager file on them because I couldn't bring myself to eliminate them with a sterile stab of my computer's delete key. To do so would have truly relegated them, once and for all, to the realm of the forgotten.

All this is by way of saying that I have included some women in this volume, despite a lack of substantive information on them, because they were genuine pioneers. Others showed enough promise that one might reasonably expect the continuation of their scientific careers. And some simply struck a resonant chord within me. To make these women more readily identifiable to the reader, I have starred (*) their names. It is my hope that someone, somewhere, may know something about at least one—or maybe more—of these women. I would greatly appreciate any information you, the reader, might have. Readers can write to me or forward information to me through Indiana University Press, or in care of the American Studies Department, State University of New York, College at Old Westbury, Old Westbury, New York.

Recent scholarship has done much to document the experiences of women in science, but there has been little comparable scholarship on Black Americans' experiences and contributions, and even less on Black women scientists. In preparation for this volume, I completed an extensive review of the literature dealing with the history of science in the United States. I do not think, however, that this present volume is the proper place in which to undertake comment on the larger issue of the underrepresentation of Black women in science. That issue will be discussed in my forthcoming book on the history of Black women in science.

Historians of American science, while occasionally paying lip service to the Black participants, have failed to integrate their contributions and participation into historical studies. While it has become politically correct to mention the occasional Black scientist in works on American science, these individuals still remain largely invisible in mainstream narratives, after a mere drop of their names. White scholars routinely point to the so-called marginal participation of Blacks in science as the typical justification for their neglect in histories of American science. And neglected they are. White historians speak about the hardships Blacks have faced, but they make little effort to identify or discuss those who did participate.[2]

One reason given for the neglect of science-trained Blacks in the standard works is the idea that they didn't achieve greatness in science and didn't work at major universities or in national labs.[3] There are two problems with this view: first, it is incorrect; and second, it presumes that science is done only by great scientists. This latter assumption is often referred to as "The Father Of" type of science history; the problem is that, as recent scholarship has shown, science is also carried forward by those working in the middle levels. Blacks working outside the mainstream as well as within it have made significant contributions to science.[4]

Recent studies indicate that by the early 1980s Black females had outstripped Black males in the attainment of doctorates in disciplines other than the sciences. Despite this "female advantage," the numbers of Black women obtaining advanced degrees in the sciences did not increase.[5] Considering the lack of historical work done on Black women scientists to date, this information should not come as a surprise. In relatively recent works that include Black women scientists, scholars have generally relied heavily on Harry W. Greene's *Holders of Doctorates among American Negroes*,[6] and James M. Jay's *Negroes in Science: Natural Science Doctorates, 1876–1969*;[7] both books are useful, but terribly dated. Additionally, since both authors relied heavily on self-identification by Black doctorate holders, the lists they compiled are necessarily incomplete.[8] More current work generally relies on Vivian Ovelton Sammons's *Blacks in Science and Medicine*.[9]

When historian Darlene Clark Hine wrote about the experiences of minorities and women in the medical professions in an earlier book, she offered some especially pertinent comments:

> If white women, black men, and poor whites, as many scholars argue, were outsiders in medicine, then black women, belonging as they did to two subordinate groups, surely inhabited the most distant perimeters of the professions. Yet it is precisely because of this dual—sexual and racial—marginality that any

examination of their lives and careers bears the possibility of shedding new light
on many conventional interpretations in American medical history.[10]

Hine's comments about Black women in the medical professions applies equally
well, perhaps even more so, to Black women in the sciences. More information
on those who did achieve would help to allay the feelings of "differentness" for
those who are still aspiring. And certainly, as my own research has shown, the
experiences of the women in this volume can shed new light on interpretations
of the American scientific enterprise.

Despite efforts to increase the participation of minorities and women in the
sciences during the last twenty-five years, recent reports indicate that the number of science doctorates awarded to Blacks, particularly to Black women, has
declined.[11] A plethora of reasons are given for this decline, but one reason is rarely
mentioned: the dearth of narratives or biographies of science-trained Blacks, particularly Black women. In those rare instances where the importance of generat-
ing narratives on science-trained Black women has been mentioned, little has
been done to follow up the suggestion. In 1981, for example, Paula Quick Hall of
the American Association for the Advancement of Science reported on remedies
suggested by a group of fifty-two minority women surveyed by the Office of
Opportunities in Science:

> Increased use of minority and women scientists as role models in the media was
> urged by some who expressed the need for motivating, encouraging and
> providing suitable models for minority students at every level. . . . Young women
> may view a career as a [scientist or a] mathematician as a desirable and an
> attainable goal if they can identify in some way with a person [from the same
> race and background] who has attained a similar goal.[12]

In recent years, major efforts have been made to illuminate the history of white
women in the sciences. An honest attempt to illuminate the experiences of Black
women scientists, particularly in terms of their struggles, strategies, successes, and
scientific work, could make a significant contribution—both to efforts aimed at
increasing the participation of minorities in the sciences and to the history of
science in America. Until such studies are made and integrated into the historical
mainstream, efforts to increase the participation of women of color in science will
have only limited success, and the history of science in America will remain incomplete.

Identification and Research

In the biographical sketches included in this volume, I have examined the lives
and career paths of more than one hundred Black women scientists. These scientists include women in the fields of anatomy, anthropology, astronautics and
space science, biochemistry, biology, botany, chemistry, geology, marine biology,
mathematics, medicine, nutrition, pharmacology, physics, psychology, and zoology. The majority of these women possessed the doctorate (Ph.D.), but I have also
included some women who had medical degrees (M.D.), a number of whom had
engaged in research.[13] Although I concentrated primarily on women who had

earned doctorates in the sciences and mathematics, I also included a number of women who did not have advanced degrees but whose contributions to a scientific field, either as researchers or as educators, were significant. I identified these women in a variety of ways. I examined various published lists and biographical directories regarding Blacks and women in science or science-related fields,[14] and I made an extensive inventory of Black women scientists whose names appeared in articles or announcements in various Black popular publications and, in some cases, in popular publications on women in general. I also attended a number of conferences whose themes pertained to the recruitment and retention of minorities and women in the sciences.

When I began the research, my intention was to rely heavily upon primary sources—extant personal accounts; interviews, conducted either in person, by mail, or by telephone; personal papers, either in a woman's own possession or housed in an archive, including publications, correspondence, curricula vitae, and other memorabilia; family papers and photographs; and the records housed at schools and other institutions. To a significant extent, I have accomplished that goal. In some cases, however, primary source materials have proven elusive. Despite increased interest in both African American and women's studies, materials on Black women scientists seem to have fallen through the cracks. Within the framework of Black or African American Studies, for example, there might be material available on a woman scientist's civil rights activities, yet not a scrap of information on her science. Or there might be material available on Black male scientists, but little on their female counterparts, even those working at the same school and department. The same is true with regard to information on women of color in recent works on the history of women in American science.[15]

My most important resource in identifying the women has been networking with the women themselves. As I came to know a few Black women scientists personally, they directed me to other women, often giving me permission to use their names in my correspondence. I sent those women I thought were still living a written questionnaire, along with a request for a curriculum vitae and copies of any articles in their possession on themselves and/or other Black women scientists. In those instances when time and finances allowed, I followed up on as many women as I could with personal interviews. In addition, I interviewed a number of women by telephone or via email; others faxed materials to me. Followup correspondence with many of the women often provided additional information on both themselves and on others.

To obtain substantive information on deceased women, I visited the archival collections at a number of historically Black colleges and universities—sometimes without success. In many cases, however, I accidentally located information on "Woman X" in the archival collection of "Woman Y." In many cases, when secondary sources listed a woman's papers as being housed in a particular archival collection, visits to those archives often resulted in disappointment—sometimes the secondary sources were simply wrong and the archivists had no record of the materials, and sometimes the materials were uncataloged to the extent that the archivists could not determine, with any degree of certainty, what they might have. This was in no way the fault of the various archivists I worked with—all of whom were most helpful; the situation results from a severe lack of the funds needed to accomplish these tasks at historically Black institutions. Nevertheless, each of the archivists I worked with managed to help me find some bits of information, occasionally on women whose names were new to me.

My visits to a number of historically Black colleges and universities were particularly fruitful in other ways: younger women scientists on the faculties at these institutions frequently helped me gather data—often from personal or departmental files. In some cases, I was able to interview women who had known some of the women who had died and were thus able to provide me with additional perspectives. A number of administrators and senior male faculty members at historically Black colleges were also supportive in this regard and often helped me acquire important data.

With regard to the living women who did not respond to my queries, and the deceased women who did not leave papers behind in archival collections, I have had to rely on secondary information, culled from a wide variety of sources. This secondary information occasionally presented some problems, as I detail below. In these instances, school records—faculty bulletins, yearbooks, and correspondence with college registrar offices—were useful, particularly when I had discovered errors or inconsistencies in the secondary sources. The fact that women marry and change their names occasionally caused difficulty in locating information. As my research has indicated, a significant number of the women scientists married more than once, often discarding names along with the husbands.

Another problem involved the way some women described themselves. A significant number of women, despite strong careers in research, described themselves (or were labeled by others) as educators rather than as scientists. This is understandable for those women who spent the major portion of their careers as teachers rather than as research scientists. There may be any number of reasons for choosing or assigning the appellation "educator" rather than "scientist." Among the more compelling reasons is the fact that, historically, the Black community has placed a tremendous cachet on educators, which according to scientist Gloria Long Anderson, was one of the time-honored ways for a Black to "move up." Another reason for choosing or assigning the appellation "educator" rather than "scientist" is that the latter may have been tainted, in some cases, by negative views of science in the Black community, while the former carried a certain status. Finally, some women used their science backgrounds as springboards to high-level administrative posts, and rather than calling themselves "college president" or "university dean," may have chosen instead to simply call themselves "educator." These problems necessitated a great deal of cross-referenced research, which often proved fruitless.

In the instances where a woman never married, using her last name did not present a problem. However, in the cases where a woman was married, I have referred to her by her married name throughout. In those cases where a woman was married more than once, I have referred to her by her most recent married name, or by the name she chose to retain if that marriage, too, ended in divorce. In all instances, I have included at least one mention of each woman's entire list of surnames. In the appendix, each woman's publications are listed under the name she was using when the article was published; to avoid confusion, each woman's entire name appears at the head of her list of publications.

A Note on Language

In some quarters the term "African American" is currently in vogue. This is not universal, however, and more recently the term "Black" is increasingly used. During the interviews, I asked each of the women which term she preferred. The overwhelming response was "Black"—with a capital "B." I have thus chosen to use the term "Black" rather than "African American." This was an arbitrary decision on my part, as was the decision not to capitalize the word "white" as it refers to Caucasians. The terms "negro" and "colored" have been employed in the past, but I will use them only where faithful citation demands. The terms "person of color" and "women of color" will be used where appropriate, and refer to non-white but not necessarily Black individuals.

Author's Note

Because I am a white woman writing about Black women, I have spent some time considering the thoughts expressed by Peter Novick on historical separatism in "Every Group Its Own Historian."[16] I have discussed the issue at a number of symposia attended by women scientists of color in the last six years, and I have also talked at length about it with the women I have interviewed, who generally expressed satisfaction that someone was (finally) interested in studying Black women scientists. Clearly, this volume is just a beginning. Others must attempt to speak about the deeper experiences, struggles, and contributions of other women scientists of color as well. I hope my effort will spur further research in this area.

NOTES

1. Wini Mary Edwina Warren, "Hearts and Minds: Black Women Scientists in the United States, 1900–1960" (Ph.D. dissertation, Department of History and Philosophy of Science, Indiana University 1997, available on University Microfilm). This volume is an expansion of my dissertation research, which looked primarily at women who had performed significant scientific work in the period from around 1900 through the 1960s. I defined significant work as research published in recognized scientific or medical journals; research published in-house by various government agencies or government-sponsored laboratories; research funded by government agencies, recognized philanthropic organizations, or academic institutions; research performed as employees of private industry, with in-house publications; a teaching career in science, medicine, or mathematics that produced students (regardless of race or gender) who pursued advanced degrees in the sciences and related fields; or a career as a science facilitator or administrator in which the primary focus was to enhance or increase opportunities for women and/or minorities in science, including efforts to enhance the science facilities and faculty credentials at historically Black colleges and universities.

2. For a fuller discussion of this issue, please see Warren, "Hearts and Minds."

3. Stephen G. Brush, "Women in Science and Engineering," American Scientist 79 (September/October 1991): 404–419.

4. Examples abound; here are two.

Between 1908 and 1923, Charles H. Turner, who earned a Ph.D. at the University of

Chicago in 1903, managed to publish 49 papers on invertebrate biology while working as a high school teacher in St. Louis, Missouri. An authority on the behavioral patterns of ants and spiders, Turner was the first scientist to prove that insects can distinguish tonal pitch. See Vivian Ovelton Sammons, *Blacks in Science and Medicine* (New York: Hemisphere, 1990), p. 235, and Aaron E. Klein, *The Hidden Contributors: Black Scientists and Inventors in America* (Garden City, N.Y.: Doubleday, 1971), p. 172. Turner's lab was a backyard shed, equipped with materials scrounged from junkyards and discards from high school labs (Klein, ibid.).

Charles Richard Drew (1904–1950) had more mainstream work. He was a research physician and a pioneer in the field of blood preservation, whose work led to the founding of blood banks. Drew directed the Plasma for Britain program during World War II, and became director of the first American Red Cross blood bank program. It is important to note that this was at a time when the blood of Blacks was routinely segregated from that of whites. As head of the surgery department at the Howard University School of Medicine between 1936 and 1950, he also helped train a considerable number of Black surgeons. See Emma Gelders Sterne, *Blood Brothers: Four Men of Science* (New York: Knopf, 1959), and Hamilton Bims, "Charles Drew's 'Other' Medical Revolution," *Ebony*, February 1974, pp. 88–92. (Also see Sammons, *Blacks in Science and Medicine*, pp. 78–79.)

5. Shirley Malcolm, "Increasing the Participation of Black Women in Science and Technology," in *The "Racial" Economy of Science: Toward a Democratic Future*, ed. Sandra Harding (Bloomington: Indiana University Press, 1993), pp. 250–251.

6. Harry W. Greene, *Holders of Doctorates among American Negroes: An Educational and Social Study of Negroes Who Have Earned Doctoral Degrees in Course, 1876–1943* (Boston: Meador Publishing, 1946). Of the 275 pages, only a small number focus on Black scientists.

7. James M. Jay, *Negroes in Science: Natural Science Doctorates, 1876–1969* (Detroit: Belamp, 1971). Jay is a biologist. His estimation that approximately 650 Black Americans had obtained doctorates in the natural sciences through 1969 was based on questionnaires he sent out; 587 people either responded or were mentioned by others as having obtained doctorates. His study was based, in large part but not exclusively, on those responses.

Although Jay's book is basically a demographic study, he does identify some Black scientists by name. What is most useful in this book are the tables, which trace the geographic, undergraduate, and graduate origins of African American scientists. Also of interest are breakdowns of scientists according to natural science discipline. We find, for example, that only two Blacks held doctorates in astronomy during the period studied: Harvey W. Bankes (Georgetown, 1961), and Benjamin Peery (University of Michigan, 1962)—most provocative when compared with the very large number who pursued careers in chemistry. Jay's chapter on Black women scientists is especially interesting. Here, he surveyed the career paths of fifty-eight women and found that (as of 1969) thirty-three were employed at predominantly "Negro" academic institutions; eleven were employed at (presumably) white academic institutions; four were employed in government; four in industry; three were housewives; and the rest were unaccounted for.

Jay notes the first Blacks to receive the doctorate in each natural science discipline. He also shows comparisons between the number of males and females earning undergraduate degrees; the figures indicate that while women outstripped their male counterparts in obtaining the bachelor's degree, they largely fell by the wayside in terms of graduate study.

The book's major flaw—one it shares with many others — is that it fails to give any information on the quality or even the direction of the subjects' research interests. Although we know, for example, that fifty-two of the women scientists were employed in venues where they might have done research, we do not know if that is what they did; they may have been simply educators or administrators.

Like Jay's *Negroes in Science*, Klein's *The Hidden Contributors: Black Scientists and*

Inventors in America is frequently cited in scholarly works. Presumably written for young adults, the book focuses more on inventors than on scientists—there are only three doctorate holders included (two Ph.D.s and one M.D.). It is, nevertheless, a useful book, although it has no bibliography. Klein does a nice job of integrating the sciences and inventions into the personal histories, and is also able to show in each case what the standards were at the time each innovation or discovery was made. Particularly well done are the chapters on Banneker, Just, and Drew.

Klein, a self-described white Southern Jew, is a scientist and educator turned historian. His other books include *Threads of Life: Genetics from Aristotle to DNA* (Garden City, N.Y.: Natural History Press, 1970), *Test Tubes and Beakers: Chemistry for Young Experimenters* (Garden City, N.Y.: Doubleday, 1971), and *Transistors and Circuits: Electronics for Young Experimenters* (Garden City, N.Y.: Doubleday, 1971). As some of the titles indicate, he writes primarily for young and young-adult readers.

Editor Julius H. Taylor's *The Negro in Science* (Baltimore: Morgan State College Press, 1955) appears to be a collective effort. (Clyde R. Dillard and Nathaniel K. Proctor are listed as the editorial board.) After its widely quoted Herman R. Branson introduction, the book is divided into chapters based on disciplines—Biology, Chemistry, Mathematics, and so on. Each chapter begins with a survey detailing such items as training, age at attainment of degrees, and publications (by number, not by name of scientist). The survey sections are followed by articles that were published by various Black scientists in disciplinary journals. Curiously, no publications are included for any women.

Although Taylor does not indicate a time frame in the text, it appears that the scientific papers reproduced represent the research of then-current scientists working actively in their various fields; all the papers were originally published between 1938 and 1954. No attempt is made, however, to indicate why a particular article might have been a contribution to a discipline, or whether it constituted anything more than a journeyman's publication credit. The overall effect on the reader is, therefore, one of simply showing that Blacks publish scientific papers, too. There is no attempt to integrate the collection of papers into a scheme vis-à-vis the history of science, and there is no bibliography (except in the individual scientific papers). (None of the contributors to the Taylor book are historians—all are either scientists or science educators.) There are, however, 133 very short biographical sketches at the end of the volume that do indicate areas of scientific interest, broken down according to discipline.

Hattie Carwell's *Blacks in Science: Astrophysicist to Zoologist* (Hicksville, N.Y.: Exposition Press, 1977) is basically just another listing. There is little substance, no analysis of the science, and no useful bibliography.

8. As I mentioned earlier, accurate records were not kept, so these volumes are important.

9. Sammons's book is both a huge disappointment and a treasure trove for the interested historian. There are over 1,500 names listed here, in alphabetical order. The title is deceptive—only 475 of the listings are holders of Ph.D.s in science or related disciplines, 65 of whom are female. It is a huge list of everyone who is or was anyone and Black, including nurses, midwives, military personnel (with no connection to science), astronauts (with some connection to science), an early American explorer, the first Black to obtain a private pilot's license, physicians, dentists, veterinarians, and even a woman who is listed as a "consultant" at a commercial relaxation-hypnotherapy clinic. The book identifies a large number of Blacks in science, going all the way back to Banneker; however, once one finds a significant individual, the amount of detail with regard to his or her *science* is little more than minimal. Sammons gives educational background, title of dissertation (when available), area of interest (but not degree of activity), and an occasional mention (no more than two per person) of scientific publications. In the case of Ernest Everett Just, for example, she lists two citations and then she simply writes "published more than 50 papers in this field between 1912 and 1937" (p. 141). Still, the book is an improvement over earlier efforts. Sammons also lists the names by discipline in the back of the book, and there is a useful bibliography of Black publications.

10. Darlene Clark Hine, "Co-Laborers in the Work of the Lord: Nineteenth Century Black Women Physicians," in Ruth J. Abrams, ed., *Send Us a Lady Physician: Women Doctors in America, 1835–1920* (New York: W. W. Norton, 1985), pp. 249–253.

11. "Current Lack of Minority Doctoral Candidates Will Lead to Future Minority Faculty Shortage," *Black Issues in Higher Education* 4, no. 23 (1988); *Changing America: The New Face of Science and Engineering*, final report of the Federal Task Force on Women, Minorities, and the Handicapped in Science and Technology (Washington, D.C., 1988); Eileen M O'Brien, "Without More Minorities, Women, Disabled, U.S. Scientific Failure Certain, Fed Study Says," *Black Issues in Higher Education* 6, no. 20 (December 1989), reprinted in Harding, *"Racial" Economy of Science*. See also Malcolm, "Increasing the Participation of Black Women" (in Harding, op. cit.); and "The Status of Blacks in Graduate and Professional Schools: A Report by the National Advisory Committee on Black Higher Education and Black Colleges and Universities," *The Black Collegian*, October/November 1981, pp. 86–87.

12. Sylvia T. Bozeman, "Black Women Mathematicians: In Short Supply," *Sage* 6 (fall 1989): 18–23.

13. Work has already been done, and continues to be done, on Black women physicians and other health care professionals. Because the majority of these worked primarily in a clinical rather than a research setting, their experiences are not pertinent to this study. For example, see the following works. Leslie L. Alexander, "Early Medical Heroes: Susan Smith McKinney Steward, M.D., 1847–1918: First Afro-American Woman Physician in New York State," *Journal of the National Medical Association* 67 (1975): 173–175. Bettina Aptheker, "Quest for Dignity: Black Women in the Professions, 1885–1900," in *Woman's Legacy: Essays on Race, Sex and Class in American History* (Amherst: University of Massachusetts Press, 1982). Sara W. Brown, "Colored Women Physicians," *Southern Workman* 52 (1923): 586. Penina Magdal Glazer and Miriam Slater, *Unequal Colleagues: The Entrance of Women into the Professions, 1890–1940* (New Brunswick, N.J.: Rutgers University Press, 1987). Basil C. H. Harvey, "Provision for Training Colored Medical Students," *Journal of the American Medical Association* 94 (1930): 1415. Darlene C. Hine, "Co-Laborers in the Work of the Lord." Hine, *Black Women in the Nursing Profession: A Documentary History* (New York: Garland, 1985). Hine, ed., *Black Women's History: Theory and Practice* (Brooklyn, N.Y.: Carlson Publishing, 1990). Hine, ed., *The State of Afro-American History: Past, Present, and Future* (Baton Rouge: Louisiana State University Press, 1986). John A. Kennedy, *The Negro in Medicine* (Tuskegee, Ala.: Tuskegee Institute, 1912). Cynthia Neverdon-Morton, *Afro-American Women of the South and the Advancement of the Race, 1895–1925* (Knoxville: University of Tennessee Press, 1989). Gerald A. Spencer, *A Medical Symphony: A Study of the Contributions of the Negro to Medical Progress in New York* (New York: Arlain Printing, 1947). Ronald T. Takaki, "Aesculapius Was a White Man: Race and the Cult of True Womanhood," in Harding, *"Racial" Economy of Science*.

14. James M. Jay, *Negroes in Science*. James M. Jay, "Michigan Negro Science Doctorates," in *Michigan Challenge*, June 1968. James M. Jay, *Minority Groups among United States Doctorate-Level Scientists, Engineers, and Scholars* (Washington, D.C.: Commission on Human Resources of the National Research Council, National Academy of Sciences, 1974). Frank Lincoln Mather, ed., *Who's Who of the Colored Race: A General Biographical Dictionary of Men and Women of African Descent* (Chicago, Ill., 1915; reprint, Detroit: Gale Research, 1976). Sammons, *Blacks in Science and Medicine*. Vivian Ovelton Sammons and Denise Dempsey, comps., *Blacks in Science and Related Disciplines* (Washington, D.C.: Science Reference Section, Science and Technology Division, Library of Congress, 1989). Greene, *Holders of Doctorates among American Negroes*. Edgar Allen Toppin, *A Biographical History of Blacks in America since 1528* (New York: McKay, 1971). *Who's Who among Black Americans* (Northbrook, Ill.: Who's Who among Black Americans Publishing, 1975–1985; Lake Forest, Ill.: Educational Communications, 1985–1989; Detroit: Gale Research, 1990–). *Who's Who in Colored America: A Biographical Dictionary of Notable Living Persons of Negro Descent in America*

(New York: Who's Who in Colored America, 1927–). Also used for cross-referencing were *American Men of Science*, 1st–11th eds. ([New York: Bowker], 1906–1968); and *American Men and Women of Science: The Physical and Biological Sciences* (New York: Bowker, 1971–), a continuation of *American Men of Science*, beginning with 12th edition.

15. Again, see Warren, "Hearts and Minds," for a fuller discussion of this issue.

16. Peter Novick, *That Noble Dream: The "Objectivity Question" and the American Historical Profession* (Cambridge, Mass.: Cambridge University Press, 1988, 1990), pp. 469–491.

BLACK WOMEN SCIENTISTS IN THE UNITED STATES

GLORIA LONG ANDERSON

Acknowledging a Debt to Martin Luther King Jr.

Gloria Anderson

Physical organic chemist Gloria Long Anderson made a conscious decision in 1968 to pursue her career at a small, historically Black college, despite her belief that her research would have been easier to conduct at a large, mainstream university. As she recalled,

> I was at the University of Chicago when Martin Luther King Jr. was killed. I was totally stunned by this. But in the aftermath of the event I made a commitment: my contribution to civil rights would involve my efforts to do science at an historically Black college/university.
>
> But my friends told me—and those were my *Black* friends, not white ones—that to do so was to commit professional suicide.
>
> You see, as a scientist at a predominantly white college you can get everything you need to do your research, at a Black college you can't. But I felt I had an obligation. A little more than ten years ago, one of my former students came back here, said I had paid my dues, and asked why I didn't move on to a mainstream college. . . .
>
> I guess I was naive enough, crazy enough, to believe I could come here and if I wanted to do research badly enough, and if I were good enough, it wouldn't

make a difference. That, essentially, I could do the same type and amount of science here as I could have done at a predominantly white school. But that hasn't been the case. . . . I have not been able to do the research here that I could have done at a white university.[1]

Anderson was born in Altheimer, Arkansas, in 1938, and raised in a mixed-race, segregated, farm community of fewer than one thousand people. Anderson's parents were sharecroppers; her father had a third grade education, and her mother had completed tenth grade.[2] All six children—three older brothers, Anderson, and two younger brothers—were expected to do their share of the farm work. "It was a farm family," Anderson said, "we all worked on the farm. . . . In those days we didn't know we were living in poverty or in a ghetto."[3] She learned to read before starting elementary school at age four, although she has no idea how she learned:

> I read before I entered grade school. . . . When I went to school at age four— they didn't check very carefully in those days—I could already read, although I have no memory of ever seeing any reading material at home. In grammar school, whatever little books they had in the classroom I finished in no time. I read every book in the library, but they didn't have many. I don't know if they had science books, but I read everything I could get my hands on. I *loved* books. I often wonder now what my life would have been like if I'd had more books to read.[4]

She attended rural, segregated schools, and thrived on the attention of her teachers, as she recalled:

> Arkansas had segregated schooling, but they didn't call it that. They called the Black high school a training school, which was a euphemism for Black. . . . We had all Black teachers. They had a kind of interest in us you don't see today. One of the disadvantages of desegregation is that you no longer have all Black schools where the teachers really pushed the students. The pushed us to succeed, to excel, because they had a vested interest in turning out successful, well-educated students. They cared, intimately, about our doing well.[5]

She and her five brothers made straight As in school, and this often entailed family sacrifice. As she explained,

> Even though they didn't have much education, our parents realized quite well the importance and value of our getting an education. They did not keep us out of school to work on the farm, which was a tremendous sacrifice on their part, which is not to say we did not help out. We did—it was survival.[6]

Two of her three older brothers completed high school. Her two younger brothers finished high school and went on to college.[7] She graduated from Altheimer Training (High) School in 1954, at age sixteen.

Despite her grades, Anderson said she had never really planned for a career in science:

> I guess my early life is a good example of how to succeed by not doing career planning. When I went to college I didn't know what to major in. I knew I didn't want to stay in Altheimer, knew that I didn't want that kind of limited life and future—as a domestic in (as we say it) some white woman's house, or on the farm.

> There were some few little stores, and a small bank, but those jobs were
> pretty much reserved for whites. So in the sense of planning a future, or a career,
> it was more a question of what I was running *from* than what I was running *to*.[8]

She thought about the possibility of majoring in physical education because she
had liked basketball in high school, but her mother vetoed that idea. Interior dec-
orating was her next choice, but her mother did not like that either. As Anderson
explained, "There were certain things people in small towns knew about—like
doctors, lawyers, and teachers—so I thought of maybe taking pre-med in college."[9] There was, however, little choice about which college she would attend;
it had to be Arkansas A & M Normal College because, as Anderson said, "It was
the only Black state school, it was close to home, and it was almost affordable."[10]
Because of her excellent high school grades, she received a small, partial scholarship for her freshman year—her parents had to pay the balance, although Anderson said it was not too expensive. At the end of her freshman year, she had
earned As in all her classes and had the highest average in the freshman class. Arkansas A & M gave her a full scholarship for the sophomore year, on the condition that she maintain the highest class average. As luck would have it, Anderson
got the only two Bs in her entire life during her sophomore year, and lost the
scholarship. She made only a B in "Negro" history, and a B in chemistry, which
was not surprising since her high school had not offered chemistry courses. Anderson recalled that she had only signed up for the chemistry course on a dare:

> There were four of us in the freshman class who thought we were smart. Some
> upperclassmen told us not to major in chemistry—it's the hardest major, they
> said. When they said that, we just signed up for it, just for the challenge, you
> know.[11]

The following year, however, she was awarded a Rockefeller Scholarship,
which she held until graduation.[12] She earned the baccalaureate in chemistry in
1958, graduating first in a class of 237.[13] The chemistry major notwithstanding,
Anderson still had little idea of a career in science:

> I didn't know I was supposed to go and teach after the B.A.—teacher or preacher
> those were the professions open [to Blacks in those days]. I had a 2.96 GPA out
> of a possible 3.0. Yet in the fall of 1958 I was the only person in the class without
> a job on the horizon. I didn't want to teach so I applied to graduate school at
> Stanford—they accepted me, but there was no funding, so of course it was
> impossible.[14]

She next decided to apply for a job at the Ralston Purina Company:

> I got the application package practically by return mail—very fast. But then I
> guess when they looked at the actual application, it was pretty clear to them that
> I was Black. I didn't hear from them and didn't hear, and finally I wrote and they
> wrote back and said the position had been filled by somebody else. Actually,
> they did me a favor by not hiring me, but I didn't realize it at the time.[15]

Teaching was the only option, so Anderson taught seventh grade for six months,
and disliked dealing with the discipline problems. Then, in the wake of the Soviet
Union's Sputnik launch, Kimuel Alonzo Huggins, chair of the chemistry depart-

ment at Atlanta University, contacted her and suggested she apply to the master's program there—he even arranged funding. The timing was fortuitous:

> Sputnik had a similar impact on us [Blacks and women in science] as World War II. When the Russians [launched Sputnik in 1957], it turned the attention of politicians and even the general public on science. That's when the government started to put lots of money in science. The program at Atlanta University was one of those funded by the government. And by 1958 lots of programs were speeding up. After the first year, the program at Atlanta got funded by the NSF [National Science Foundation], and this was directly a result of Sputnik. [The program] was for high school math and science teachers. Technically, [because I had taught seventh grade] I qualified as a high school teacher. We [school teachers] were the most serious students in the program.[16]

Anderson entered the master's program at Atlanta University in 1959; then in June of 1960 she married and decided over the summer that she was not going to return to the program. Despite the NSF funding, which paid tuition, finances were the problem. Huggins, who was set to be Anderson's thesis advisor, intervened once again, as she recalled:

> Dr. Huggins was a mentor of the old school. At Atlanta University, it was a given that it took two years to get the master's degree. I got married in June of 1960 and during the summer I decided I wasn't going to return to school. Huggins took me aside and talked to me, as if I were his own daughter. He promised to help with the search for scholarship money, and was just very supportive. If that's not a mentor, I don't know what is. He arranged for scholarships and loans. If he hadn't talked to me I would have gone with my then husband, who was a high school teacher. It was not a continuous mentor/student relationship, but it was crucial for me.[17]

Anderson completed the master's degree in chemistry in 1961, and then taught for a year at South Carolina State University (1961–1962). After earning the master's degree she said she "never thought about medical school again," but realized that if she intended to continue teaching at the college level she ought to get a doctorate: "I didn't get any counseling, it was all rather haphazard."[18] Then another mentor entered Anderson's life.

By all accounts Henry Cecil Ransom McBay single-handedly trained the lion's share of Black chemists in the United States in the years between 1936 and 1994.[19] It was his practice to seek out talented Blacks with chemistry degrees, and he sought out Anderson. While she was in the master's program at Atlanta University, he was at nearby Morehouse College. Her master's thesis, done under Huggins' direction, had described a new three-step synthesis of butadiene using infrared spectrometry to confirm the structures; McBay was aware of the work. After her year at South Carolina State, he arranged for Anderson to come to Morehouse, both as a chemistry instructor and research assistant. During her two years at Morehouse (1962–1964), he continually encouraged her to go on for the doctorate. In the fall of 1965, she enrolled at McBay's alma mater, the University of Chicago, which awarded Anderson a research and teaching assistantship.[20]

Despite the University of Chicago's excellent record of producing Black science doctorates, Anderson does not recall it as an especially "friendly" place:

Although there was no organized harassment at the University of Chicago, racism came in with some of the faculty, though I'm not sure they were even conscious of it. The attitude was, "We let you people in and you don't *do* anything." Some of it probably had to do with society's view of Blacks in general.[21]

She said there was neither a mentor nor a support program at the University of Chicago, but she did find a role model—Thomas Cole, a young Black teaching assistant in the instrumentals course. Cole's ability with the instruments made a tremendous impression on her, particularly in terms of offering "proof" that a Black person could do that type of work.[22]

During her first year of graduate school, Anderson began to work with white female undergraduates, tutoring them in chemistry: "I had taught it for a few years, and mechanistic organic chemistry is not easy if you didn't have it as an undergrad, so I'd help. But after the first year I stopped because I sort of realized that they had all the advantages I'd never had."[23] She recalled that there were a few Black students in the chemistry department, and generally the environment consisted of white males who would not study with her. Eventually, she did find two study partners, Sarah Woods and Loretta Smith Jones.[24] The study partnership between Anderson and Jones, who was white, developed into a friendship:

Our first year together as organic chemistry majors, she was very helpful. The white kids did not study with us—in fact, they had better access to friends among the upperclassmen who would help them out by providing copies of old tests, and so forth. Loretta was the only white student I felt really comfortable with—I didn't even think of her as white. She was one of the few people who was truly a friend—not only to me but to other Black students, as well. No one knew it, but she had a Black husband, yet I don't think that had anything to do with why she was so nice—she just wasn't prejudiced![25]

At some point, Jones became disenchanted with the program and dropped out, which left Anderson pretty much on her own. She decided to "tough it out."[26] Her dissertation project was a study of fluorine–19 nuclear magnetic resonance (NMR) substrate chemical shifts and CF infrared frequency shifts, done under the direction of Leon M. Stock. The research generated at least one paper, which was published before she had completed her dissertation.[27] "In my dissertation," she said, "I had explained away anomalous phenomena but I had no real data and so my advisor said I should look at stretching frequencies of the fluorine bond as a way to get the data."[28] At the present time, according to Anderson, there are two other groups in the United States doing work with fluorine–19, but when she started working with it during her graduate studies at the University of Chicago it was very new.

During Anderson's last year of graduate school, the American Chemical Society held its annual meeting in Chicago, which Henry McBay attended. On a walk after one of the sessions, McBay strongly suggested that Anderson apply for the chairmanship of the chemistry department at Morris Brown College, in Atlanta. "I never would have thought of applying for a chairmanship, certainly not right out of graduate school," Anderson said, "but Dr. McBay said to call the school and to mention his name."[29] She completed the doctorate in 1968, and was immediately named an associate professor and department chair at Morris Brown.[30] Anderson also did postdoctoral work, "Studies on the Mechanism of Epoxidation," under Charles L. Liotta, at the Georgia Institute of Technology.[31]

Anderson's next project was designed to acquire experimental data on the transmission mechanism of chemical substituents, because the substituent effect gives information on reactivity and mechanism, which she uses frequently in her synthetic work. As she explains:

> In organic chemistry one part is organic synthetic chemistry, where we make stuff, another part is when you try and figure out how the components got together—the more theoretical aspects. . . .
>
> If a hydrogen group is replaced with another chemical group, for example, how does it transmit its effect to the reaction site? It has to transmit its effect to the site, but the real interesting question is *how*. We are looking for an understanding of the *mechanism* of the reaction, the chemical reaction, and there are several ways of doing this, one of which is using fluorine–19 as a probe and looking at the resulting NMR spectrum.
>
> If you have an oxygen molecule, for example, and you put fluorine in it, it will give a certain signal in the NMR, and if you place a hydrogen in it, it will cause that signal to shift its position. The shift is supposed to tell you something about how electron density is changed in the molecule. The molecule is the crucial component of the study.[32]

The major portion of Anderson's research over the years has involved synthetic organic chemistry, including work on fluorine–19, which she uses to understand how chemical substituents combine at the molecular level; synthetic techniques for developing solid-fuel rocket propellants; synthesis of antiviral drugs and medicinal compounds containing fluorine; structural studies on substituted amantadines; and studies on the mechanism of epoxidation.

Although she has received a considerable number of grants over the years, Anderson believes they have not been commensurate with her level of performance and expertise.[33] Her quest for funding has been frustrating, and at times she has attributed her problems to both racism and sexism:

> I've spent the last fifteen years trying to get research money and I can't. In the first case—or first, came a National Science Foundation grant. I applied for a mainstream grant because I was doing mainstream work and I was recommended for the funding—they *approved* it. But then the money didn't arrive. They sat on it, kept the money for a whole year. Then I found out that they had taken the money for my *mainstream* grant out of the *minority* budget. And this was even though my research was mainstream.
>
> What I was told—after the minority program was set up—was that they funded the mainstream grant for me through the minority budget. They didn't give me the money until the new minority funds came through. Then when I applied for renewal, NSF cut me off entirely, even though they usually give "phase-out" funds when they decide not to re-fund. They cut me off completely. When I finally got in touch with someone in the program office, he told me he didn't know why my funding had been cut and it was he who told me that NSF always gave phase-out grants.[34]

Anderson said that later, because she had gotten that first NSF mainstream grant, she was not eligible for other grants designed for women and minorities:

> I tried for several years after that to get funding under any category, but the mainstream grant put me in a category of someone who didn't *need* minority support and I've been stuck in that Catch–22 ever since. In my opinion, NSF is the most racist, most elitist funding organization in Washington. It may not be that way now—in fact things should be better now because Shirley Malcolm is on the NSF board and she's a fighter, a go-getter—but it was then.[35]

She also said her success in obtaining alternative funds hurt her chances of securing NSF funding:

> Later [in the 1980s when I reapplied to NSF], I wasn't eligible because I had had an industrial fellowship from Lockheed. They penalized me because I had managed to find little ways to keep up my research. I need to document the rejection letters [which clearly show the] pattern of rejections.[36]

She has tried every avenue she knows of to obtain the funds necessary to continue her research: "I attended a meeting for women in science here in Atlanta some years ago, on how to get through the funding maze. And I told them, 'everything you've said, I have already done—and it didn't work.'"[37] She recounted similar problems in her efforts to obtain funding from the Army:

> With the Army, I sent samples to be tested, they claimed they lost them, then they found them a year and a half later—it's all part of a pattern. I think the reason why this may have happened is that the Army doesn't fund work they can't get a full patent on. And there's a good chance that some of the compounds I'm testing are antiviral agents.[38]

Although she has accomplished a great deal with her research, Anderson believes she would have been able to do much more if she had chosen a different career venue:

> At a major white university, there would be funding and equipment—things like nuclear magnetic resonance spectroscopes, graduate students to help, and other faculty to bounce ideas off, so it's been a trade-off. I really wanted to develop my research program here, but it has been impossibly difficult. I don't have help for the research, I don't have an assistant.[39]

In an increasingly difficult research funding atmosphere, Anderson uses a sizable portion of her salary to fund her research. "I spend far more of my own money— about $40,000 per year—on my research than I should because my funding is not adequate," she said.[40] That statement—coming as it does from a scientist who had secured more than a quarter million dollars in research grants through 1985 —clearly indicates how expensive it is to do scientific research in the United States today.

Since the late 1980s, her research has focused on synthetic antiviral agents, but she notes that almost all the requests for reprints of her publications come from abroad. She has not submitted for publication since 1989 because she suspects her work is patentable, although she continues to present her findings at symposia.[41] In recent years, she has been a consultant for BioSPECS, a chemical and pharmaceutical research firm in the Netherlands, which has purchased a number of her structures. "If I took the time and sent them other structures," she said, "I'm sure they would want them too."[42] She does not have enough time, however, because her teaching load is so heavy. Yet Anderson made a conscious decision to carve out her career at an historically Black college: "When I was a student, I would have been grateful for a professor/scientist such as myself."[43]

She is currently the Fuller E. Callaway Professor of Chemistry at Morris Brown College. The Callaway Professorship, a partially endowed chair sponsored by the Callaway Trust, is designed to raise the salaries of selected professors at four-year colleges in Georgia to levels commensurate with national averages.[44] Anderson says she is grateful for the endowment because it gives her the ability to spend her own money for research:

> The Callaway Foundation is particularly helpful to schools such as Morris Brown and other small schools. The base salaries are so low that this puts you where you should be anyway. The Foundation awards me an additional one-half of my regular Morris Brown salary, which is about $54,000, so I make a total of about $80,000 a year, which is pretty good—or close to what I would make at a mainstream university.[45]

Despite her funding problems, Anderson is determined to remain at an historically Black college, and has, in fact, turned down offers from majority white universities.[46] A tireless grant writer, between 1969 and 1992 Anderson also secured nearly $1,000,000 in grants for college and pre-college science improvement and faculty development programs at Morris Brown,[47] and more than $5,000,000 in grants for other pre-college and college programs in the Atlanta area.[48] During her tenure as chair, she upgraded the chemistry department's faculty to five doctorates; revised the curriculum to make it consistent with criteria developed by the American Chemical Society; and introduced modern instrumentation, acquired through numerous institutional grants.[49]

A longtime advocate of increasing opportunities in science and science education for women and minorities, Anderson has also been active in efforts to encourage positive portrayals of minorities in public broadcasting. In 1972, she was appointed by President Richard M. Nixon to a six-year term on the Board of Directors of the Corporation for Public Broadcasting (CPB), chaired the CPB National Task Force on Minorities on Public Broadcasting (1976–1978), and was appointed to the Georgia Public Telecommunications Task Force in 1980.[50] Divorced in 1977, Anderson is single and the mother of one adult son.

NOTES

1. Gloria Long Anderson, interviews by author, Morris Brown College, Atlanta, Ga., 10 and 15 November 1995.

2. Anderson, interviews. Gloria Long Anderson, author questionnaire, 10 November 1995. Gloria Long Anderson, curriculum vitae (1995). Gloria Long Anderson, assorted personal papers provided to author (1995). See also *American Men and Women of Science: The Physical and Biological Sciences*, 15th ed. (New York: Bowker, 1971–), p. 117. *Biography Index*, vol. 11, September 1976 and August 1979. "Physical Organic Chemist," in *Contributions of Black Women to America*, ed. Marianna W. Davis, vol. 2 (Colum-

bia, S.C.: Kenday Press, 1982), pp. 438–439. *Dictionary of International Biography* (Cambridge, England: International Biographical Centre, 1986), p. 29. *Who's Who among Black Americans,* 1980–1981 ed. (Northbrook, Ill.: Who's Who Among Black Americans Publishing, 1975–1985), p. 18. *Who's Who in America,* 47th ed. (Chicago: Marquis Who's Who, 1992–1993). *Who's Who in Science and Technology,* 1st ed. (Chicago: Marquis Who's Who, 1991). *Who's Who in Engineering* (New York: American Association of Engineering Societies, 1971–). *Who's Who among U.S. Executives,* 1989. *International Who's Who of Intellectuals,* August 1984.

3. Anderson, interviews; Anderson, author questionnaire. Vivian Ovelton Sammons in *Blacks in Science and Medicine* (New York: Hemisphere, 1990), p. 12, erroneously states that Anderson grew up in Arizona.

4. Anderson, interviews.

5. Ibid.

6. Ibid.

7. Anderson says that she has "two younger brothers and three older brothers, the eldest of whom was my half-brother. My two younger brothers graduated from high school and both graduated from college. One of them has an M.B.A. The other studied oceanography at MIT [Massachusetts Institute of Technology], although he didn't finish the master's, and works for the government. The one with the M.B.A. started in chemistry and math, but he couldn't stand the mess of being in the labs. . . . My three older brothers—the oldest, my half-brother, his father lived in Chicago, and he moved there. He may have finished high school there, I don't know. My next eldest brother didn't finish high school, he got married. And the brother just older than me, went into the Korean War, but after he went to college for two years" (ibid.).

8. Ibid.

9. Ibid.

10. Ibid. She added that, at that point, her parents were in a better financial position. Her father had "moved up" and begun working at the Pine Bluff Arsenal as a janitor.

11. According to Anderson, the other students were "Barbara Wallace, Lawrence Davis Jr., and I can't remember the other fellow's name. I knew Lawrence before college; we had met when we attended the college's laboratory high school during the summer. . . . Barbara and the other fellow eventually dropped out, although they were very good. Lawrence Jr. and I finished" (ibid.).

12. Ibid.

13. Ibid. According to Anderson, "Some music major, whose name I don't remember was number 2, and Lawrence [Davis Jr.] finished number 3. Lawrence's father was then president of the college, and today Lawrence is the chancellor of the University of Arkansas at Pine Bluff" (ibid.). See also Anderson, author questionnaire.

14. Anderson, interviews; Anderson, author questionnaire.

15. Anderson, interviews.

16. Ibid.

17. Ibid. Chemist Kimuel Alonzo Huggins earned the B.A. at Morehouse College, 1923; the M.S. and Ph.D. at the University of Chicago in 1929 and 1937, respectively; and was chair of the chemistry department at Atlanta University from 1933 to 1968 (Sammons, *Blacks in Science and Medicine,* p. 125).

18. Anderson, interviews.

19. Chemist Henry Cecil McBay earned the B.S. at Wiley College in 1934; the M.S. at Atlanta University in 1936; and the Ph.D. at the University of Chicago in 1945. He spent the bulk of his career at Morehouse College, where he was chair of the chemistry department from 1960 to 1981; he then moved to Atlanta University (Sammons, *Blacks in Science and Medicine,* p. 167). He trained entire cohorts of Black chemists, according

to Dr. Prince Rivers of the Atlanta University consortium. McBay died in June 1995, at age 81. According to Rivers, people came from around the world to attend the memorial service (Prince Rivers, personal interview at Atlanta University, 10 November 1995). Dr. Gloria Long Anderson of Morris Brown College, also in Atlanta, stressed McBay's contributions (Anderson, interviews).

20. Anderson, curriculum vitae.

21. Anderson, interviews.

22. Ibid.

23. Ibid.

24. Sarah Woods has a Ph.D. in chemistry (ibid.).

25. Ibid.

26. Ibid.

27. See the appendix.

28. Anderson, interviews.

29. Ibid.

30. Ibid. See also Anderson, curriculum vitae.

31. The postdoctoral work involved studies of the mechanism of epoxidation of 1, 3-diphenyl-1,3-propanedione with *m*-chloroperbenzoic acid using gas chromatography (Anderson, curriculum vitae, 1995).

32. Anderson, interviews.

33. Between 1969 and 1985, Anderson received $271,614 in grants for various research projects, provided by organizations as diverse as the U.S. Air Force Office of Scientific Research, the United Negro College Fund, the National Institutes of Health, and the National Science Foundation. The grants were as follows: "New Synthetic Techniques for Solid Rocket Propellants: Nucleophilic Triflate Displacement Reactions," SCEEE (Southeastern Center for Electrical Engineering Education) Air Force Office of Scientific Research, Research Initiation Grant ($12,000), 1985; "Synthesis of Some Potential Antiviral Drugs: Amantadine Analogs," United Negro College Fund Distinguished Scholar Award ($35,000), 1985; "Structure Activity Studies on Some Substituted Amantadines," Minority Biomedical Support Program, National Institutes of Health ($45,996), AY 1979–1980; "Structure Activity Studies on Some Substituted Amantadines," MBRS Program, National Institutes of Health ($32,818), AY 1978–1979; "$^{19}F$ Chemical, Shifts and C-F Stretching Frequencies of Bridgehead Fluorides," Office of Naval Research ($20,000), 1973–1975; "$^{19}F$ Chemical, Shifts and Infrared C-F Stretching Frequencies," National Science Foundation ($66,500), 1972–1974; "The Synthesis of Medicinal Compounds Containing Fluorine," National Institutes of Health ($52,000), 1972; "$^{19}F$ Chemical, Shifts and Infrared C-F Stretching Frequencies for 3-Substituted-1-Fluoroadamantanes," Atlanta University Center Research Committee ($1,600), 1971; "The Preparation and Decomposition of Pyridinium Salts," Atlanta University Center Research Committee ($1,850), 1969; "Studies on the Mechanism of Epoxidation," National Science Foundation ($1,850), 1969; "Studies on the Mechanism of Epoxidation," National Science Foundation ($2,000), 1969 (Anderson, curriculum vitae).

34. Anderson, interviews.

35. Ibid. Also see the entry "Shirley M. Malcolm" in this volume.

36. Ibid. Anderson was referring to two Lockheed Corporation fellowships: a 1982 summer research consultancy at Lockheed Georgia Corporation, Marietta, Georgia, and a 1981 summer research fellowship (which was actually funded by the NSF) also at Lockheed. There was also a 1984 Summer SCREEE Faculty Research Fellowship at Edwards Air Force Base, California, in the Air Force Rocket Propulsion Laboratory (Anderson, interviews; see also Anderson, curriculum vitae).

37. Anderson, interviews.

38. Ibid.

39. Ibid.

40. Ibid.

41. See the appendix to this book.

42. Anderson, interviews.

43. Ibid.

44. The Callaway Professorships were established by the Callaway Foundation in 1968 with a $10 million trust fund to provide senior Georgia colleges and universities with a means of retaining and adding faculty members. The Callaway family is involved in the Georgia textile industry. The Foundation endows chairs in each four-year college in the state of Georgia, enabling colleges whose base salaries tend to be lower than the national average to attract minorities and other highly qualified faculty members. The foundation pays 50% of a chairholder's base salary as a supplement (Anderson, interviews). See also "Physical Organic Chemist," pp. 438–439.

45. Anderson, interviews. When she served as dean and interim president, Anderson had to give up her Callaway stipend, but it was reinstated when she stepped down.

46. Out of deference to the universities, Anderson did not wish to have the universities' names published here.

47. Anderson's grants for science improvement programs ($725,000): Science Improvement Grant Program, David and Lucile Packard Foundation, California, 1991, 1992 ($200,000); Science Improvement Program Grant, Pew Charitable Trusts, Philadelphia, 1988–1991 ($500,000); Instructional Equipment Grant for Chemistry, Title VI, Department of Health, Education and Welfare, Washington, D.C., 1969 ($25,000).

Anderson's grants for pre-college science programs ($58,824): Pre-College Science and Engineering Program Grant, Minority Institutions Science Improvement Program, U.S. Department of Education, Washington, D.C., 1986 ($44,824); Pre-Freshman Engineering Program, U.S. Department of Energy, Washington, D.C., 1985 ($10,000).

Anderson's grants for faculty development programs ($220,000): Faculty Development Program Grant, Bush Foundation, St. Paul, Minn., 1988–1991 ($210,000); Faculty Development Planning Grant, Bush Foundation, St. Paul, Minn., 1987 ($10,000) (Anderson, curriculum vitae).

48. Anderson's grants for other pre-college/college programs ($5,708,317): Upward Bound Project Grant, U.S. Department of Education, Washington, D.C., 1989–1990 ($313,664—funded for 3 years, estimated total award, $950,000); Student Support Services Grant, U.S. Department of Education, Washington, D.C., 1987–1990 ($317,218); Educational Talent Search Grant, U.S. Department of Education, Washington, D.C., 1988–1990 ($164,189—funded for 3 years, estimated total award $250,000); Educational Talent Search 7th And 8th Grade Initiative Grant, U.S. Department of Education, Washington, D.C., 1989–1990 ($50,000—funded for 2 years, estimated total award, $100,000); Summer Youth Employment and Training Program Grant, Fulton County Private Industry Council, Atlanta, Ga., 1989 ($71,906); Summer Youth Employment and Training Program Grant, Metropolitan Atlanta Private Industry Council, Atlanta, Ga., 1988 ($150,000); Upward Bound Project Grant, U.S. Department of Education, Washington, D.C., 1986–1989 ($834,017); Student Support Services Grant, U.S. Department of Education, Washington, D.C., 1984–1987 ($270,784); Educational Talent Search Grant, U.S. Department of Education, Washington, D.C. ($226,231); Upward Bound Project Grant, U.S. Department of Education, Washington, D.C., 1983–1986 ($731,380); Summer Youth Employment and Training Program Grant, Metropolitan Atlanta Private Industry Council, Atlanta, Ga., 1987 ($184,000); Summer Youth Employ-

ment and Training Program Grant, Dekalb County Private Industry Council, Decatur, Ga., 1987 ($61,936); Summer Youth Employment and Training Program Grant, Metropolitan Atlanta Private Industry Council, Atlanta, Ga., 1986 ($176,210); Summer Youth Employment and Training Program Grant, Metropolitan Atlanta Private Industry Council, Atlanta, Ga., 1985 ($114,184); Summer Youth Employment and Training Program Grant, Dekalb County Private Industry Council, Decatur, Ga., 1985 ($55,426); Student Support Services Grant, U.S. Department of Education, Washington, D.C., 1983–1984 ($80,952); Special Student Services/Upward Bound Grants, U.S. Department of Education, Washington, D.C., 1980–1983 ($913,000); Special Student Services/Upward Bound Grants, Office of Education/Department of Health Education and Welfare, Washington, D.C., 1977–1980 ($753,000); Special Student Services Grant, Office of Education, Department of Health, Education and Welfare, Washington, D.C., 1970–1973 ($240,000) (Anderson, curriculum vitae).

49. Among the instruments she has secured for Morris Brown are infrared, ultraviolet, and nuclear magnetic resonance spectroscopes and a gas chromatography unit (Anderson, curriculum vitae; Anderson, interviews).

50. The appointment was confirmed by the United States Senate; Anderson served as vice-chairman of the board from 1977 to 1979. The CPB National Task Force on Minorities on Public Broadcasting produced the report "A Formula for Change" in 1978. The recommendations of the Public Telecommunications Task Force formed the basis of the Georgia Public Telecommunications Act passed by the state legislature. She also served on the CPB National Task Force on Women in Public Broadcasting (1975) and the CPB National Advisory Panel on Essentials for Effective Minority Programming (1973–1974). Georgia governor George Busbee appointed Anderson to the Georgia Telecommunications Task Force, which was charged with making recommendations on the future direction of the state's public telecommunications (Anderson, interviews).

JUNE BACON-BERCEY*

In 1979, June Bacon-Bercey became chief administrator for Television Weather Activities at the National Oceanic and Atmospheric Administration (NOAA) in Rockville, Maryland.[1] She was then one of only a few Black women with degrees in meteorology (B.S., University of Kansas; M.S., University of California at Los Angeles).[2] Before joining NOAA, Bacon-Bercey had been employed as an engineer at the Sperry Rand Corporation, as a weather analyst at the National Meteorological Center and at the Aviation Branch of the National Weather Service, and as a consultant with the U.S. Atomic Energy Commission.[3]

NOTES

1. *Black Women: Achievements against the Odds.* Washington, D.C.: Anacostia Museum, Smithsonian Institution, 1984, p. 39. See Vivian Ovelton Sammons, *Blacks in Science and Medicine* (New York: Hemisphere, 1990), p. 17. Note that Sammons lists her as Berey.

2. Marianna W. Davis, ed., *Contributions of Black Women to America*, vol. 2 (Columbia, S.C.: Kenday Press, 1982), p. 451. No degree dates were given. Note that Davis lists her as Bercey.

3. Ibid.

PATRICIA ERNA BATH
Eye Surgeon

When Patricia Erna Bath was only a teenager, she seemed likely to become a scientist, as she was already doing research on cancer. Bath had been interested in science all of her life:

> I was tinkering with toy chemistry sets as a child. One of the gifts my parents bought for me was a microscope set from Macy's. General Science was my favorite subject when I was attending Julia Ward Howe Junior High School. Later, at Charles Evans Hughes High, I took two years of biology and advanced chemistry.[1]

In particular, she had become interested in cancer research:

> [C]ancer offers such a challenge. Medical science has conquered so many diseases that used to make shambles of human lives. But this one still has to be understood—and curbed. . . .
>   Our people [Blacks] are hit hardest by cancer—particularly by certain types such as cancer of the breast, lungs, and digestive organs. Fewer of them will die when we can find the general causes of the disease and end discrimination in treatment facilities at hospitals and clinics.[2]

In 1961, she was featured in a *Sepia* magazine article by Harold Preece—at that time, Bath and her research collaborator were only seventeen years old:

> Today a pair of pathfinders happen to be teenagers. Each is from a minority group whose respective members have had much trouble in getting medical training. One is a Jewish boy—Arnold Lentnek, the son of a businessman in suburban Far Rockaway. The other is a Negro girl—Patricia Erna Bath, the daughter of a seaman living in the heart of Harlem. Already she is being acclaimed as one of the most promising and talented young scientists in America.[3]

The style and focus of the Preece article provide interesting historical insight: Racial integration—the value of people of different races working together—was as important as the achievements of the two talented youngsters. Their "interracial collaboration is in the deepest interest of all humanity since not even the legislature of Mississippi might pass an enforceable law to segregate cancer."[4]

Bath and Lentnek's research on the relationship between cancer, nutrition, and stress factors began in July 1959, when both were fifteen-year-old high school sophomores. Because of their scientific promise, they had won grants from the National Science Foundation to Yeshiva University's Summer Institute in Bio-Medical Science. The two students had never met each other until they began their experiments at New York's Jewish-supported—but completely integrated—Yeshiva University.[5]

Bath was interested in Canadian physician Hans Selye's stress theory of disease, which posited that cancer would be a gradual and total breakdown of healthy tissues:

> Lentnek found that paramecii [*sic*] showed marked growth when he fed them a substance called dried streptomycin beer. . . . But the way that the minuscule organisms thrived on it upset something else that had been proved in cancer

therapy. Something that was well known to Dr. Moses D. Tendler, associate professor of biology at Yeshiva and [then] director of the program.

By an involved mathematical equation, [Bath] struck a balance between the processes that build up and those that tear down. Her colleagues in the project found . . . that they could anticipate the corresponding growth response of a paramecium in the streptomycin beer and the reduction response in a cancer patient, [a comparison] "much sharper than medical science ever believed."[6]

In September 1960, a research paper co-written by Bath and Lentnek was read at the Fifth International Congress on Nutrition in Washington, D.C.:

[Their research had] thrown new and surprising light on the nature of cancer. Shown better how to provide nourishment for patients wasting away from the disease as it spread through their bodies. Provided experimental support for a new theory steadily gaining ground among cancer researchers that cancer, like that other killer, arteriosclerosis, is a general disease of the whole human system, manifested by arrested normal growth plus total breakdown of the physical structure.[7]

At that time, Bath, whom Preece described as "one of two of the most brilliant women I've ever interviewed during a lifetime as a reporter" (at times he felt as though he was "covering Einstein"),[8] was enrolled as a sophomore at the then female-only Hunter College, part of the City University of New York (CUNY). She intended to enter medical school and become a physician. She called her experience in the Yeshiva program a "milestone in my life. It really gave me a chance to do something. I had ideas before but no adequate facilities, no close, highly-trained supervision, and little contact with extremely bright contemporaries. I got all these through the Yeshiva program."[9]

Bath completed the baccalaureate degree at Hunter College in 1964, and completed the M.D. degree at Howard University, in 1968.[10] Between her medical school graduation and 1974, she returned to New York where she took additional classes at both Columbia University and New York University and completed an internship at Harlem Hospital.[11]

Patricia Erna Bath fulfilled both her dream and her potential, but in a somewhat different area of medicine and research than she had originally intended. She is not a cancer specialist, but rather is an internationally known ophthalmologist and surgeon who specializes in the treatment of glaucoma and cataracts. Now 53, Bath is known as an inventor as well as an eye surgeon—she created and patented a surgical instrument, the Laserphaco probe, used in cataract surgery. Although lasers were used for treating secondary-level cataracts prior to 1986, the then common method for removing primary cataracts was a drill-like tool that ground them away. Bath's instrument, which she developed during research in Germany and received a U.S. patent for in 1988, combines an optical fiber to conduct laser radiation with irrigation and aspiration lines, vaporizing a cataract in minutes. "The ability to restore vision is the ultimate reward," Bath said, and added, "It is a really great joy to remove eye patches the day after surgery and the patient can see again."[12] The Laserphaco probe is currently undergoing testing by the U.S. Food and Drug Administration.[13]

Bath was both a surgeon and a faculty member at the University of California at Los Angeles (UCLA) Medical Center, where she co-founded an ophthalmology training program at the Martin Luther King Jr. and Charles R. Drew Medical Center.[14] She was the keynote speaker at the annual Women Pioneers in Science and Technology luncheon held at Akron, Ohio, for the National Inventors Hall of Fame Induction Weekend in 1998.[15] She is currently the director of Telemedicine Services at Howard University Hospital in Washington, D.C., the Telemedicine SIG chairman of the Washington Metropolitan Distance Learning Association,[16] and the president of the American Institute for the Prevention of Blindness in Los Angeles.[17]

NOTES

1. Harold Preece, "The Team Who Challenges Cancer: These Two New York Teenagers Work Together in Fighting America's Number One Killer; Their Success Might Well Mean Thousands of Human Lives Saved Yearly," *Sepia*, January 1961, p. 55.

2. Ibid.

3. Ibid., p. 54.

4. Ibid.

5. Ibid.

6. Ibid., p. 56.

7. Ibid., p. 54.

8. Ibid., p. 54. Preece said the other brilliant woman was Maria Telkes, "the internationally known authority on solar energy."

9. Ibid., p. 57.

10. "Profile on Inventors: The Laserphaco Probe," *TDP Newsletter* [web page] 1 March 1999; http://www.3dpublishing.com/learningcenter/bath.htm [accessed 25 March 1999]. Also Joe Nazel, "Cataract Surgery Inventor," 3D Publishing [web site] 7 December 1998; http://www.erols.com/tdpedu [accessed 20 December 1989]. A fuller version of Nazel's article appeared in the *Wave Community Newspaper* (213–290–3000 ext. 239).

11. "Profile on Inventors."

12. Bath, quoted in Nazel, "Cataract Surgery Inventor."

13. Ibid.

14. Ibid., p. 1. Bath is now professor emerita of ophthalmology at UCLA Medical Center.

15. "Annual National Inventors Hall of Fame Induction Week Schedule of Events," p. 1 [web page]: http://www.cl.akron.oh.us/Temporary_Pages/InductionWeekInventure Place1998.html [accessed 20 December 1998].

16. Washington Metropolitan Distance Learning Association [web site]: http://www.wmdla.org/chapter.htm [accessed March 1999]. Bath's daughter is currently studying at the Howard University Medical School (Nazel, "Cataract Surgery Inventor," p. 2).

17. "Patricia Bath, MD, Gives Sixth White Coat Address," St. George's University (Grenada) Latest News: Student Life, 2 February 1999 [web page]: http://www.sgu.edusgunews.nsf/504ca249c786e20f85256284006da7ab9bcc5e31880a94718525670c006 db8b1?OpenDocument [accessed 20 March 1999].

MATILENE BERRYMAN*

In 1982, oceanographer Matilene Berryman was a professor of marine science and head of the Department of Environmental Science at Washington Technical Institute of the University of the District of Columbia, in Washington, D.C.[1]

Born and raised in Prince Edward County, Virginia, Berryman was an avid childhood reader who developed a love for mathematics during a high school geometry course taught by Mrs. Dorothy Vaughn: "I loved her [Vaughn], for she made me love the subject she taught."[2]

Berryman earned a baccalaureate degree in mathematics, and a master's degree in marine affairs, concentrating in oceanography and sonar engineering. She then joined the U. S. Naval Oceanographic Office in Maryland, where she found "absolute fulfillment as far as a job is concerned. Physics, oceanography, meteorology, and mathematics all welded together in a nice, easy picture of how nature operated this world." Berryman is a strong advocate of education: "I am a believer that education is the key that could truly spell the difference between abject poverty and the extreme wealth of that one percent of the population."[3] Berryman continued her own education and eventually earned a law degree at Howard University.

NOTES

1. "Oceanographer," in *Contributions of Black Women to America,* ed. Marianna W. Davis, vol. 2 (Columbia, S.C.: Kenday Press, 1982), pp. 449–450.

2. Ibid.

3. Ibid.

SYLVIA TRIMBLE BOZEMAN

Fostering Mathematics at Spelman College

Sylvia Bozeman
Photo courtesy of Spelman College Archives, Atlanta, Georgia

Mathematician Sylvia Trimble Bozeman, a recipient of both a White House Initiative Faculty Award for Excellence in Science and Technology and a Tenneco United Negro College Fund Award for Excellence in Teaching, grew up and attended segregated schools in Camp Hill, Alabama.[1] Receiving encouragement from her parents, Horace and Robbie Jones Trimble, and inspiration from Frank Holley, her high school geometry teacher, Bozeman earned the baccalaureate degree in mathematics at Alabama A & M University in 1968, graduating second in a class of 200.[2] Awarded funding for graduate study from the Southern Fellowship Fund, Bozeman began work on her master's degree at Vanderbilt University in Nashville, Tennessee—the first non-segregated school she had attended. She quickly discovered that her white classmates had benefited from undergraduate mathematics courses that had not been among the course offerings at Alabama A & M. Bozeman persevered and earned the master's degree in mathematics in January 1970.[3]

During this period she married Robert E. Bozeman, also a mathematician, and bore two children, a son and daughter. Juggling household chores and motherhood, she taught part time at Vanderbilt and then in the Upward Bound Program at Tennessee State University.[4] In 1973, her husband joined the mathematics faculty at the historically Black, all-male Morehouse College in Atlanta, Georgia; the following year, 1974, she joined the mathematics faculty at the historically Black, all-female Spelman College, also in Atlanta. The move to Spelman was fortuitous, particularly for a young mathematician.

According to mathematician Etta Zuber Falconer, in the early 1960s the science faculty at Spelman had already begun to question "the low production of science graduates and the 'perceived' low status of science at the College."[5] Falconer believed the "situation would have continued if the science faculty, under the leadership of mathematician/chemist Shirley Mathis McBay, had not presented a strong case" for improving the science and mathematics offerings.[6] Falconer, McBay, and Gladys Glass collaborated to build a reputable mathematics program at Spelman.[7] In 1972, Spelman created a Division of Natural Sciences, with McBay as its chairman; its goal was to develop and implement plans for increasing the science and mathematics emphasis at the college.[8] Bozeman arrived in the midst of this crucial reorganization of the science and mathematics initiative.

In 1976, two years after she had arrived at Spelman, the college encouraged Bozeman to take a study leave, whereupon she entered a doctoral program in mathematics at Emory University. She studied with John Neuberger, a specialist in functional analysis, whom Bozeman has described as providing strong encouragement to aspiring women mathematicians. Bozeman has also said that her husband, Robert, provided strong, steady support throughout her career.[9] She completed the doctorate in 1980, and returned to her teaching duties at Spelman.[10] She continued her own research work—funded by the Army Research Office, the National Science Foundation, and the National Aeronautics and Space Administration—and has produced publications on functional analysis and research on image processing.

One legacy of the combined leadership of Falconer, McBay, Bozeman, and others was that by 1993 the all-female Spelman had acquired a reputation as having one of the best undergraduate science and mathematics education programs in the country. Between 1987 and 1992, more than 37 percent of Spelman's students majored in science and mathematics.[11] According to President Johnnetta B. Cole, Spelman College ranks among the top ten United States institutions in the number of Blacks admitted to medical school. Between 1989 and 1993, Spelman science major alumnae were selected for fellowships by the National Science Foundation, the National Consortium for the Physical Sciences, and the Watson Research Fellowship. The goal of Spelman's $22.5 million "Initiatives for the 90s" science campaign is to create the state-of-the-art learning and experimental facilities necessary to educate future generations of doctors and scientists.[12] In 1993, the W. K. Kellogg Foundation, one of the nation's largest private philanthropic organizations, identified Spelman as one of the ten premier historically Black colleges and awarded the school a $3 million grant to establish a Center for Scientific Applications of Mathematics. The effort to secure the grant and initiate the center was driven by Falconer, by then Associate Provost for Science Programs and Policy, and Bozeman, who had by then become chair of the mathematics department.[13]

Bozeman has served on the Board of Governors of the Mathematical Association of America, has co-chaired the organization's Committee on Minority Participation in Mathematics, and is a past vice-president of the National Association of Mathematicians.[14]

NOTES

1. Bozeman was born in 1947; she received both awards in 1988. Patricia C. Kenschaft, "Black Women in Mathematics in the United States," *Journal of African Civilizations* 4 (1982): 63–83; see especially pp. 79–80. Patricia C. Kenschaft, "Sylvia Trimble Bozeman," in *Black Women in America: An Historical Encyclopedia*, ed. Darlene Clark Hine, Elsa Barkley Brown, and Rosalyn Terborg-Penn, vol. 1 (Brooklyn, N.Y.: Carlson, 1993; Bloomington: Indiana University Press, 1994), p. 159.

2. Kenschaft, "Black Women in Mathematics."

3. Ibid.

4. Ibid.

5. Etta Z. Falconer, "A Story of Success: The Sciences at Spelman College," *Sage* 6 (fall 1989): p. 36. According to Falconer,

> There was an abundance of evidence to support the premise the Spelman students were not seriously encouraged to pursue the sciences: the science building was dark and uninviting; there was a lack of emphasis on science and health careers and little recognition of scientists and their contributions to society; the role of women in science and engineering was not discussed in classes, or presented in College publications; major classes in the sciences beyond the freshman level had tiny enrollments; major classes were offered only in biology and mathematics—the only chemistry course was a service course for majors in home economics and physical educations—the only physics course had been deleted from the curriculum before 1960; [and] science was perceived by the general student population as difficult and uninteresting. (Ibid., p. 36)

6. Ibid.

7. Ibid.

8. Ibid.
9. Kenschaft, "Black Women in Mathematics."
10. Bozeman's dissertation was "Representations of Generalized Inverses of Fredholm Operators."
11. Based on a report on the top 100 degree producers conducted by the journal *Black Issues in Higher Education's* (May 1993). See "$3 Million Grant Establishes Math Center," *Inside Spelman*, 1993.
12. Johnnetta B. Cole, "Cole's Corner," *Inside Spelman*, 1993. Cole was president of Spelman College.
13. *Inside Spelman*, ibid.
14. See Kenschaft, "Black Women in Mathematics"; and Kenschaft, "Sylvia Trimble Bozeman."

DOROTHY LAVINIA BROWN

From Orphan to Surgeon to Teacher

Surgeon Dorothy Lavinia Brown was not a research scientist, but I have included her here for three reasons. Her quest for a medical education illustrates the tremendous struggle Black women have undertaken to pursue careers in the sciences and related fields. During her more than forty-year career as a professor of surgery at Meharry Medical College—until recently one of only two historically Black medical colleges in the country—she helped train a significant number of Black surgeons. And, although she had no family, she managed to succeed despite this handicap—her life illustrates the importance of community support, which in her case took the place of a biological family.[1]

Born to an unwed mother in 1919, at age five months Brown was placed in an orphanage in Troy, New York, where she remained for the next thirteen years.[2] According to her autobiography, Brown was one of only a "handful of Black children in a sea of white faces" at the orphanage, and one of the few who never received a visitor.[3] A very bright child who regularly took top academic awards at the orphanage school, she was befriended by Herbert Hunn, the superintendent of the orphanage. Her interest in medicine stemmed from a tonsillectomy that she underwent at age five—clearly becoming a doctor was an unusual dream for a Black orphan in the 1920s. By the time she was nine years old, she had apparently became depressed at her lack of visitors, and Hunn arranged for a local family, the Frank Coffeens, who were members of the Presbyterian Church, to visit with her. A relationship developed, and the Coffeen family and Hunn apparently offered her a certain amount of encouragement.[4]

When Brown was thirteen years old, however, her mother reentered her life and took her out of the orphanage. During the next year or so, Brown ran away frequently—each time returning to the orphanage—until her mother finally placed her in domestic service.[5] She became a mother's helper at the home of the Jarrett family in Albany. According to Brown, she discussed her dream of becoming a doctor with the family, and they were encouraging—Jarrett had a considerable home library where she read Latin, algebra, and chemistry textbooks.

By the time Brown was fifteen, she had saved a few hundred dollars, and the Jarretts agreed to allow her to return to Troy. The efforts of Herbert Hunn brought the now homeless Brown to the attention of the high school principal in Troy. Hunn also arranged for her to live in the home of Samuel Wesley and Lola (Cannon) Redmon, who became her foster family.

After graduating from high school, Brown again entered domestic service, but she soon met Mrs. Charles E. Smart, who was active in the Troy chapter of the Women's Division of Christian Service of the Methodist Church. The group was interested in sending a Black female to college, and in 1937 they arranged for Brown to receive a four-year scholarship to Bennett College in Greensboro, North Carolina.[6] At Bennett, Brown was advised to study for a career as a school-teacher, yet she managed to take the kinds of courses that would prepare her for medical school. Even though the administrators at Bennett College had told the Methodist church women at the end of Brown's first year that she was not "Bennett material," she graduated second in her class in 1941 and returned to Troy.[7]

Brown again met with the Methodist church women and told them she intended to work and save money for medical school—apparently, up until this point, no one (other than Brown herself) had actually taken her plans for a medical career seriously.[8] At that time, World War II had not yet begun, and the country was still in the waning days of the Depression. For Blacks, finding employment was difficult, particularly in Troy, which had a very small Black population. Despite her college degree, Brown could not find any type of professional-level job and wound up working as a presser in a large commercial laundry.[9] Within a few months, however, the United States had entered World War II, and this worked to Brown's advantage. She got a job as a civilian inspector of Army ordnance.[10]

Brown was often the only Black woman working in the upstate New York ammunition and bomb factories. According to historian Karen Tucker Anderson, white women workers in defense plants during this period expressed a great deal of prejudice toward Black women: "For women workers the desire to maintain social distance, rather than a wish to safeguard economic prerogatives seemed to be the dominant motivation for [the prejudice]. White female workers frequently objected to working closely with black women or sharing [toilet] facilities . . . because they feared that blacks were dirty or diseased."[11]

By 1944, not only had Brown saved a considerable amount of money, but once again the Methodist church women came to her aid. She applied to both Howard and Meharry medical schools, choosing Meharry because it would be more economical to live in Tennessee than in Washington, D.C.[12] During her time in medical school, Brown returned to Troy each summer to work. Along with the support of the church women, Herbert Hunn, her old friend from the orphanage, arranged for her to receive an additional $700 in support.[13]

Brown completed her medical degree in 1948, and interned at Harlem Hospital in New York City. She then applied for a surgical residency at Harlem Hospital but was turned down, apparently on the basis of gender. According to Brown, the popular view in those days was that women were not fit—neither physically, intellectually, nor emotionally—to perform surgery: "Men said that a

woman wasn't able to stand up to five years of training in surgery. . . . I tried to be
. . . not hard, but durable. . . . I am a fighter who learned how to get along with the
male ego."[14]

She returned to Meharry and begged the then Chief of Surgery Matthew
Walker to allow her into the surgical residency program: "Dr. Walker was a brave
man," she said, "because despite advice from his staff that a woman couldn't with-
stand the rigors of surgery, he accepted me into the program."[15] Brown recalled
that in her attempt to prove the male surgeons wrong, she worked so hard they
called her "Mule."[16] The only female surgical resident, she completed her residency at Meharry's affiliated Hubbard Hospital in 1954 and became the first
Black female surgeon in the South. At that time, she also became one of the few
Blacks elected a Fellow in the American College of Surgeons.[17]

Accepted on the surgical staff at Riverside Hospital in Nashville and appointed
chief of surgery in 1957, Brown held the position until the hospital closed in
1983.[18] In conjunction with her faculty position at Meharry, she also served as
Educational Director of the Riverside-Meharry Clinical Rotation Program.[19] In
1983, she became an attending surgeon at George W. Hubbard Hospital, while
continuing as a professor of surgery at Meharry and maintaining a private practice.[20] She also served on the National Advisory Board of the National Institutes
of Health, the American Medical Association's Joint Committee on Opportunities for Women in Medicine, the Board of Trustees of Bennett College, and was
very active in the Methodist Church, both locally and at the national level.[21]

In 1966, Brown became the first Black (male or female) elected to the Tennessee State Legislature. Due in part to her clinical experience, she sponsored a
bill to change what she has described as the state's "archaic" abortion laws. The
bill was defeated, and Brown resigned her seat.[22] Her 1973 article "History and
Evolution of Abortion Laws in the United States" remains a compelling argument for change.[23]

Brown has been a strong supporter of historically Black colleges and universities, which she views as having been threatened by desegregation. "We still need
Meharry Medical College and we still need Howard University," she said in a
1986 interview. "Why should we throw away our heritage?"[24] In 1982, the Board
of Education of the United Methodist Church produced a film on Brown's life,
Run to Live: A Day in the Life of Dr. "D" Brown.[25] The film continues to be shown
at fund-raisers for twelve historically black colleges and universities maintained
by the church.[26] Brown has said that her basic philosophy of life encompasses "the
belief that we are here for a purpose—each of us being endowed with multiple
talents; our charge is to develop one or as many of these talents as possible and to
use these talents and the days of our living to glorify God. Therefore I must 'Run
to Live,' and I must seek to serve in as many different areas of endeavor as I can."[27]

Brown believes that "surgery is a natural [field] for women, because after all,
[they do] the cutting and sewing at home," and that lack of opportunity could be
"overcome by the motivation and the will to do what you want to do."[28] Although
no one ever encouraged her to become a doctor when she was a youngster, Brown
said, "No one actually said I couldn't do it because [I was] black, [a] woman and
. . . poor."[29]

NOTES

1. Howard University College of Medicine is the other school. Although there are now a few other Black medical schools—one at Morehouse College in Atlanta, and the Charles Richard Drew Medical College in California, among others—these have been relatively recent efforts. Howard and Meharry have been the mainstays for training Black physicians and surgeons.

2. Dorothy L. Brown, "Thus Would I Live," autobiography, papers of Dorothy L. Brown, Special Collections, Fisk University Library, Nashville, Tenn. See also Lois J. Dunn, "Dorothy L. Brown," in *Notable Black American Women*, ed. Jessie Carney Smith, vol. 1 (Detroit: Gale Research, 1992), pp. 114–116. Elizabeth Levy and Mara Miller, "Dorothy Brown: A Doctor for the People," in *Doctors for the People: Profiles of Six Who Serve* (New York: Knopf, 1977), reprinted from *Ms.*, May 1978, pp. 65–68. "Outstanding Women Doctors: They Make Their Mark in Medicine," *Ebony*, May 1964, p. 74. *Ebony Success Library*, vol. 1 (Chicago: Johnson Publishing, 1973), p. 43. Marianna W. Davis, ed., *Contributions of Black Women to America*, vol. 2 (Columbia, S.C.: Kenday Press, 1982), p. 385. Ronnie Miller, "Spotlight: Dorothy Brown: Strong Will Overcomes Obstacles to Career," *Nashville Banner* (July 28, 1986), pp. B3–7. Christine A. Lunardini, "Dorothy Lavinia Brown," in *Black Women in America: An Historical Encyclopedia*, ed. Darlene Clark Hine, Elsa Barkley Brown, and Rosalyn Terborg-Penn (New York: Carlson, 1993; Bloomington: Indiana University Press, 1994), 1: 175. "Bachelor Mother: Unmarried Meharry Surgeon Adopts a Child." *Ebony*, September 1958, pp. 92–96. Doris F. Innes, and Julianna Wu, eds., *Profiles in Black: Biographical Sketches of 100 Living Black Unsung Heroes* (New York: CORE Publications, 1976). "Meharry Gets Woman Chief of Surgery," *Afro American*, 25 July 1953. Claude H. Organ Jr., and Margaret M. Kosiba, eds., *A Century of Black Surgeons: The USA Experience* (Norman, Okla.: Transcript Press, 1987). "A Surgeon Goes to Legislature," *Washington Post*, 4 June 1967. Vivian Ovelton Sammons, *Blacks in Science and Medicine* (New York: Hemisphere, 1990), p. 38. Note that the Troy Orphanage was later named Vanderhyden Hall (Dunn, "Dorothy L. Brown").

3. "Bachelor Mother."

4. Brown, "Thus Would I Live"; see also Dunn, "Dorothy L. Brown."

5. Lunardini, "Dorothy Lavinia Brown."

6. Ibid.

7. Ibid.; Brown, "Thus Would I Live"; Dunn, "Dorothy L. Brown," p. 115.

8. Brown, ibid.

9. Miller, "Spotlight: Dorothy Brown." Brown considered herself lucky to get the job, which was arranged through a friend of the Redmon family.

10. Ibid.

11. Karen Tucker Anderson, "Last Hired, First Fired: Black Women Workers during World War II," *Black Woman and U.S. History* 5 (1982): 17–32. "More than 2,000 white women employed at the U. S. Rubber plant in Detroit walked off the job in March 1943, demanding separate bathroom facilities. A similar walkout occurred at the Western electric plant in Baltimore in the summer of 1943," and another strike at a Dan River plant—a traditional female-employing industry—in 1944 (ibid.). According to Anderson, this co-worker bias was, among other things, based on the belief that "Negroes couldn't pass Wasserman tests" (ibid.). Citing Emory S. Bogardus, Anderson said this attitude among white women workers could be "attributed to the infrequency of social contacts experienced by women as a result of their confinement to homemaking, to the fact that women's contacts tend to be more personal while men's are more professional, and to the customary constraints on women's interracial contacts" (ibid.).

12. Miller, "Spotlight: Dorothy Brown."

13. Brown, "Thus Would I Live"; Dunn, "Dorothy L. Brown," p. 115.

14. "Meharry Gets Woman Chief of Surgery."

15. Brown, "Thus Would I Live"; see also Miller, "Spotlight: Dorothy Brown."

16. Brown, "Thus Would I Live"; Miller, "Spotlight: Dorothy Brown."

17. Dunn, "Dorothy L. Brown," p. 115.

18. Lunardini, "Dorothy Lavinia Brown."

19. Dunn, "Dorothy L. Brown," p. 115. The hospital was run by the Seventh Day Adventist Church.

20. Dunn, "Dorothy L. Brown," p. 116.

21. Sammons, *Blacks in Science and Medicine,* p. 38.

22. Dorothy L. Brown, "History and Evolution of Abortion Laws in the United States," *Southern Medicine* 61 (August 1973): 11–14. See also *Ebony Success Library.*

23. The article appeared in the August 1973 edition of *Southern Medicine.*

24. Miller, "Spotlight: Dorothy Brown," p. B7.

25. United Methodist Communications. *Run to Live,* 16 mm, 22 min. (Nashville, Tenn.: United Methodist Film Service, 1981). Revised edition of *Run to Live,* 16 mm, 30 min. (Nashville, Tenn.: United Negro College Fund/United Methodist Board of Higher Education, 1976).

26. Dunn, "Dorothy L. Brown," p. 116. Brown also wrote a narrative essay, "My Journey through Reason in Search of an Answer," which was presented in 1982 at the Fifteenth Quadrennial Assembly of United Methodist Women, in Philadelphia, Pennsylvania (ibid.).

27. Brown, "Thus Would I Live"; see also Dunn, "Dorothy L. Brown."

28. Miller, "Spotlight: Dorothy Brown," p. B7.

29. Brown, quoted in Miller, "Spotlight: Dorothy Brown," p. B7.

MARJORIE LEE BROWNE

Teacher and Mentor

Mathematician Geraldine Darden has said that the confidence she needed to enter doctoral studies was directly attributable to the influence of mathematician Marjorie Lee Browne.[1] In 1949, Browne was one of the first two Black women to earn a doctorate in mathematics,[2] and she became a teacher, mentor, and grant writer, as well as a tireless advocate for promoting the study of higher mathematics among Black students. She proved very effective in using a series of National Science Foundation teaching grants to improve the mathematics education of Blacks at North Carolina Central University.

Browne was born in Memphis, Tennessee, in 1914, to Lawrence Johnson Lee and Mary (Taylor) Lee.[3] Her father, a railway postal clerk, had, according to Kenschaft, "attended college for two years (which was extremely unusual at the beginning of this century) and was known as a 'whiz' at mental arithmetic."[4] Her mother died before she was two, and her father later remarried; her stepmother was equally supportive of Browne's education, and taught her arithmetic and reading before she began school.[5] Browne attended the segregated public schools of Memphis, before being sent to LeMoyne High School, a private all-Black school which had an interracial faculty. During the worst days of the Depression, Browne entered Howard University. According to Kenschaft, "College funding was difficult . . . but some combination of scholarships, working, and borrowing [on the part of her father] took her through."[6] Browne graduated, cum laude, with a bachelor's degree in mathematics in 1935.[7]

Browne began her teaching career at Gilbert Academy, a Methodist secondary school for Blacks in New Orleans. According to Kenschaft, Browne immediately began to think of graduate education, but finances were a problem. On a trip home to Memphis, she spoke with a neighbor who had attended the University of Michigan. He apparently told her it was a good school and that the fees were affordable. She left the job at Gilbert Academy and enrolled in the master's program at Michigan, paying the tuition herself, again with continuing support from her father.[8]

Browne earned the M.S. in mathematics in 1939, and joined the faculty of Wiley College in Marshall, Texas. During her time at Wiley, Browne returned to the University of Michigan during the summers to work on her doctorate. In 1947, Michigan awarded her a teaching fellowship; she took a leave from Wiley, and attended Michigan full time. She was elected to the Sigma Xi honor society in 1948, and was nominated for membership in the American Mathematical Society.[9] She completed her dissertation, "On the One Parameter Subgroups in Certain Topological and Matrix Groups," in 1949.[10]

Upon completion of the doctorate, Browne secured a faculty position at North Carolina Central University, where she remained until her retirement in 1979, serving as department chair from 1951 to 1970.[11] For the twenty-five years from 1949 to 1974, she was the only person in the her department with a doctorate in mathematics.[12] As is the case in many historically Black colleges and universities, Browne's teaching load was heavy—she taught fifteen hours a week, both undergraduate and graduate courses.[13] Nevertheless, she managed to do some research and to write grant applications. In 1955, she published the paper "A Note on the Classical Groups."[14] During the academic year 1952–1953, she was awarded a Ford Foundation Fellowship to study combinatorial topology at Cambridge University in England. In the academic year 1958–1959, she was a National Science Foundation Faculty Fellow at the University of California at Los Angeles, where she studied numerical analysis and computing. In 1960, she was the principal writer on a $60,000 grant awarded by the IBM Corporation to fund the installation of the first electronic digital computer laboratory at North Carolina Central. She received a second NSF Faculty Fellowship in 1965–1966 to study differential topology at Columbia University, and in 1969, she obtained a Shell Oil Company grant that for the next ten years provided awards to outstanding students in her department.[15] In the years 1966, 1967, and 1973, Browne served on the advisory panel of the Undergraduate Scientific Equipment Program of the National Science Foundation, and from 1968 to 1969, she was a faculty consultant in mathematics for the Ford Foundation.[16]

During Browne's tenure, North Carolina Central University was awarded thirteen yearly grants by National Science Foundation to conduct summer institutes for secondary teachers of mathematics. Browne also wrote four sets of lecture notes for the summer institutes: "Sets, Logic, and Mathematical Thought" in 1957, "Introduction to Linear Algebra" in 1959, "Elementary Matrix Algebra" in 1969, and "Algebraic Structures" in 1964.[17] She served as graduate advisor for ten master's degree theses,[18] and one of the students she mentored was William T. Fletcher, who earned a doctorate in mathematics at the University of Idaho and succeeded her as department chair at North Carolina Central University.[19]

NOTES

1. Geraldine Darden, quoted in Patricia Clark Kenschaft, ed., "Excerpts from the Association for Women in Mathematics Panel on Black Women in Mathematics," Atlanta, 7 January 1978. My thanks to Dr. Kenschaft for providing me with a copy of this paper. Also see the entry for Geraldine Darden in this volume.

2. See my entry on the other woman, Evelyn Boyd Granville, in this volume.

3. Patricia Clark Kenschaft, "Marjorie Lee Browne: In Memoriam," *Association for Women in Mathematics Newsletter* 10, no. 5 (September/October 1980): 8–11. According to Kenschaft, "LeMoyne High School [was] a private A.M.A. school started after the Civil War by the Methodist and Congregational Churches to educate Negroes. . . . [Browne] credited much of her later success to her excellent preparation there" (pp. 9–10).

4. Ibid.

5. Patricia C. Kenschaft, "Blacks and Women in Mathematics," paper presented at the Science and Technology Seminar, The City University of New York, 3 March 1983, in *Educational Policy Seminar Papers* (New York: New City University of New York, Office of Special Programs, 1986), pp. 22–23. (I am indebted to Dr. Kenschaft for providing me with a copy of this paper, as well as others she has written.)

6. Ibid., p. 10.

7. Ibid., p. 8.

8. Ibid., p. 9.

9. Ibid., pp. 8–11.

10. Ibid., p. 10.

11. Ibid., p. 10.

12. Ibid., p. 10.

13. Kenschaft, "Marjorie Lee Browne: In Memoriam," p. 10.

14. Marjorie Lee Browne, "A Note on the Classical Groups," *American Mathematical Monthly*, August 1955. Cited in Patricia C. Kenschaft, "Black Women in Mathematics in the United States." *Journal of African Civilizations* 4 (April 1982): 63–83.

15. Kenschaft, "Marjorie Lee Browne: In Memoriam," p. 9.

16. Ibid., p. 11.

17. Ibid., p. 8.

18. Patricia C. Kenschaft, "Marjorie Lee Browne," in *Black Women in America: An Historical Encyclopedia*, ed. Darlene Clark Hine, Elsa Barkley Brown, and Rosalyn Terborg-Penn, vol. 1 (New York: Carlson, 1993; Bloomington: Indiana University Press, 1994), p. 186. See also Kenschaft, "Marjorie Lee Browne: In Memoriam," p. 10.

19. Kenschaft, "Marjorie Lee Browne: In Memoriam," p. 9.

CAROLYN CANNON-ALFRED*

Born in Tyler, Texas, in 1934, pharmacologist Carolyn Cannon-Alfred attended the University of Toledo, graduating with the baccalaureate degree in 1954. She earned the M.S. degree at Howard University in 1957. From 1957 to 1959 she was both an assistant pharmacologist and an instructor of pharmacology at Howard, before earning the Ph.D. from Georgetown University in 1961. Immediately after earning her doctorate, Cannon-Alfred became a senior pharmacologist at Riker Laboratories, where she remained until 1962, when she became a research associate at California Riverside Hospital—a position she held until 1964. In 1964, she received a two-year grant from the National Heart Institute, and took a position as an assistant professor of pharmacology at the School of Medicine at Southern California University.[1] She is the co-author of *Medical Handbook for the Layman.*[2]

NOTES

1. Vivian Ovelton Sammons, *Blacks in Science and Medicine* (New York: Hemisphere, 1990), p. 7. See also *American Men and Women of Science: The Physical and Biological Sciences,* 11th ed. (New York: Bowker, 1971–), supplement 1, p. 8.

2. Joseph Tyrone Alfred and C. Cannon-Alfred, *Medical Handbook for the Layman* ([Los Angeles]: n.p., [1969]).

MAY EDWARD CHINN
From Slave's Daughter to Physician

In a 1977 interview, physician May Edward Chinn recalled that her mother had been the crucial element in her attending college. Her father, "who had been a slave, opposed her even going to college," but her mother "scrubbed floors and hired out as a cook" to support her education.[1] Her father, William Lafayette Chinn, had escaped from slavery on a Virginia plantation when he was just eleven years old. Her mother, Lula Ann (Evans) Chinn, had been born on the Chickahominy Indian Reservation, near Norfolk, Virginia, and was of mixed Black and Native American descent. William Chinn was an alcoholic; when May Chinn was three years old, she and her mother moved to New York. At age five, she was sent off to board at the Bordentown Manual Training and Industrial School.

Chinn developed osteomyelitis, and had to leave school. At that time, her mother was employed on the Tiffany Estate, in Tarrytown, and she joined her there. Although the Tiffany family was white, Chinn later recalled that she "was raised as one of the Tiffany children: dining with them, studying the classics with them; attending concerts with them as would any child in the house."[2] A few years later, when the estate was sold, Chinn and her mother moved back to New York City. Chinn then attended grammar school and took piano lessons. By the time Chinn dropped out of high school, she was an accomplished enough musician to give piano lessons to younger children.[3]

Although Chinn did not complete high school, she passed a series of tests to gain admission to Columbia University Teachers College, and entered in 1917 as a music major. While in college, she sang and played in concerts around New York, and even accompanied the Black opera singer Paul Robeson. Despite her musical talent, one of her teachers at Columbia, Jean Broadhurst, persuaded her to change her major to science.[4] She graduated in 1921, and applied to medical school.[5]

Chinn later recalled that during the interview at Bellevue Medical Center, the assistant dean had asked her if she knew Paul Robeson.[6] She replied that not only did she know Robeson, but that she had accompanied him for several years. The remainder of her interview was spent talking about Robeson, and she believed her relationship with him played a part in her acceptance.[7] When she entered in 1922, she became the first Black, male or female, to attend that medical school. She earned the M.D. in 1926, and completed her internship at Harlem Hospital (again as the first Black) in 1928.[8] Chinn entered private practice, and became the physician to the Franciscan Sisters of Mary, a Black Catholic convent in New York City.

When Chinn began her practice, it was difficult for Black doctors to obtain hospital privileges—most of the hospitals in New York City were still predominantly white, as she discussed in her autobiography:

> We doctors in Harlem had many problems in common in the late 1930s and early 1940s. . . . Chief among them was that Negro doctors were denied any hospital connection whatever. There was not a City Hospital in New York City where we could attend an Out-patient Clinic or a Ward Service for [the] study and observation of the newer diseases and the effects of the newer drugs. [The] Negro doctor, including myself, was forced to practice medicine, surgery, and obstetrics [just] as the old-fashioned "family doctor" did one hundred years ago in the deep rural South—with the nearest hospital being fifty miles away; for us, [it] was so far away that we did not know "how far." Even if a hospital was around the "bend of the road" it was useless to us who were denied any privilege whatsoever of its facilities. We managed the best we could.[9]

In 1928, she joined a number of other Black doctors who practiced at the Edgecombe Sanitarium, which was established as an alternative to the then predominantly segregated New York hospital system.

Although she did not restrict her practice, Chinn generally focused on the care of females, babies, and small children. She also saw more than what she considered an "average" number of cancers in her practice. By her second year in private practice, she had determined that she would need to continue her education, with an eye toward cancer diagnosis and treatment:

> After seeing such misery in suspected cancer patients, I kept saying to myself—there must be some way of having this disease diagnosed earlier. Certainly there must be some way of slowing its process down until newer methods are found to control it or . . . perhaps to cure it.[10]

As a result, Chinn entered Columbia University's School of Public Health, where she earned an M.S. in 1933. This was the start of her long (although not always continuous) relationship with George Papanicolaou, the developer of the Pap smear test, who was developing cytological methods for the diagnosis of cancer. When Papanicolaou moved from the Columbia University School of Public Health to Cornell University Medical College, Chinn continued to work with him, especially between 1948 and 1955, when she studied exfoliative cytology.[11]

She credited Papanicolaou's support, as well as her own clinical observations of patients with advanced cancers, as having spurred her research interest in developing methods for early cancer detection. In 1957, she received a citation from the New York City Cancer Committee of the American Cancer Society.[12]

In addition to her private practice, Chinn held a number of posts over the years: examining physician for the Juvenile Aid Bureau in New York City (1937); staff member at the Strang Clinic of the Memorial Hospital in New York City (1945–1974); staff member at the New York Infirmary for Women and Children (1945, 1960–1965); and clinician with the Department of Health Day Care Centers (1960–1977).[13] In 1975, Chinn was one of the founders of the Susan Smith McKinney Steward Medical Society, which was designed to aid Black women in medical school and to document the achievements of Black women in medicine. In 1977, she became a consultant to the Phelps-Stokes Fund, an educational foundation that brings medical students from Africa and other parts of the world to universities in the United States. In 1980, she was awarded an honorary doctorate of science degree by New York University for her pioneer work in early cancer detection, and she was honored by the New York Urban League. May Edward Chinn died in 1980, at age 84.[14]

NOTES

1. Charlayne Hunter-Gault, "Black Women MDs," *New York Times*, November 16, 1977.

2. Gloria Long Anderson, "May Edward Chinn," in *Notable Black American Women*, ed. Jessie Carney Smith, vol. 1 (Detroit: Gale Research, 1992), pp. 183–185. See also Susan Shifrin, "May Edward Chinn," in *Black Women in America: An Historical Encyclopedia*, ed. Darlene Clark Hine, Elsa Barkley Brown, and Rosalyn Terborg-Penn, vol. 1 (New York: Carlson, 1993; Bloomington: Indiana University Press, 1994), pp. 235–236.

3. Anderson, "May Edward Chinn."

4. Ibid.

5. Anderson, "May Edward Chinn," p. 183. See also "Medical Profession in Harlem Gets New Addition," *New York Amsterdam News*, 16 June 1926. "Her Heart Is Strong," *New York Times*, 6 January 1980.

6. May Edward Chinn, autobiography (May Edward Chinn Collection, Schomberg Center for Research in Black Culture, New York Public Library). The reason why the dean asked her that question was not made clear, but apparently it started out as "small talk." (Note: pages in the autobiography are not numbered.) See also May Edward Chinn, interview by Ellen Craft Dammond, Black Women's Oral History Project, Schlesinger Library, Radcliffe College (the complete transcript can be accessed at the Fisk University Library, Nashville, Tenn.).

7. Ibid.

8. Shifrin, "May Edward Chinn," p. 236.

9. Chinn, autobiography.

10. Ibid.

11. Anderson, "May Edward Chinn," p. 184.

12. Ibid.

13. Chinn, autobiography.

14. Thomas W. Ennis, "Dr. May Edward Chinn, 84, Long a Harlem Physician" [obituary], *The New York Times*, 3 December 1980.

GLORIA TWINE CHISUM
Using Psychology to Protect Pilots

Born in Muskogee, Oklahoma, in 1930, experimental psychologist Gloria Twine Chisum earned her bachelor's and master's degrees at Howard University in 1951 and 1953, respectively. Apparently, she received a graduate fellowship from the University of Pennsylvania in 1958, and earned the Ph.D. in experimental psychology in 1960 with her dissertation, "Transposition as a Function of the Number of Test Trials."[1] Immediately after earning the doctorate, Chisum began her career as a research psychologist at the Naval Air Warfare Development Center (NAWDC) in Pennsylvania;[2] concurrent with that position, she continued as a psychology instructor at the University of Pennsylvania until 1968.

In 1965, Chisum was named director of the Vision Research Laboratory of the NAWDC, and remained in that position until 1980. At the Center, her research focused on methods for protecting jet pilots against loss of vision during sharp turns, and the development of protective goggles that automatically darken to shield pilots' eyes from sudden bright flashes of light from sources such as lightning or nuclear explosions. In 1979, Chisum was awarded the Raymond F. Longacre Award of the Aerospace Medical Association for her research contributions. In 1980, she was named director of the Center's Environmental Physiology Research Team.[3]

NOTES

1. Vivian Ovelton Sammons, *Blacks in Science and Medicine* (New York: Hemisphere, 1990), p. 24. *Who's Who among Black Americans*, 5th ed. (Northbrook, Ill.: Who's Who among Black Americans Publishing, 1985), p. 154. "Vison Lab Director," *Ebony*, September 1970, p. 7. See also *American Men and Women of Science: The Physical and Biological Sciences*, 13th ed. (New York: Bowker, 1971–), p. 211. *Who's Who of American Women*, 5th and 6th eds. (Chicago: Marquis Who's Who, 1968–1971). It seems apparent that she had funding for doctoral study: all sources list her as a lecturer/teaching assistant at the University of Pennsylvania beginning in 1958.

2. The U.S. Naval Air Warfare Center Aircraft Division (NAWCAD) is located at Warminster, Pennsylvania. The *Ebony* article, "Vison Lab Director," erroneously cites its location as being near Johnsville.

3. Sammons, *Blacks in Science and Medicine*, p. 24; *Who's Who among Black Americans*, p. 154.

MAMIE PHIPPS CLARK
Science as a Tool to Effect Legal and Social Change

Psychologist Mamie Phipps Clark's work on the negative self-image of Black children as a result of segregation and racism led directly to legal and social change; most notably it was cited as expert testimony in the 1954 U.S. Supreme Court decision in the case of *Brown v. Board of Education of Topeka*. Clark's own self-image was put to the test both during her graduate school days at Columbia University—where her advisor, Henry E. Garrett, made no secret of his view that Blacks and whites had different mental abilities—and later when her husband, Kenneth Clark, received the lion's share of the credit for what had originally been her work.

Born in Hot Springs, Arkansas, in 1917 to physician Harold H. and Katie F. Phipps, Mamie Clark attended segregated public schools, and graduated from Langston High School in 1934.[1] Offered a number of scholarships, she accepted one from Howard University, where she intended to major in mathematics and minor in physics. By her own account, at that time Howard University housed a number of important scholars:

> Many of the great teachers—Negroes not yet accepted into the major white educational institutions—were at Howard University: Alain Locke in philosophy; E. Franklin Frazier in sociology; Ralph Bunche in political science; Benjamin Brawley and Sterling Brown in English and literature; Francis C. Sumner in psychology; and many others who served as role models of academic excellence.[2]

By the end of her first year, however, she had become disenchanted with the idea of a mathematics major, due largely to what she termed "the detached and impersonal approach of the mathematics teachers (and I believe particularly toward female students)."[3] She had already met her future husband, Kenneth B. Clark, who was then studying for his master's degree in psychology at Howard, and he had a strong influence on her choice of career.[4] He suggested she change her major to psychology, describing it as an "extremely stimulating" field, which offered "good job opportunities" and fit in with her already well-developed interest in children.[5] Kenneth Clark steered her toward supportive faculty members, introducing her to Francis Sumner, then head of the psychology department, and also to Max Meenes, with whom she would later do research. Both men were, according to Mamie Clark, "warm, friendly and eager to have me in their courses."[6] Howard University agreed to have her scholarship allotment transferred to the psychology department. Nevertheless, Mamie Clark was acutely aware that "there were no Negro women on the psychology staff. . . . the almost total absence of Negro females with advanced degrees in psychology at Howard was in itself a 'silent' challenge."[7]

By Mamie Clark's junior year at Howard, Kenneth Clark had finished his master's degree and had moved on to pursue doctoral studies at Columbia University, where he worked as a research assistant to Gunnar Myrdal.[8] A year later, in her senior year, Kenneth and Mamie Clark were married. In 1938, Clark earned her bachelor's degree in psychology, magna cum laude, then she obtained a graduate fellowship from Howard University and immediately began study for the master's degree. During this period, her interest in developmental psychology intensified. Again, Kenneth Clark offered a suggestion: she should go to New York and meet with Ruth and Gene Horowitz (who later changed their names to Ruth and Gene Hartley), who were doing developmental studies on self-identification with preschool children. The Hartleys had already determined the need to expand their studies, and particularly to involve many more "Negro" children in the sample.[9] After meeting with the Hartleys in New York, Clark embarked upon her ground-breaking research on the self-image of Black nursery school children. Through her connection to an all-Black nursery school in Washington, D.C., Clark had access to the numbers of Black children the Hartleys required; with their encouragement, she embarked upon her master's research project. The research involved subjecting the children to two already established tests: a coloring test and one using dolls. This work culminated in her thesis, "The Development of Consciousness of Self in Negro Pre-School Children."[10] She completed her master's degree in 1939.[11]

Clark's major findings were that Black children became aware of their racial identity at about age three, and—simultaneously with their awareness of racial identity—acquired a negative self-image. Her subsequent research, a substantial portion of which was produced during the next year (1939–1940), was conducted in collaboration with her husband. According to Clark's own account, her husband was "interested and excited" by her research.[12] Together, during the next year, they published three major articles on racial identification among Black children.[13] Kenneth Clark completed his doctorate in psychology at Columbia University in 1940; his dissertation was titled "Some Factors Influencing the Remembering of Prose Material."[14]

The Clarks set about preparing a funding proposal to develop and expand their research on racial identification in Black children, based on their having developed newer versions of the standard coloring and doll tests. They submitted their proposal to the Julius Rosenwald Fund in 1939, and were awarded a Rosenwald Fellowship in 1940, with renewals in the second and third years.[15] The Rosenwald grant made it possible for Mamie Clark to enter the doctoral program at Columbia, mere months after the birth of her first child. Immediately after receiving the Rosenwald research grant, Kenneth Clark embarked on a series of extended trips to selected southern and northern states to gather research data. As he was often away on research trips, Mamie Clark was the principal care-giver for their small baby, in addition to her doctoral studies.

Clark completed her studies with her dissertation "Changes in Primary Mental Abilities with Age," and she was awarded the doctorate in 1944.[16] In 1943—the year she completed her doctoral work—Clark also gave birth to her second child. In addition to her duties as a wife and mother, she was employed as a research psychologist for the American Public Health Association (1944–1945), and a research psychologist for the United States Armed Forces Institute (1945–1946); she served a concurrent term as a psychologist at the Riverdale Children's Association.

Meanwhile, the pair continued to publish.[17] The Rosenwald Grant, which the Clarks held for three years, was awarded to them *jointly* but was an outgrowth of Mamie Clark's master's thesis research. By her own account, "My husband, Kenneth Clark, shared my interest and excitement in this research, and together we published articles."[18] Their findings established that the racial self identification of Black children—and their resulting negative self-image—was determined by the larger society's negative views toward Blacks. Their work was viewed with a high degree of interest in the academic world, and their findings were later repeated and verified by others.[19]

A Career Begun at the Start of the Modern Civil Rights Era

In the late 1930s, Mamie Clark first came in contact with Thurgood Marshall and other activists, who would eventually use the Clarks' data to support a legal challenge of the constitutionality of the 1896 Supreme Court decision *Plessy v. Ferguson.* Immediately after her graduation from Howard University in June 1938, she had obtained a summer job as a secretary in the law office of Washington, D.C. attorney William Houston, which she has described as the "veritable 'hub' of early planning for the civil rights cases which challenged the laws requiring or permitting racial segregation."[20] According to Clark, the experience remained in her mind as an "enormously instructive and revealing one in relation to my own identity as a 'Negro'":

> This opportunity to learn, not only about the plan for the eventual repeal of the
> *Plessy v. Ferguson* case, but also to observe firsthand the "giants" who were
> preparing these cases, made a deep impression on me. Frequently there would
> be much excitement in the office when the Washington-based lawyers would be
> joined by William Hastie, Thurgood Marshall, and others from the original
> group of legal activists.[21]

Later, in 1953, Clark's husband organized a group of social scientists to prepare a social science brief summarizing the major research findings on the effects of racial segregation, which they submitted to the National Association for the Advancement of Colored People (NAACP) lawyers as a supplement to their legal brief.[22] This social science supplement was alluded to in the 1954 U.S. Supreme Court decision in *Brown v. The Board of Education of Topeka:*

> To separate [Negro children] from others of similar age and qualifications solely
> because of their race generates a feeling of inferiority as to their status in the
> community that may affect their hearts and minds in a way unlikely ever to be
> undone.[23]

For his contribution to the civil rights effort, Kenneth Clark was awarded NAACP's Spingarn Medal in 1961.[24]

Although Mamie Clark was the one who had initiated the research on the self-image of Black children, Kenneth Clark is most often cited as the source of the ground-breaking work that was later cited in the Supreme Court decision.[25] For example, the biographical essay on Kenneth Clark in the *Ebony Success Library* mentions only that Kenneth Clark's wife was "also a psychologist."[26] The essay goes on to specify that "nationally prominent in psychological research, [Kenneth Clark] contributed data cited by the United States Supreme Court in its landmark ruling in 1954, outlawing segregated schools."[27] And, although biographical essays on both Mamie and Kenneth Clark are provided in Guthrie's book *Even the Rat Was White,* the entry for Kenneth Clark makes no mention of Mamie Clark's research contribution to the Supreme Court's decision: "The

U.S. Supreme Court cited Dr. Clark's work on the harmful effects of segregation in its 1954 decision, *Brown versus Board of Education.*"[28]

The seeming neglect of Mamie Clark's contribution may be due to sexism, or to a combination of events. The way in which the Clarks' joint research was presented—and by whom—may be responsible. Although both of the Clarks provided testimony in the Virginia trial, Kenneth Clark alone made the presentations in the majority of court appearances, as in the South Carolina and Delaware school desegregation trials.[29]

Meaningful Clinical Work in the Face of Entrenched Prejudice

Despite Mamie Clark's ground-breaking research in psychology during the 1940s, she experienced difficulty in securing a position commensurate with her training. During her doctoral studies at Columbia University, Henry E. Garrett was her sponsoring professor and guided her through her dissertation research on the development of primary mental abilities with age. Garrett, whom Clark later described as "not by any means a liberal on racial matters" (an understatement), suggested in no uncertain terms that her advanced training in psychology was nothing more than preparation for a career teaching Black high school students in the South.[30] Furthermore, Garrett firmly believed that Blacks and whites had different mental abilities, a view later shown by his testimony in the Virginia school desegregation case.[31] Given his bias, it is all the more remarkable that Clark succeeded as well as she did under his direction.

Nor was Garrett alone in his ideas concerning Clark's lack of career potential. After Clark received her doctorate from Columbia in 1944, she wrote that "it soon became apparent that a black female with a Ph.D. in psychology was an unwanted anomaly in New York City."[32] When she applied for a research position at a major broadcasting company, she was "heartbroken" to learn "that a number of white men and women with far less qualifications were hired at relatively high salaries," whereas she was rejected without explanation.[33] Once again, her husband provided a useful contact: A colleague of his at the City College of New York helped her obtain a position at the American Public Health Association, where from 1944 to 1945 she analyzed research data gathered on American nurses. The only other person in the office with a doctorate was the director, and Clark was the only Black person. She found it a "humiliating and distasteful first employment experience."[34]

Because of the wartime research being conducted by the armed forces, she was able to obtain her next position, as a research psychologist, at the United States Armed Forces Institute from 1945 to 1946. Although this position was better than her former job at the American Public Health Association, Clark considered herself to be in a "holding pattern."[35]

Then a "career milestone" occurred: Clark secured a position as a psychologist at the Riverdale Home for Children, a private agency for homeless Black girls. Not only did the position offer an opportunity for professional growth, but according to Clark it also "afforded new insights into the enormous lack of psychological services for black and minority children in New York City."[36] There were almost no resources for referring these girls for the plethora of services, particularly mental health services, that they needed. Again, Mamie Clark's experience would provide the basis for a turning point in both her own and her husband's careers.

In the 1940s, psychological and psychiatric services were generally available to white New York City children through a network of social services agencies, but Clark observed that the "minority children of Harlem generally did not have access to these programs at that time or, at the least, were not encouraged to apply."[37] Seeing the great need for such services among minority children, she and her husband tried to enlist the support of a number of social service agencies and religious organizations, and encourage them to extend their scope to Harlem; the Clarks' proposals were universally rejected on the grounds that such services were not needed by Black children. Thus, in 1946 the Clarks established their own service, the Northside Testing and Consultation Center (later the Northside Center for Child Development). Mamie Clark served as the executive director, and her family provided the money to cover the rent and furnish the basement space for the new center. The center, which was staffed by friends and by volunteer psychiatrists, psychologists, and social workers, was the first full-time child guidance center offering psychiatric, psychological, and social services to children and their families in the Harlem area.[38]

Initially, the local community did not look with favor on the Clarks' efforts:

> In the mid 1940s, the climate of acceptance of any psychiatric of psychological program in the Harlem community was very poor. Many families felt that such services would place a stigma on their children; they did not want to have their children called "crazy." This attitude reflected the prevailing tendency of the larger society to reject the concept of emotional disturbance. [However,] about this time the movie *The Snake Pit* began to have enormous impact on attitudes toward the mentally ill throughout the United States.[39]

Also crucial, according to Clark, was "the Roosevelt Administration's awareness of needs of people on all levels, and the federal advocacy of social programs resulting in attention to health and mental health needs of families."[40]

Despite their traditional antipathy toward psychology and psychiatry, a few frustrated Black parents turned to the Northside Center for assistance in dealing with the public schools—their children were being incorrectly placed in classes for the mentally retarded, often without their parents' permission.[41] Clark observed that such placements were "rampant in schools located in minority and deprived areas," and were probably used as a means of ensuring de facto segregation.[42] The Center provided outside evaluations, and found that most of the children who had been labeled "retarded" by the city schools were in fact above the intelligence level for placement in those classes. Soon other parents seized upon the opportunity to bring in their children, and were advised that the former designations were incorrect—and that using them to "track" minority children into special classes was illegal. By the end of its first year, the Northside Center had widened its scope, and taken on an active advocacy role on behalf of its clients.

Clark's findings "highlighted the high degree of educational retardation, which was either not being recognized or corrected in the public schools." To counter this trend, she and the staff added a remedial reading and arithmetic program to supplement the Center's psychiatric and psychological services. At that time, 1946 to 1947, an educational component in a child guidance center was an innovation. And, as Clark later said, "many times educational remediation was the treatment of choice."[43]

Clark was executive director at Northside from 1946 to 1979, and often served as chief psychologist. From 1946 to 1966, Kenneth Clark served as the research director. In addition to her work at the Center, Clark served on the board of directors of several companies and institutions: Teachers College at Columbia University, the American Broadcasting Company, Mount Sinai Medical Center, Union Dime Savings Bank, the Museum of Modern Art, the Phelps Stokes Fund, the New York City Mission Society, and the New York Public Library. She also participated in several advisory groups, including Harlem Youth Opportunities Unlimited (HARYOU), which was organized in the 1960s by Kenneth Clark, and the National Head Start Program Planning Committee.[44] Kenneth Clark was for many years a professor of psychology at City College of the City University of New York; at various times during his career he was also a visiting professor at Columbia University, the University of California at Berkeley, and Harvard University; and he was a member of the New York State Board of Regents, the Board of Trustees of the University of Chicago, and the New York Urban Development Corporation. In 1967, he founded the Metropolitan Applied Research Center, which was to serve as a catalyst for change and an advocate for America's urban poor.[45] He also served as president of the American Psychological Association from 1970 to 1971, the first Black to hold that office.[46]

Mamie Clark saw the psychologist's role in society as one of research, prevention, and advocacy for those in need, particularly children. As of 1983, she felt that her profession had yet to make as great a contribution to American society as it could have. In particular, she decried the predominance of clinical psychologists (42% of APA members) as an "attempt to be competitive with psychiatrists, both professionally and monetarily":

> This flight into the field of clinical psychology has been at the expense of urgently needed research into effective treatment of emotional disturbance in children and, most importantly, the *prevention* of emotional disturbance in children. Unless this trend away from theoretical and research psychology is slowed, I believe that children, and most particularly minority children, will suffer. Indeed, the flow of federal and private foundation funds for research into issues of child mental health will decrease in exact relationship to the lessening of an advocacy role for these children. We will hope that a steady flow of younger psychologists—male and female—into the field will bring more responsiveness to the need for psychological research in general, and [of] research oriented toward prevention of emotional illness in particular.[47]

Clark recalled that when she was attending Howard University in the 1930s, in what she described as "the security of an almost completely segregated student body," she never considered how a Black woman would fare in the nearly all-white and male field of psychology, or how a female psychologist could satisfy her interest in working with children in a society that offered services mainly to white children.[48] As her career illustrates, she not only fared well, but she made an important contribution to science and to society.

NOTES

1. Mamie Phipps Clark, personal account quoted in *Models of Achievement: Reflections of Eminent Women in Psychology*, ed. Agnes N. O'Connell and Nancy Felipe Russo (New York: Columbia University Press, 1983), pp. 266–278. See also Robert V. Guthrie, *Even the Rat Was White: A Historical View of Psychology* (New York: Harper & Row, 1976), pp. 166–167.

2. Clark, personal account, pp. 226–278.

3. Ibid., pp. 266–267.

4. Kenneth Clark was born in 1914, in Panama, the Canal Zone. After becoming a naturalized citizen in 1931, he graduated from Howard University with a bachelor's degree in psychology in 1933. He was awarded a master's degree in 1936, and earned the Ph.D. at Columbia University in 1940. See the *Ebony Success Library*, vol. 1 (Chicago: Johnson Publishing, 1973), p. 67. See also Guthrie, *Even the Rat Was White*, pp. 150–151.

5. Clark, personal account, pp. 226–278.

6. Ibid.

7. Ibid.

8. *Ebony Success Library*, p. 167. Gunnar Myrdal was working on a comprehensive study of American social problems; he published this work as *An American Dilemma: The Negro Problem and Modern Democracy* (New York: Harper & Brothers, [1944]).

9. Clark, personal account, pp. 226–278. See R. E. Horowitz, "Racial Aspects of Self-Identification in Nursery School Children," *Journal of Psychology* 7 (1939): 91–99. The word "negro" was the proper term at the time; I have placed it in quotation marks.

10. Clark, personal account, pp. 268–269. See Mamie Phipps Clark, "The Development of Consciousness of Self in Negro Pre-School Children," *Archives of Psychology* (Washington, D.C.: Howard University, 1939).

11. Guthrie, *Even the Rat Was White*, p. 166.

12. Clark, personal account, pp. 269–270.

13. Mamie Phipps Clark and Kenneth B. Clark, "The Development of Consciousness of Self and the Emergence of Racial Identification in Negro Preschool Children," *Journal of Social Psychology* 10 (1939): 591–599; Clark and Clark, "Segregation as a Factor in the Racial Identification of Negro Pre-School Children," *Journal of Experimental Education* 8 (1939): 161–165; Clark and Clark, "Skin Color as a Factor in Racial Identification of Negro Preschool Children," *Journal of Social Psychology* 11 (1940): 159–169.

14. Vivian Overton Sammons, *Blacks in Science and Medicine* (New York: Hemisphere, 1990), p. 54.

15. Clark, personal account, pp. 270–271.

16. Guthrie, *Even the Rat Was White*, p. 166. See also Stella Chess, "Very Gifted and Black: Mamie Phipps Clark," in *The Women of Psychology*, ed. Gwendolyn Stevens and Sheldon Gardner, vol. 2 (Cambridge, Mass.: Schenkman, 1982), p. 191.

17. See the appendix.

18. Clark, personal account.

19. Ibid. Also see Mary Ellen Goodman, "Evidence Concerning the Genesis of Interracial Attitudes," *American Anthropologist* 48 (1946): 624–630; Goodman, *Race Aware-*

ness in Young Children (Cambridge, Mass.: Addison-Wesley, 1952). On the scholarly attention the work received see Chess, "Very Gifted and Black," p. 191.

20. Clark, personal account.

21. Ibid.

22. See Clark, personal account, pp. 166–184.

23. Clark, personal account.

24. Guthrie, *Even the Rat Was White*, pp. 150–151.

25. See, for example, the section on Kenneth Clark in *Ebony Success Library*. It is interesting to note that there is no comparable entry on Mamie P. Clark—in fact, there is no entry at all. See also the section on Kenneth Clark in Guthrie's *Even the Rat Was White*, which, again, seems to indicate that Kenneth Clark earned the lion's share of the recognition.

26. *Ebony Success Library*, p. 67.

27. Ibid.

28. Guthrie, *Even the Rat Was White*, pp. 150–151.

29. Clark, personal account.

30. Ibid.

31. According to Clark, "Some years later my husband and I were to meet Henry Garrett in a federal courtroom where the school desegregation case regarding Prince Edward County in Virginia was being tried. We were testifying on opposite sides of the issue—Henry Garrett opposed the desegregation of public schools. He testified to the effect that black and white children had different talents and abilities which presumably justified segregated schools" (ibid.).

32. Ibid.

33. Ibid.

34. Ibid.

35. Ibid.

36. Ibid.

37. Ibid.

38. Ibid.

39. Ibid. *The Snake Pit* (Twentieth Century Fox, 1948) was a film adaptation of the best-selling novel by Mary Jane Ward (New York: Random House, 1946).

40. Ibid.

41. Ibid. The New York City Public Schools called these "classes for children of retarded mental development."

42. Ibid.

43. Ibid.

44. Chess, "Very Gifted and Black," p. 191.

45. Guthrie, *Even the Rat Was White*, pp. 150–151.

46. Ibid.

47. Clark, personal account.

48. Ibid.

YVONNE CLARK

Engineer/space scientist Yvonne Clark was the first female to earn a degree in mechanical engineering from Howard University, as well as the first Black woman engineer in Nashville, Tennessee.[1] Clark was born in Houston, Texas, in

1925, and was raised in Louisville, Kentucky, the daughter of a physician/surgeon father and librarian/journalist mother. In a 1996 interview, she recalled,

> I have always used my hands to build things. When I was a child you could put something in front of me and say, "Put this together," and I loved it. I had Erector sets and things like that when I was a kid. And I fixed things around the house.[2]

In high school, Clark took an aeronautics class as part of her science requirements:

> In aeronautics class, we'd build planes, fly them off the fire escape at school, and let them crash.
>   I don't remember whether I was the only girl in those classes; I don't think so. But I do remember I couldn't take mechanical drawing because I was a girl.[3]

She also joined the high school's Civil Air Patrol, where she learned the rudiments of flying in a Link trainer, a mechanism that simulated actual flight conditions. She graduated from high school in 1945 at age sixteen, but because her parents thought she was too young for college, Clark spent the next two years studying in Boston, where she lived in the home of family friends.

Because race restrictions barred her from attending the University of Louisville, the State of Kentucky paid for her tuition at Howard University, where she was the only female in mechanical engineering. As Clark recalled, she pursued studies in mechanical engineering, "a curriculum which was not especially encouraging to Black women, and [invaded] a section of academia where it was necessary to overcome the obstacle of being female."[4] Clark has described mechanical engineering as a (then) stigmatized field for women:

> They did have female graduates in electrical engineering before I got [to Howard]. Electrical engineering and chemical engineering didn't have the stigma that mechanical engineering did—that you're out there getting dirty.[5]

Her size and physique were often cited in an effort to discourage her from completing her area of study: "As a senior [at Howard] I interviewed with companies that came to campus, one interviewer said . . . 'Well, we can't hire you.' I asked why not and he said I didn't have the muscles to break down the machinery."[6]

After Clark completed her B.S. degree at Howard in 1951, she continued to seek employment in her field, but discovered that "the engineering job market wasn't very receptive to women, particularly women of color."[7] Three months after her graduation, Clark finally landed a job at the Frankford Arsenal-Gage Laboratories in Philadelphia, then quickly moved on to a job at the RCA Corporation's Tube Division in Harrison, New Jersey.

In 1955, Clark left RCA, married William F. Clark Jr., then a biochemistry instructor at Meharry Medical College, and moved to Nashville, Tennessee. Clark said she turned to teaching at that point in her career because there were no industry jobs available to her in mid-1950s Nashville.[8] She did manage to secure a job as an instructor in College of Engineering and Technology at Tennessee State University (TSU), where she moved up to the level of associate professor, and served as department head in mechanical engineering (1965–1970,

1977–1988). In 1970, she took a two-year leave of absence to study for her master's degree in engineering management at Vanderbilt University; she completed the degree in 1972.

During the summers, Clark worked in industry and government, always on the lookout for new techniques and information to bring back to the classroom. She spent several summers at the Frankford Arsenal, working on recoilless weapons; another summer with the National Aeronautics and Space Administration (NASA) division in Huntsville, Alabama, where she investigated Saturn 5 engines for hot spots; a summer at the NASA Manned Spacecraft Center in Houston, where she worked on containers for returning moon samples to earth; and a summer at Westinghouse's Defense and Space Center in Baltimore, Maryland, where she worked in an interdisciplinary team seeking methods for revitalizing and modernizing a section of the inner city.[9]

Clark's research currently focuses on refrigerants. She is the principal investigator for a research project, "Experimental Evaluation of the Performance of Alternative Refrigerants in Heat Pump Cycles," funded by a grant from the Department of Energy's Oak Ridge National Laboratory in Tennessee.[10] She is also the student division team leader for the Center for Automated Space Science, a NASA-funded project at TSU. Clark is now in her fourth decade of teaching; since 1988, when AT&T donated a lab to the TSU mechanical engineering department, she has also taught computer graphics courses.[11]

During her tenure at TSU, Clark has seen the female representation in the university's engineering fields move from two individuals—herself and a secretary—to a ratio of one female out of every four students.[12] She has learned how to pick and choose her battles, and offers the following advice:

> The important thing to remember is that you have to make your own openings. You also have to discern which problems are yours and which problems belong to the other person. Don't wear another person's problems.[13]

Clark has been recognized by Howard University for outstanding achievement in the engineering profession, and for leadership and distinguished service by the Society of Women Engineers.[14]

NOTES

1. "Engineers and Space Technologists," in *Contributions of Black Women to America*, ed. Marianna W. Davis, vol. 2. (Columbia, S.C.: Kenday Press, 1982), pp. 459–460.

2. "Yvonne Y. Clark," in Susan A. Ambrose, Kristin L. Dunkle, Barbara B. Lazarus, Indira Nair, and Deborah A. Harkus, *Journeys of Women in Science and Engineering: No Universal Constants* (Philadelphia: Temple University Press, 1997), p. 63.

3. Ibid.

4. "Engineers and Space Technologists," pp. 459–460. See also Ambrose, *Journeys*.

5. Ibid., pp. 63–64.

6. Ibid.

7. "Opening Doors for Women and Minorities in Engineering: TSU's Clark Earns National Engineering Award," Tennessee State University Office of Public Relations News Releases [web page]: http://www.tnstate.edu/opr/yyclark.htm [accessed 15 December 1998].

8. Ibid. William F. Clark Jr. died in 1994.
9. Ambrose, *Journeys*, p. 65.
10. "Opening Doors."
11. Ambrose, *Journeys*, p. 67.
12. "Opening Doors."
13. Ibid.
14. "Engineers and Space Technologists"; the latter was awarded in 1974.

JEWEL PLUMMER COBB

World-Renowned Scientist and Path-Breaker

Born in Chicago, Illinois, in 1924, cell biologist/physiologist Jewel Plummer Cobb was, like her colleague and sometime collaborator, Jane Cooke Wright, a product of America's "Black elite." Her father, Frank V. Plummer, was a physician and her mother, Carribel (Cole) Plummer, was a physical education teacher. Her mother's sister was also a physical education teacher, and her paternal grandfather was a graduate pharmacist.[1] In the Plummer home, there was always great emphasis on the concerns and accomplishments of Black Americans. Frank Plummer, who according to Cobb, "specialized in dermatology to the degree that Black doctors could in the 1920s," was on the staff at Provident Hospital, which was affiliated with the University of Chicago and had been established just before the turn of the century:

> [Provident Hospital handled] Black patients . . . the hospital where Black
> medical students and interns from the University of Chicago Medical School
> and other medical schools could be sent to care for and observe patients. It was
> unthinkable then that a Black medic would touch a white patient, especially a
> Black male medic.[2]

According to Dona Irvin, "from her earliest memory [Cobb] heard discussions of racial matters—the hopes and frustrations of her family and their associates. She became familiar with the aspirations, successes, and talents of black people."[3] Cobb recalled that the lack of recognition given to the accomplishments of Black Americans like Daniel Hale Williams, who had performed the first open-heart surgery in 1893, was a prominent topic of conversation.[4] Cobb's mother was a friend of historian Carter G. Woodson and writer Arna Bontemps. Allison Davis, the famous Black anthropologist, and Alpha White, the director of the YWCA, lived in their apartment building; other important black American artists and professionals lived in the vicinity, although the neighborhood was predominantly white.[5] According to Irvin, Cobb also had the advantage of her father's home library, which contained "a comprehensive collection of material about black Americans, scientific journals and magazines, and periodicals of current events."[6]

Cobb began her schooling at the predominantly white Sexton Elementary School. However, in 1929–1930, the Chicago Board of Education gerrymandered the school districts so that fewer Black children were eligible to attend Sexton; Cobb was transferred to the "overcrowded, old and dilapidated Betsy Ross

Elementary School."[7] During this time, Martin Jenkins, then a graduate student at Northwestern University, began his research project "A Study of 100 Gifted Negro Children"—and Cobb was one of those designated as gifted and talented. As Cobb recalls,

> The research format involved selecting 100 Black gifted children in grammar school and giving us a battery of standard intelligence tests for several days. The tests were followed by an in-depth study of the social, intellectual, personal and family environment of each child tested. These same 100 children, from all over Chicago and in different grammar schools, were again tested four to six years later when attending high school.[8]

By the time Cobb had entered Englewood High School, the Chicago Board of Education had developed a new gerrymandering scheme "for purely racial reasons," to place the high schools on double shifts: "So I began Englewood High under a double shift scheme, starting at 11:45 A.M. and ending at 5:15 P.M, often walking home in the dark in the winter months."[9]

Englewood High had a special honors track which, among other aspects, allowed selected students take five years of science, which Cobb did:

> It was in Ms. Hyman's sophomore year biology class that I was given a microscope to view an entirely new world beyond my normal viewing capacity. . . . Having been inspired by that sophomore year biology class, I then signed up for a second year of biology. . . . Again I remember my teacher, Ms. Mardoff, and the classes in botany. We learned how to identify a tree's genus and [its] common name using only a winter twig without its leaves. Following my two years of biology, I decided then and there that I would be a biology teacher. . . . I also had lectures and labs in physics, algebra, geometry and chemistry, but biology was my favorite subject. . . . As I became a senior with advanced honors classes in English and five years of science, I naturally decided to major in biology in college.[10]

By this time, she had also amassed an insect collection "mounted on Bohemian steel needles in mothproof boxes,"[11] and been further inspired by reading Paul DeKruif's *The Microbe Hunters*.[12] Both her schoolwork and her interest in science were strongly encouraged by her parents, as well as by family friends. There was never any question but that she would attend college.

The choice of which college Cobb would attend was made for social as well as academic reasons. The Plummer family had always spent the summer months at Idlewild, a resort in northern Michigan where wealthy Blacks maintained homes around a lake: "Year after year, [Cobb] met friends there, peers who were sons and daughters of the privileged class of black Americans. These friends were typical of the young people who were her friends in public school and in college."[13] Cobb recalled the importance of those summers in her choosing the University of Michigan:

> This decision was based on the long relationships I'd had with my teen-age friends . . . while spending summers in Idlewild, a community of vacation homes owned by Black folks from the Middle West, including Ohio, Pennsylvania, Indiana, and Illinois.

The choice of Michigan socially . . . was also a good one. . . . at that time over 200 Black students (over half of whom were graduate students) attended the university. Many were from the South where they could not, as Black students, be admitted to their state university or study for a professional degree. So those southern states paid the tuition and other fees for these students to study in the North.[14]

Cobb entered the University of Michigan in the fall of 1941, and it was an excellent academic choice, particularly in terms of the science courses she took there. However, in terms of how the Black students were treated, it was not what she and her parents had expected—in fact, it was a tremendous disappointment:

The choice of the University of Michigan . . . was a disaster for Black students . . . in terms of dormitory living arrangements. [When I was] a Black incoming freshman student, the Dean of Women at the University of Michigan wrote my parents in January 1941 saying that as a mid-year freshman, "your daughter has been assigned to Benjamin House at 1102 East Ann Street, Ann Arbor." This assignment to a League House (a Black "official" residence) was given only to Black students because we were not allowed to live in the dormitories! Also the popular grills and the famous Pretzel Bell Tavern did not welcome Black students. And so I was never allowed in the mainstream of social life on campus. No Black students belonged to the big fraternities or sororities. This situation prompted my parents and me to consider leaving Michigan.[15]

In addition to feeling excluded from the mainstream of campus social life, Cobb also felt somewhat isolated academically, as none of her Black friends were science majors.

As luck would have it, Hilda Davis, then Dean of Women at Talladega College, a historically Black college in Alabama, was doing graduate study at the University of Chicago during the summer of 1942. Apparently Davis was a Plummer family friend, and convinced Cobb and her parents that she should transfer to Talladega, which had a strong science program.[16] Talladega did not accept transfer credits, but it did allow selected students to take qualifying examinations to satisfy requirements without actually enrolling in a course. Cobb entered an accelerated program in which she took a certain number of these examinations, attended summer courses, and took private courses in addition to her regular semester work.[17]

Although there were no lab assistants at Talladega, and students prepared their own media and plates, Cobb thrived there: "My friend and mentor . . . was my bacteriology professor, known affectionately as 'Captain Jack,' [who] suggested that I apply to New York University's Graduate School of Arts and Sciences in biology."[18] Because of World War II, the student body at Talladega was predominantly female during Cobb's time there.[19] Despite not having transferred in any credits, she completed her course work in two years, and graduated in 1944 with a major in biology. She was accepted at New York University, and entered in the fall of 1944. Although she was initially turned down for a teaching fellowship, Cobb requested a review of the decision soon after her arrival; the faculty reconsidered and she was later awarded funding, which she held for the next five years.[20]

Cobb's original plan had been to earn a master's degree and become a high school biology teacher, but she abandoned the idea when, as a substitute teacher, she "discovered that I didn't like the discipline problems I encountered in the classroom at New York City's Julia Richmond Girls High School."[21] As in the case of so many of the women, a mentor's support proved crucial:

> Gradually, I became involved in my graduate courses and after being offered a teaching fellowship in my own department by the chair, Harry Charipper, I decided I would continue past the master's and go on to a Ph.D. My second mentor was M. J. Kopac, professor of biochemistry and involved in micrurgy (microsurgery) research. He was my advisor and supporter throughout graduate school. I never experienced a time when he was not willing to talk about science and share many ideas.[22]

Cobb worked toward the master's degree in cell physiology for three years (1944–1947). Her thesis work, an "original laboratory project using the intricate Warburg respirometer apparatus," dealt with a series of organic molecules, aromatic amidines, and their effect on the respiration of yeast cells.[23] She completed the master of science degree in 1947; her thesis was "The Effect of Several Aromatic Amidines on the Respiration and Aerobic Metabolism of Yeast Cells."

In 1949, Cobb spent the summer as an independent investigator at the Marine Biological Laboratory at Woods Hole, Massachusetts, and the following year (1950) she earned her Ph.D. in cell physiology.[24] Her dissertation, "Mechanisms of Pigment Formation," was done under the supervision of biochemist M. J. Kopak.[25] In it, she examined the way melanin pigment granules could be formed in vitro using the enzyme tyrosinase, which is needed for melanin pigment synthesis. Various substrates of the enzyme were also tested as model systems for pigment formations.[26] According to Cobb,

> My advisor and I worked together for months so that we both were satisfied that I had a "perfect" dissertation, capable of passing muster by the severest critic. Then finally, it was scrutinized by several readers and I came before the committee for my oral defense. I was, of course, extremely nervous. Despite that, I was comforted deep down inside with the awareness that I knew much more about my subject, "Mechanisms of Pigment Granule Formation," than anyone else in the world.[27]

In July of 1950, Cobb began a two-year postdoctoral fellowship to study various factors influencing normal and abnormal pigment cell growth. Her studies, sponsored by the National Cancer Institute of the National Institute of Health, were based at the Harlem Hospital Cancer Research Foundation, headed by chief of surgery Louis Tompkins Wright.[28] With this work she embarked on the type of research that would become her specialty: the factors influencing the growth, morphology, and genetic expression of normal and neoplastic pigment cells, the in vitro growth of mammalian neoplastic cells, and the changes produced in vitro by cancer chemotherapeutic agents, hormones, and other agents known to disrupt cell division.[29]

Cobb designed new experiments for growing human tumor tissue in vitro, using specimens acquired from autopsies and from patients undergoing surgery.

These experiments were designed to complement the in vivo studies being conducted by Jane Cooke Wright:[30]

> As part of the cancer research team at Harlem Hospital, I undertook tissue
> culture (in vitro) studies of the tumors of cancer patients. Clinical studies
> in human patients treated (in vivo) with (a) the newly synthesized agent,
> triethylene melamine (TEM), as a radiomimetic compound, or (b) aureomycin,
> an antibiotic, or (c) some 4-amino derivatives of folic acid (folic acid is part of a
> molecule needed in cellular growth) were also done.[31]

In addition to working independently and with the Wrights, Cobb also worked in conjunction with another young investigator, Grace Antikajian, to find the modes of action of the various compounds on cancer cells in vitro and to discover which compounds had the most damaging effects on which types of cells:

> We described, in 1951, certain cytotoxic changes. TEM [triethylene melamine]
> in vitro at lower doses prevented human tumor cells from migrating along the
> flask surfaces. Amethopterin at significantly low doses caused cell nuclei to
> enlarge while aureomycin at similar concentrations had no effect.[32]

She participated in further studies that compared the in vivo effects of other chemotherapeutic agents with the in vitro effects of the same tissue obtained from a patient.[33] According to Cobb,

> A preliminary study of 18 cases revealed a meaningful relationship between the
> effect of a chemotherapeutic agent used clinically in the same patient with
> advanced neoplastic disease. I then continued extensive research identifying a
> spectrum of in vitro cellular changes in tissue culture from a variety of sources.
> They included human tumors of various types, and also fetal mouse and chick
> embryo tissue.[34]

Between 1950 and 1952, Cobb also spent time in Margaret Murray's tissue culture laboratory at Columbia University's College of Physicians and Surgeons, where she learned new techniques for growing and analyzing nerve cells in vitro.[35]

In 1952, the National Cancer Institute (of the National Institutes of Health) awarded Cobb her first grant as an independent investigator.[36] She moved to the University of Illinois Medical School, where she was named an instructor in the anatomy department. More importantly, she established the first tissue culture research laboratory at the medical school (an accomplishment she would repeat at other institutions throughout her career).[37] At Illinois, Cobb developed a new research program that was an extension of her graduate research: cytological studies of normal and malignant pigment cells.[38] During her time at Illinois she also published data on human bladder cancer cell growth in tissue culture.[39] Except for her two years at Harlem Hospital (1950–1952), a major portion of Cobb's research has focused on pigment cell research—the role of melanin in skin cancers, as well as its potential for shielding human skin from cancer-causing ultraviolet rays.

Cobb was married to Roy Raul Cobb in 1954, and, with another grant from the National Cancer Institute, she returned to the Harlem Hospital Cancer Research Foundation, directed by Jane Cooke Wright.[40] The following year, 1955, the entire laboratory of the Cancer Research Foundation joined the Fifth Division Surgery of New York University's Post-Graduate Medical School; Cobb was appointed an instructor of research surgery. She was promoted to the level of assistant professor the following year.

At New York University–Bellevue Hospital Medical Center, Cobb designed and established a new tissue culture research laboratory. Here, she "entered a most exciting phase of basic cell research in close coordination with [Wright's] clinical cancer chemotherapy program."[41] She began working with animal models — specifically, with a type of JAX mice bred for a tendency to express Cloudman-S91 tumors.[42] The research on mouse melanomas was designed to discover whether

> the resistance to radium and x-ray therapy observed in patients with melanoma was a function of pigment density. Since human experimentation was not practicable, I attempted to elucidate this resistance phenomenon using tissue slices of densely pigmented and very pale areas of the same mouse tumor specimen. I exposed them to varying doses of x-rays and then implanted them immediately in untreated host mice or *grew* them in vitro. Pale melanoma slices versus pigmented slices exposed to equal x-ray dosages displayed a greater sensitivity to radiation. Pigmented tissue slices survived in vitro radiation at doses that prevented growth of pale tissue when implanted to mice. Microscopic analyses determined that there were quantitatively more pigmented cells per area in the pigmented tissue than in pale tissue slices while the number of mitoses was the same. Melanin protected the cells from x-ray damage.[43]

During her tenure at New York University (1955–1960), Cobb was continually funded by the National Cancer Institute for in vitro work on the production of melanin using tissue cultures of amelanotic and melanotic tumors.

Cobb also produced significant work on the exposure of living cells in vitro to the antibiotic actinomycin D; the resulting paper contained the first published data on the ability of actinomycin D to cause a reduction of nucleoli (a complex nucleotide) in the nucleus of human normal and malignant cells. Her research was reinforced by others, whose later findings explained the changes taking place at the molecular level.[44]

Along with her New York University research assistant and colleague Dorothy Walker Jones,[45] Cobb continued studies on breast and other cancerous tumors treated in vivo and in vitro with nitrogen mustard derivatives,[46] the antibiotic puromycin derivative, and the folic acid antagonist A-methopterin (methotrexate), as well as actinomycin D.[47] By her own account, her most interesting work during this period was an in-depth analysis of Thio-Tepa, a nitrogen mustard derivative; she and Walker conducted a cinematographic analysis of cell division movement behavior, photographed through a special phase contrast microscope. The paper resulting from these experiments was presented in 1962, at the Eighth International Cancer Congress in Moscow.[48]

In 1960, Cobb left New York University to assume a full professorship in the biology department at Sarah Lawrence, a women's college in Bronxville, New York. She continued her tissue culture studies, and once again established and became director of a tissue culture laboratory. Again, with funding from the National Can-

cer Institute, she studied hormonal factors influencing the growth and melaniza-
tion in vitro and in vivo of the Cloudman-S91 mouse melanoma. She worked to
determine the optimal nutritive requirements of Cloudman-S91 melanoma in
organ culture, and conducted studies of induced tolerance of melanoma in a
foreign mouse strain.[49] With additional funding from the Undergraduate Research Participation Program of the National Science Foundation, Cobb's laboratory at Sarah Lawrence was closely tied to helping students do research with
mouse melanomas. Two of her former students from this program now have
Ph.D.s and are, according to Cobb, at the forefront of research in their respective
fields. During this period, Cobb's work in the lab also involved "perfecting new
techniques for organ cultures in sealed chambers placed in the peritoneal cavity
of mice or in mini-dishes incubated at 37 degrees Celsius."[50]

Between February and September of 1967, Cobb took a seven-month leave
from Sarah Lawrence to accept a research fellowship from the National Institutes
of Health to study cancer polyoma viruses at the Laboratorio Internazionale di
Genetica e Biofisica in Naples, Italy.[51]

Cobb left Sarah Lawrence in 1969 to become dean of Connecticut College,
and a professor in the zoology department. At Connecticut College, Cobb built
a new laboratory. With the help of assistants, and with grants from the American
Cancer Society and the National Cancer Institute, she studied hormone action on
human and mouse melanoma in tissue culture (1971–1973). She again developed melanotic and amelanotic strains of the Cloudman-S91 mouse melanoma
in cell culture, and published a series of papers detailing the changes in the morphology and behavior of malignant pigment cells when they are exposed to the
female hormone 17B estradiol, and the male hormone testosterone. According
to Cobb,

> The changes in such a model melanoma cell system permit examination of the
> direct effect of added agents thought to cause pigment changes in vivo. Changes
> in pigment granule density, their distribution and relationships to cell division
> provide changes that can give us useful clues in cell biology. For example, the
> pituitary hormone melanocyte-stimulating hormone (MSH) causes increased
> pigment production. Our work described significant changes in melanin
> intensity while reducing cell division in MSH-treated cells. When combined
> with cytochalasin B, giant multi-nucleated pigment cells formed. These data
> gave me new insight into the dynamics of the pigment cell division cycle and
> the relationship of that cycle to the time of pigment formation.[52]

During her tenure at Connecticut College, Cobb's administrative duties began to encroach steadily on her laboratory time, so she devised a strategy of
spending early morning hours in the laboratory and acquired a staff of talented
assistants. In her position as dean at Connecticut College, Cobb also developed
other interests—specifically, creating opportunities for minorities in science.
With funding from the Van Ameringen and the Macy foundations, she established a new program, the Postgraduate Premedical and Predental Program for
Minority Students. The program enrolled recent minority college graduates who
had demonstrated high levels of learning skills but had decided too late in their

undergraduate schooling that they wanted to enter graduate school in a health field. Under the program, such students would attend Connecticut College for one year to complete concentrated studies in the sciences—biology, physics, and organic and inorganic chemistry—with an eye toward making them more viable applicants for graduate school.[53]

In 1974, Cobb became a member of the National Science Board, the policy body of the National Science Foundation, the governmental agency involved in guiding science policy by shaping the direction of basic science research in the United States. She became involved in the original formation of the NSF's Committee on Women and Minorities in Science.[54] In 1976, Cobb left Connecticut College to become the dean at Douglass College, the women's division of Rutgers University. Although she was also a professor of biological sciences, at this point she ceased being an active researcher. During Cobb's more than twenty-five years in research, her work had been continuously supported by grants from the National Cancer Institute, the Damon Runyon Cancer Fund, and the American Cancer Society.

Cobb remained at Douglass College until 1981, then left to become president of California State University at Fullerton, a position she held until her retirement in 1990. She continues to serve as head of the NSF's Committee on Women and Minorities in Science.

NOTES

1. Jewel Plummer Cobb, personal communications, including curriculum vitae and correspondence. See also Jewel Plummer Cobb, "A Life in Science: Research and Service," *Sage* 6 (fall 1989): 39–43. "Cell Physiologist, Researcher, College Dean," in *Contributions of Black Women to America*, ed. Marianna W. Davis, vol. 2 (Columbia, S.C.: Kenday Press, 1982), pp. 426–428. *Ebony Success Library*, vol. 1 (Chicago: Johnson Publishing, 1973), p. 72. Rosalyn Mitchell Patterson, "Black Women in the Biological Sciences," *Sage* 6 (fall 1989): 3–13. "Shaper of Young Minds," *Ebony*, August 1982, pp. 97–98, 100. *Who's Who among Black Americans*, 4th ed. (Northbrook, Ill.: Who's Who among Black Americans Publishing, 1985), p. 165. Dona L. Irvin, "Jewel Plummer Cobb," in *Notable Black American Women*, ed. Jessie Carney Smith, vol. 1 (Detroit: Gale Research, 1992), pp. 195–198. Gaynelle Evans, "The 'Crown Jewel' of California State University, Fullerton," *Black Issues in Higher Education*, September 1988, p. 5. *American Men and Women of Science: The Physical and Biological Sciences*, 15th ed. (New York: Bowker, 1971–), p. 266. *Who's Who 1985: An Annual Biographical Dictionary*, 136th ed. (London: A. & C. Black, 1985), pp. 165–166. *Who's Who in America* (Chicago: A. N. Marquis, 1986–1987), p. 527. Vivian Ovelton Sammons, *Blacks in Science and Medicine* (New York: Hemisphere, 1990), pp. 56–57. James M. Jay, *Negroes in Science: Natural Science Doctorates, 1876–1969* (Detroit: Belamp, 1971). *Who's Who among Black Americans*, 1980–1981 ed. (Northbrook, Ill.: Who's Who among Black Americans Publishing, 1975–1988), p. 156. James H. Kessler, J. S. Kidd, Renee A. Kidd, and Katherine A. Morin, *Distinguished African American Scientists of the Twentieth Century* (Phoenix, Ariz.: Oryx Press, 1996), pp. 49–53. Margaret W. Rossiter, *Women Scientists in America: Before Affirmative Action, 1940–1972* (Baltimore: Johns Hopkins University Press, 1995), p. 74.

Cobb's father was a graduate of Cornell University and Rush Medical School (1923), in Chicago. Her mother, also an interpretive dancer, attended Sargeants College, a physical education college then affiliated with Harvard University (Cobb, "A Life in Science";

Kessler, *Distinguished African American Scientists*, p. 50; Irvin, "Jewel Plummer Cobb," p. 195).

2. Cobb, "A Life in Science," pp. 39–40.

3. Irvin, "Jewel Plummer Cobb," p. 195.

4. Cobb, "A Life in Science." Daniel Hale Williams (born 1856; graduated Hare's Classical Academy, 1877; received M.D., Chicago Medical College, Northwestern Medical School, 1883) founded the Provident Hospital and Training School for Nurses in 1891, which would later be known as Provident Hospital. Provident is the oldest free-standing Black-owned hospital in the United States. Williams performed the "first open-heart surgery in 1893 by removing a knife from the heart of a stab victim and sewing up the pericardium—the victim recovered and lived several years afterwards" (Sammons, *Blacks in Science and Medicine*, pp. 251–252).

5. Irvin, "Jewel Plummer Cobb," p. 195.

6. Ibid.

7. Cobb, "A Life in Science," p. 39.

8. Ibid. Martin Jenkins later earned his Ph.D. and eventually became president of Morgan State College in Baltimore, Maryland.

9. Ibid. Cobb said that "despite these attempts to redistrict students, the teachers at Englewood High were excellent. The racial mix was approximately 60% to 70% white and 40% to 30% Black."

10. Ibid.

11. Ibid., p. 40.

12. Irvin, "Jewel Plummer Cobb," p. 196. Paul DeKruif, *The Microbe Hunters* (New York: Harcourt, Brace, 1926).

13. Irvin, "Jewel Plummer Cobb," p. 196.

14. Cobb, "A Life in Science," pp. 40–41.

15. Ibid. In the first quoted line, Cobb also included "as well as Illinois and other Big Ten universities."

16. Ibid., p. 41.

17. Ibid. See also Irvin, "Jewel Plummer Cobb," pp. 196–197.

18. Cobb, "A Life in Science," p. 42. "Captain Jack" was James R. Hayden, Ph.D. (Jewel Plummer Cobb, personal communication, 27 June 1996).

19. According to Irvin, of the thirty-two students in Cobb's graduating class only four were males ("Jewel Plummer Cobb," p. 197).

20. Cobb, curriculum vitae. Cobb taught biology at Washington Square College of NYU.

21. Cobb, "A Life in Science," p. 41.

22. Ibid.

23. Ibid. See also Cobb, curriculum vitae.

24. Cobb, curriculum vitae.

25. Ibid. See also Cobb, "A Life in Science," p. 41.

26. Cobb, "A Life in Science." The Warburg apparatus was again used to study the biochemical reactions. In her dissertation research, Cobb also tested tyrosine and various other substrates as model systems for pigment formation.

27. Ibid.

28. Cobb, curriculum vitae. See the entry for Jane Cooke Wright in this book for more information on Louis T. Wright.

29. Cobb, curriculum vitae.

30. Ibid. See also Cobb, "A Life in Science," p. 41, and "Cell Physiologist," p. 426. See the entry for Jane Cooke Wright in this volume.

31. Cobb, "A Life in Science," p. 41.

32. Ibid.

33. According to Cobb, "The experimental design for an in vitro test of a chemotherapeutic agent on the patients' cells followed by a comparison of the in vivo effect on the same patient was modeled on bacterial sensitivity tests that were used for selecting the antibiotic of choice for clinical use" (ibid.).

34. Ibid., p. 42. See also Cobb, curriculum vitae.

35. Ibid.

36. Between 1955 and 1960 Cobb, as principal investigator, had a series of renewable research grants from the National Cancer Institute of the National Institute of Health to study the production of melanin in vitro using tissue cultures of amelanotic and melanotic tumors (Cobb, curriculum vitae).

37. Cobb, "A Life in Science," pp. 42–43. See also Cobb, curriculum vitae; Irvin, "Jewel Plummer Cobb"; "Cell Physiologist," pp. 426–427; and *Ebony Success Library*.

38. According to Cobb: "Melanin is a pigment found throughout the animal and plant kingdom, including in human skin. In humans, the density of melanin-bearing cells in the skin determines the hue and darkness of the skin as in Negroid skin. Cells which contain melanin . . . may change into fast-growing and fast-spreading tumors called melanomas" (Cobb, "A Life in Science," pp. 41–42).

39. Ibid.

40. The Cobbs were divorced in 1976; they had one son. She has not remarried.

41. Ibid., p. 41.

42. For information on Cloudman-S91 tumors I am indebted to my colleague Karen Rader. According to Rader, the Cloudman-S91 tumor was named after researcher Arthur Cloudman, who worked with C. C. Little at the University of Michigan Cancer Research Lab before becoming one of the founding members of the Jackson Laboratory in Bar Harbor, Maine, where the JAX mice were bred. Cloudman worked on transplantation of mammary tumors in various strains as well as on the diagnosis of multiple spontaneous tumors. The Jackson Laboratory was set up around 1931 (Rader, personal communication, March 1996). See also Karen Rader, "Making Mice: C. C. Little, the Jackson Laboratory, and the Standardization of Mus Musculus for Research," Ph.D. dissertation, Indiana University, March 1995. Cobb's team obtained samples of S91 through the Bar Harbor Laboratory (Cobb, personal communication, 27 June 1996).

43. Cobb, "A Life in Science."

44. Ibid., p. 42. Cobb notes that actinomycin D is now used as a lab tool to inhibit RNA synthesis.

45. Dorothy Walker Jones (Ph.D.) is now a professor of biology in the Graduate School of Arts and Sciences at Howard University. She was mentored by both Cobb and Jane Cooke Wright (Cobb, personal communication, 27 June 1996).

46. Thiotepa and chlorambucil (Leukeran).

47. Cobb, "A Life in Science." According to Cobb, "Direct cytotoxic changes in living cells in vitro were observed in some tumor series but no promising pattern emerged in terms of comparative studies, in vivo versus in vitro."

48. Ibid., p. 42. See also Cobb, curriculum vitae.

49. Cobb, curriculum vitae.

50. Cobb, "A Life in Science," p. 42.

51. Cobb, curriculum vitae. See also "Cell Physiologist," p. 427.

52. Cobb, "A Life in Science," pp. 42–43.

53. Ibid., p. 43.

54. Ibid. See also Cobb, curriculum vitae.

JOHNNETTA BETSCH COLE
Anthropologist, College President

Dr. Johnnetta B. Cole, President of Spelman College
Photo courtesy of Spelman College Archives, Atlanta, Georgia

In 1987, anthropologist Johnnetta Betsch Cole became the first Black woman president of Spelman College, the oldest college for Black women in the United States.[1] Cole was born in Jacksonville, Florida, in 1936, and attended school there. In 1952, at age 15, she entered Fisk University, a historically Black university in Nashville, Tennessee, under an early admissions program, but the following year she transferred to Oberlin College in Ohio.[2] During her first semester at Oberlin she took a course on racial and cultural minorities taught by George Eaton Simpson, which apparently whetted her interest in sociology and anthropology; she earned the baccalaureate degree in sociology in 1957.

Cole entered the master's program in anthropology at Northwestern University, and earned the M.S. degree in 1959.[3] Secondary sources do not reveal what Cole did during the next few years, but in 1967 she earned the Ph.D. in anthropology, at Northwestern, studying under Melville J. Herskovits and Paul J. Bohannan.[4]

After serving on the faculties of the University of California at Los Angeles and Washington State University in Pullman, Cole joined the faculty of the University of Massachusetts at Amherst in 1970. During her thirteen years at Amherst, she also held visiting scholar appointments at Oberlin College, and Williams College in Massachusetts. In 1983, she was named the Russell Sage Visiting Professor at Hunter College of the City University of New York, and later received a tenured position as a professor of anthropology and director of the Latin American and Caribbean Studies Program.

A cultural anthropologist, Cole did research on both Black and women's issues, including, according to Beverly Guy-Sheftall, studies on "systems of inequality based on race, gender, and class in the Pan-African world of the United States, the Caribbean, and Africa, [and work on] female-headed households in New York City, the lives of Caribbean women, racial and gender inequality in Cuba, and economic issues in Liberia."[5]

NOTES

1. Spelman was founded in 1881.

2. Beverly Guy-Sheftall, "Johnnetta Betsch Cole," in *Black Women in America: An Historical Encyclopedia*, edited by Darlene Clark Hine, Elsa Barkley Brown, and Rosalyn Terborg-Penn, vol. 1 (Brooklyn, N.Y.: Carlson, 1993; Bloomington: Indiana University Press, 1994), p. 261. According to Guy-Sheftall, Cole left Fisk for Oberlin because her older sister, Marvyne Betsch, was at Oberlin.

3. Ibid. See also Paula Giddings, "A Conversation with Johnnetta Betsch Cole," *Sage* (fall 1988). Susan McHenry, "Spelman College Gets Its First Sister President," *Ms* (October 1987), pp. 58–61, 98–99.

4. Guy-Sheftall, "Johnnetta Betsch Cole."

5. Ibid.

BETTY ELAINE COLLETTE*

A 1958 *Jet* article described Asheville, North Carolina, native Betty Elaine Collette as "a trim, golden-brown research scientist" conducting experiments in hypertension at Georgetown University Medical School in Washington, D.C., where she was the "only Negro in the [pathology] laboratory."[1] Pictured as using "white mice, guinea pigs, dogs, rabbits, and other animals" in her work, Collette is variously described as an "animal-lover" who keeps animal skulls as a hobby, and as a student of auto mechanics who managed to "reline the brakes on her 1951 car."[2] Although the article says Collette was educated in biology at Morgan State College and in bacteriology at Catholic University, it does not mention completion of degrees, nor does it state exactly what her position was at the medical school laboratory.

NOTES

1. "Research Scientist Has Beauty, Brains," *Jet*, 3 April 1958, p. 21.

2. Ibid.

MARGARET JAMES STRICKLAND COLLINS
The Making of a Scientist

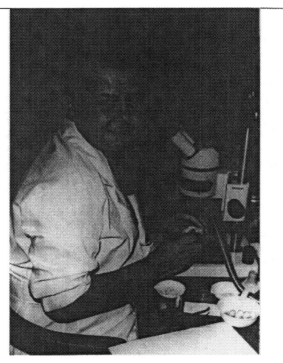

Margaret James Strickland Collins

*Because of my family and our community, my childhood was unique. I never
learned what I couldn't do—as a child, as a woman, or as a Black person.*[1]

—*Margaret Collins*

Zoologist Margaret James Strickland Collins was born in Institute, West Virginia, in 1922, the fourth of Rollins Walter and Luella (Bolling) James' five children.[2] She recalled that Institute was "unusual in that it was an all Black town *and*
a college town; in fact, the college *was* the town. As a result, there were a lot of
educated Black people living there."[3] Her father, Rollins James, had earned the
baccalaureate degree at West Virginia State College and the master's degree at
Tuskegee Institute. He worked for a time with George Washington Carver at Tuskegee, before returning to teach Vocational Agriculture at West Virginia State.
In addition to his college teaching duties, he was also responsible for running
the school's poultry program, teaching biology in the college's laboratory high
school, and serving as the County Agent for the United States Department of Agriculture (USDA), a program which Collins recently described as "sadly limited

in terms of the services it provided for the segregated African-American members of the county."[4] Her mother, Luella James, had wanted to become an archaeologist, but found her educational opportunities limited. She too had attended West Virginia State, but did not complete a degree. "My mother suffered from being unlike her family," Collins recalled. "She was independent, reckless, rebellious."[5] Luella James was also an avid reader, who followed the exploits of naturalist Roy Chapman Andrews.[6]

When Collins was still a toddler, she learned to read by sitting on her parents' laps each evening as they took turns reading aloud to their children: "My parents collected an impressive library for their income level, and a regular feature of the evenings would be one member reading aloud to the rest of the family—with the youngest on the lap of the reader, in my case. I learned to read by following the finger of the reader."[7] At age six she was given access to the college library and allowed to check out any book she could reach. At that point, Collins's mother had become ill, and was unable to keep her confined to the house. "I grew up in the woods and the barn," Collins said, "playing in or on the big river near the yard."[8] Two of her favorite childhood books were naturalist Ernest Thompson Seton's *Two Little Savages* and *Rolf in the Woods*, which had as their main theme the idea that a child could escape into the woods and wild places, and find wonders.[9] And that is what Collins did. Filled with curiosity, she investigated every backwater and pool, searching for wonders and collecting "interesting critters," as she called them.

With the rebellious inquisitiveness she inherited from her mother, she was always questioning things: "I ruined Christmas for everyone because I couldn't figure out how a reindeer could fly. I mean they just aren't built for flying, anyone could see that. So I had to reject either the truthfulness of adults or the conclusions of my own mind."[10] In that innocent childhood questioning was the beginning of the scientific frame of mind, and in Collins's case that proclivity was nurtured. When she began to bring home some of the little critters she'd found, her father encouraged her to try to identify them in his books: "I had access to an array of books on natural history, including the massive volumes of *The Nature Library*; my favorite books in the series were Rogers' *The Shell Book*, Ditmars' *The Reptile Book*, and Jordan and Everman's *The Fish Book*, each of which contained English technical terms as well as Latin names."[11]

Thus, Collins's educational experiences extended beyond the classroom. She recalled that youngsters in Institute were careful about how they behaved—good behavior was expected. All the children—but especially Collins, who continued to be a voracious reader—benefited from what she described as an "unusual group of neighbors." Two groups of neighbors, in particular, gave her free, unlimited access to their books, thoughts, and hospitality:

> Dr. Sinclair was the community doctor. He had all the volumes in the *Book of Knowledge* series, and I had the privilege of coming to his house and reading as long and as often as I wanted to. His mother had been a slave, and knew lots of interesting things which she liked to talk to me about, like how to make soap and the uses of medicinal plants.[12]

She also spent a great deal of time with John Matthews, a scholar of Haitian and African literature:

> He was extremely knowledgeable in history, literature, play writing, and even sports. He had gone to Liberia for two years as President Tubman's secretary. While he was in Liberia, he read every medical text he could get hold of, and then when he returned to West Virginia he would talk to me about these things. His wife was odd, but interesting. She was a spiritualist, and one of the first people in the country to become a member of the Baha'i faith.[13]

Clearly, there was something special about Margaret Collins—something everyone in her life recognized and sought to nurture.

Collins believed that her early and continuous exposure to books was one of the major influences in determining her interest in science:

> I suspect that plenty of books and early, unrestricted reading were quite important. Hearing about scientists doing adventurous things, and reading the works of naturalist William Beebe were probably important. Perhaps the biggest influence of all was contact with individuals who found the discipline of biology fulfilling—enthusiasm sometimes behaves like an infectious agent.[14]

By the time Collins was in high school, she had also discovered science fiction, thanks to her brother's interest in the genre. She particularly recalled a story by S. F. Light about a man reduced to the size of an ant. "That was my first exposure to science fiction," she said, "and it too inspired me to be a scientist."[15]

Collins's formal schooling was rather sketchy during her elementary school years, due both to upheavals in the local school system and to her own precocious intellectual talents. During her first year of school she was advanced two grades, ending up in the third grade class. In addition, for all of her elementary school years, she was the only girl in her class. She later skipped eighth grade and wound up, at age eleven, in the freshman class at West Virginia State College's Laboratory High School, one of a group of girls who were three to five years older than she.[16] As she recalled,

> One never really knows how early experiences contribute to personality development. In looking back I realize I was totally out of step with other kids in school, especially in high school when I was so much younger. My social life probably never recovered from the traumas associated with trying to compete with girls so much older, but I can't really regret this aspect of my life as it may have helped reinforce my self-sufficiency.[17]

Despite the age disparity, her high school years afforded contact with an excellent science teacher, who was firm but showed a great interest in her and provided rich experiences in both the formal courses and the science club. Collins's grades in high school were good, especially, she said, in "the things I was good at, like biology and physics. . . . The only thing I regret in my life is not having been better in mathematics and chemistry—and if it weren't for the tutoring of Angie Turner King, I would never have succeeded at all in those two subjects."[18] Collins graduated from high school in 1937.[19]

There had been about a hundred other students in school with her, and all went on to college: "In my community, if you had even a grain of sense, you went."[20] She entered West Virginia State College in 1936, at age fourteen, on a tuition scholarship. "The scholarship was very needed," she recalled, "because my family already had [the expense of] my older sister at Hampton Institute. I wanted to attend elsewhere but we couldn't afford it, so I stayed at State against my will."[21] Her first year of college proved disastrous.

Collins had intended to major in biology, but by the end of her first year her interest in science had waned. It cost her the scholarship. "You know how boring the wrong teacher can make science," she said. "My first college biology course was so stereotyped, dull, and malodorous and the teacher so gruff and frightening that I lost interest in it."[22] During her second year, her parents had to pay her tuition, but they never managed to get the money in on time. "I was always being pulled out of class, which was embarrassing," Collins said. "That's when I decided to do it myself."[23] She worked summers as a maid or household helper in nearby Huntington, and learned how to shop for bargains. Then a young biologist, Toye George Davis, came to the college — and he would provide the key for her return to science.

Although Collins had lost her interest in college biology, she still spent much of her free time near the Kanawha River, hunting for critters. One day, in a back-water near her home, she found colonies of a marine animal she had never seen before, so she immediately collected samples. Collins took her samples to Davis — the Black biologist with the Harvard Ph.D. whom she had heard about, but never before met — and asked if he could identify them.[24] When he couldn't, he sent a sample to the Marine Biological Laboratory at Woods Hole, Massachusetts. While Davis and Collins waited for the results, he made a microscope available to her, showed her how to use a taxonomic key, and introduced her to Ward and Whipple's *Fresh Water Biology*, then a standard in the field.[25] By the time the peo-ple at the Marine Biological Laboratory wrote back and identified the sample as bryozoans, Collins was once again a biology major. Davis had given her a job as a $10 a month lab assistant, and had promised to protect her from the department chair. Collins had found a mentor, and by the start of the next school year Davis had helped get her scholarship reinstated.[26]

During this time, the trickle of refugees from Hitler's purges had become a flood. European scientists and mathematicians fared well in the United States, but scholars from less "strategic" disciplines did not have an easy time. Frederick Lehner, a Jew, had been teaching German at a university in Vienna when all his possessions had suddenly been confiscated; he and his wife had barely managed to escape to the United States. Because he was not in a strategic field, he had difficulty finding a job and eventually wound up at West Virginia State.[27] Lehner's German classes were not well attended, but Collins — independent and rebel-lious like her mother — hooked up with the only white professor at the college be-cause she thought he might be "interesting."[28] Imagine the spirit it must have taken for a young Black woman in 1940 to seek out a white professor — and a foreigner at that. Then the unthinkable happened: the United States entered

World War II, and shortly after Toye Davis received his draft notice. Davis, who had already struck up a friendship with Lehner through Collins, discussed his prize pupil's future with Lehner, and asked him to take over as her mentor.[29] Meanwhile, Collins married Howard University pre-medical student Bernard E. Strickland in July 1942, and within a matter of months he too was drafted.[30] Guided first by Davis and then Lehner—the only white teacher she had ever had —Collins graduated from West Virginia State in 1943, with a major in biology and minors in physics and German.

Collins entered graduate school at the University of Chicago in 1943, not really expecting to earn an advanced degree—she had just wanted to take enough field zoology and ecology so that she could start a business collecting and selling animals to biological supply houses. During registration, however, she met an advisor who told her those courses were offered in different semesters and suggested that she might as well "fill out her dance card" and sign up for some other courses.[31] This advisor was the great zoologist Alfred Emerson, the grandson of Ralph Waldo Emerson. Collins loved Emerson's course—he was "incredibly knowledgeable" about the animals, the field trips to a wide variety of habitats were beyond anything she had ever imagined, and best of all, she began to see and touch things that she'd only read about in books. She became, as she put it, "absolutely intoxicated by the wonder of it all."[32]

But as intoxicating as that time in her life was, she was also exhausted and hungry much of the time. In those days, because Blacks were not allowed to attend the University of West Virginia, the state instead offered a $125 a year stipend toward out-of-state tuition, which Collins applied to her courses at the University of Chicago. No matter how Collins tried to stretch her state stipend and her husband's military allotment, after tuition and rent there was only enough money for ten meals a week, which meant she was often hungry. She considered herself lucky when she got a night job at a defense plant; the work was exhausting, yet she managed to keep up her grades.[33] As she recalled,

> Several of the other professors in the ecology area became interested in me, probably because of my interest in the field. Among them was W. C. Allee, who was the great animal behaviorist of the time. He taught animal geography, which I took in the winter quarter, and he gave me the option of either helping Dr. Emerson with the maps he was doing on the world distribution of termite genera, or writing a term paper. I elected to work with Dr. Emerson.[34]

Because she was clearly interested, even thrilled by the field, Emerson increasingly took notice. When he learned about her financial struggle, he offered her an assistantship and she was assigned to look after the termite collection.[35] As a perquisite, she was given access to his huge office library. During the next few years, Collins said she read every book in his library except one, a philosophical treatise on holism, which she described as "incomprehensible and dull."[36] Despite her progress, she hadn't yet selected a specialty. Then she chanced upon a children's book in Emerson's library that he had co-authored, called *Termite City*, which described the research he had conducted at a field station at Kartabo, British Guiana.[37] Perhaps because the book was written for children it reminded her of her long-ago favorites *Two Little Savages* and *Rolf in the Woods*. In any case, Collins's two childhood passions—reading and searching for interesting critters—had come full circle.

At the close of Collins's second year of graduate study, World War II ended, and her husband, Bernard Strickland, returned to his pre-medical studies at Howard University. As a result, she had to give up graduate school on a full-time basis. In 1947, she obtained an instructor's position in the biology department at Howard, a move that was necessary for both financial and personal reasons. The loss of Strickland's military allotment was clearly a factor, but she also felt she had to contribute monetary support for his return to pre-medical studies; also, because they had only spent a few months of their married life together, she wanted to be with him. She spent the next three summers in Chicago—again, with funding provided by Emerson—completing her course work and dissertation research. It was during this time that Collins had her only negative experience with Emerson:

> After the war with Japan ended, a team of scientists was needed to describe the flora and fauna of the Marshall Islands. Emerson was to lead the expedition, and I wanted to go, but he vetoed the request, saying, among other things, that young women were troublesome on expeditions. Instead, he insisted that I take my finals and complete my thesis.[38]

Despite this disappointment, Collins's relationship with Emerson was particularly fruitful. He was totally lacking in racist tendencies, and continued his strong support of her throughout her graduate years. During the 1950s he continually supplied her with termite specimens gathered on his research trips to Guyana, South America. In 1949, her marriage ended, but she completed her dissertation, "Differences in Toleration of Drying Between Species of Termites (*Reticulitermes*)." She earned the Ph.D. in 1950, becoming only the third Black woman zoologist in the country.[39]

Although Collins had received a promotion from instructor to assistant professor upon the completion of her doctorate, she was unhappy with her prospects for advancement at Howard: "They refused to promote me because they said I was too young. But it was also because I was a woman."[40] She was also unhappy with the scientific focus of the zoology department. Although Ernest Everett Just had died a number of years earlier, she believed the department still reflected his interests. "Just was an elitist. He had an interest in medicine, and therefore he had a strong medical slant in his work at Howard; the department was geared more toward medical training than I thought necessary."[41] Nor was she a fan of Howard's president at the time, Mordecai Johnson:

> Howard was a place of entrenched chauvinism. . . . Johnson was a minister, and did not like independence of thought. When I was at Howard, he said he didn't want me to wear trousers.
>
> I replied, "President Johnson, I wouldn't be so immodest as to wear a skirt on my field trips, doing the kind of work I do. What would my male colleagues think!"[42]

Perceiving that Howard was slow in promoting women faculty, she quit and, in 1951, obtained a position as a full professor at Florida A & M University, a historically Black school that was then making a strong effort to seek role models for its female students.[43] She was named chair of the biology department in 1953.[44]

Science in the Field and Civil Rights on the Back Roads

I'm a field scientist. When I started, field biology was not considered women's work. But gender is incidental if a scientist is good.[45]

—Margaret Collins

Collins always saw herself as a field scientist, but actually *being* one was often a struggle. In 1951, immediately before joining the faculty at Florida A & M, she had married Herbert L. Collins, and the couple had two sons, Herbert Louis Collins Jr. and James Joseph Collins.[46] In 1953, despite a heavy teaching and administrative load at Florida A & M, as well as the care of a husband and a toddler, Collins embarked on the field studies—largely self-funded at the time—that were to become the dominant characteristic of her career. Often accompanied by her husband and child, she began collecting specimens in Everglades National Park and Highlands State Park for an in-depth study of Florida termites.[47]

Collins's research and field work were to be interrupted, however, when she embarked upon a new and often dangerous period in her life. Collins had been invited to lecture on her research at a predominantly white university nearby.[48] When word of her lecture got out, the school received a bomb threat.[49] The administration, fearing for the safety of its students and facilities, withdrew the invitation. Then, in 1956, the president of the Florida A & M Student Council called for a boycott of the bus company in Tallahassee, and she was among those who volunteered to drive people back and forth to work.[50] She recalled being followed by local and state police, as well as the FBI.[51] When the protest organization received a tip that the police and FBI were going to raid their headquarters, it fell to Collins to transport the records containing the names, addresses, and activities of group members to safety—a midnight road trip that she said "terrified" her.[52] When one looks at Collins's list of publications, it is fairly obvious what the civil rights activities cost in terms of her science. Although her usual average had been about one or two papers a year, from 1952 though the end of 1957 she had no publications. At that time in her life, civil rights issues took precedence over science: "A lot of people opposed our civil rights efforts. I had to do what I thought was the most important thing. That's all there was to it."[53]

Throughout this time, however, she continued to maintain a strong correspondence with Emerson, who supplied her with specimens from his collection, particularly those he had gathered in Guyana.[54] Her publishing resumed in 1958, but her various activities had taken their toll on her ability to keep current with changes in biology:

> When I was at Florida A & M, I published a book, and when it came back from the publisher there was an illustration in it that I didn't understand. I had been spending so much time on teaching and administrative duties that new information in genetics had passed me by. Finally, I told [Florida A & M President

George W.] Gore that I had to have the summer off. I needed to catch up on my science. I told him if he didn't give me the time off I'd just quit.[55]

Gore agreed, and she applied for and received an National Science Foundation Summer Research Fellowship to study new developments in genetics and molecular biology at New York's Cold Spring Harbor Laboratory.[56] Collins recognized that Gore was also instrumental in the effort to see that Black professors received salaries commensurate with those of their white counterparts during his tenure as president at Florida A & M: "His attempts were eventually successful in Florida and helped set a lot of precedents with regard to salaries for Black academics."[57]

From 1961 to 1962, Collins took a leave of absence from Florida A & M. She had received a National Science Foundation grant for a year's study at the University of Minnesota, where she was a research associate, and at the Minnesota Agricultural Experimental Station, in St. Paul, where she began research with A. Glenn Richards on a series of studies of North American termites.[58] According to Marc Epstein, her friend and colleague at the Smithsonian, during Collins's time at Minnesota she was severely injured in a laboratory accident and the injury left her lungs permanently damaged.[59] Following her second divorce in 1963, she left Florida A & M, and in 1964 she returned to Howard University as a full professor in the zoology department. For the next ten years she was also a tenured professor at Federal City College (now the University of the District of Columbia).

In terms of Collins's science, the return to Howard University marked the beginning of an incredibly productive period, which she would maintain for the remainder of her life. Collins always described herself as an ecologist, whose primary research focus was on termite ecology and distribution:

> My work has helped to clarify an issue related to the evolution of glue-squirting termites. The older concept of evolution held that there were two separate evolutionary branches of termites. We proved that this was not the case.[60]

Her return to Howard also put her in contact with two students from Guyana—Theodore Bremmer and Lynette Padmore—whom she described as among her best students.[61] Due to her relationship with Emerson, Collins had long entertained the idea of focusing more of her research on the termites of Guyana, and the contact with Bremmer and Padmore heightened her interest.[62] She began exploring various possibilities to secure funding to reestablish Emerson's original field station at Kartabo, which had been closed since the late 1950s. During this period, she also began working with the staff and collections at the Smithsonian, and stepped up the pace of her research.[63]

With support from the Smithsonian and a travel grant from the Graduate School of Howard University, Collins made a field trip to western Mexico in 1968. Her trip to the Sonoran Desert in Arizona in 1972 was supported by the United States IBP Desert Biome Project and the National Science Foundation, and in 1972 she was also a visiting professor at the University of Arizona at Tucson.[64] The pace of her publications increased, and in 1974 she was named to the

Clark Lectureship at the Scripps Colleges at Claremont, California, where she discussed her research and experiences.[65]

Guyana and the Smithsonian: In the Sun, in the Lab

*I think my career and my work have functioned at the international level —
this is both a burden and a reward, because I tend to take on very large
tasks without thinking of the consequences and the demands they take
on my time.*[66]

— *Margaret Collins*

Around 1975 to 1976, Collins finally made formal contact with the government of Guyana, placing her one step closer to securing the funding necessary to reopen the research station at Kartabo. In December 1977, she traveled to Georgetown, Guyana, to address the Conservation Society in an effort to stress the country's potential as a host nation for scientific study groups. Then, in January 1978, she received official Guyanese government sponsorship, along with that of the Guyana Department of the Army.[67] With primary support from the government, augmented by travel grants from Howard and occasional funding from the Smithsonian, Collins began rebuilding the Alfred Emerson Research Station in 1979.[68] That same year, she embarked on a new area of research: defense mechanisms in termites. In collaboration with Glenn D. Prestwich, Department of Chemistry and the Program of Ecology and Evolution at the State University of New York at Stony Brook, she began studying the chemicals that termites use to defend themselves from predators.[69]

At that point, Collins was nearly 60 years old, but the pace of her field trips increased, as she continued to seek pristine research sites:

> Termites are essentially tropical animals, and I've always wanted to be where the termites are. Although North America is unique among temperate areas in the size of its fauna, most of our species seem drab and commonplace in comparison to the tropical forms. South America has been inhabited by man a much shorter length of time than Africa, and the habitats are still relatively unspoiled in some areas. I have been fortunate enough to study and collect in Mexico, Costa Rica, Colombia, Barbados, Belize, Suriname, the Cayman Islands, Guyana, Guatemala, and Panama, in addition to most parts of the United States.[70]

She made a series of research trips to Guyana in 1980, 1981, 1983, and 1984, dividing her time between collecting and instructing the Guyana Army on termites.[71] In 1982, she led an Earthwatch-sponsored expedition to Suriname, and served as president of the Entomological Society of Washington.[72] In 1983, she conducted another field trip to the research station in Guyana, where among other tasks, she collected termite samples for what she described as the "poorly thought out, poorly directed" Biosphere II, a privately funded research project. During this trip she also conducted preliminary negotiations and arrangements for the establishment of a "graveyard" test site on the Mazaruni River in Guyana for the United States Department of Agriculture's Forest Service. During the

1983–1984 Christmas break, she led another Earthwatch expedition—the Takatu Species Survey—in Guyana. Her letter to the members of her group indicates just how scientifically valuable and eclectic her field trips were:

> More than 13,000 insect specimens were taken, plus [the] ectoparasite stuff; samples of about a dozen identifiable medicinal plants, along with photographs and data on their utilization, were taken, and the plants are now in the hands of specialists at the U.S. National Museum. Among the highlights of our collective efforts that we can recognize *now*:
>
> 1. New taxonomic groups—both new species and new genera, especially in the beetles, and probably in the microlepidoptera. 250 of these [are already] pinned, and the expert on the group indicated that the collection included some forms in which he was particularly interested.
>
> 2. New distribution records were obtained, especially for the *caddis flies*; these are of value to students of the evolution of this group.
>
> 3. Species of both insects (many) and birds (42) *new* to the Smithsonian have already been recognized; more will probably be recognized as study of the collection progresses. At least two birds and several species of insects new to the world were also taken.
>
> 4. Even the chigoe fleas are contributing something besides discomfort, because [a group member] reared one, counted the eggs, and has recovered a whole specimen. Not much has been documented about these pests, but he will be able to add to what knowledge is available. The life cycle is pretty horrible, we will give you the details at the reunion. . . .
>
> 5. The scope of [the] collection of bird parasites is unique for the Neotropics. Relationships between the highly host-specific parasites and their hosts may provide important information to assist in solving some problems of bird classification.[73]

The letter also provides some interesting insights on the experience of field work:

> Let me take this opportunity to express my admiration of your flexibility, durability, tolerance, cheerfulness, and capacity for enjoying life! You were a magnificent team, and your efforts were critical to the success of this expedition. Very special accolades for (or to):
>
> 1. Martin's patience in carrying water, wood, baggage, or doing whatever needed to be done *cheerfully*.
>
> 2. Paul's skill in splitting kindling, setting up tents, erecting tarpaulins; willingness to load luggage, even in the rain; and most importantly, involving and training collectors in his aquatic insect survey.
>
> 3. Robin's quiet efficiency, courage, good humor, and dedication to the success of the expedition.
>
> 4. Fred and John, whose voices, skills, and temperaments kept me going long past my tolerance point; and for their immortal "Don't List."
>
> 5. Jim, who saw me though a bad fever and whose professionalism and skill in personal relationships consolidated our chances for continuing work in Guyana, not only in the interior, but also in the museum and zoo.
>
> 6. The stomachs of all the participants concerned, who ate and digested half a tiger, curassows, piranhas, agoutis, wild pigs, assorted sprouts, and a variety of exotic fruits and vegetables without obvious signs of protest or damage to the local environment.[74]

Collins retired from Howard University in 1983, but she soon accepted a position as an unpaid research associate at the Smithsonian Institution's National Museum of Natural History.[75] Unfortunately, during her 1983–1984 trip to Guyana she had contracted dengue fever, suffering a variety of health problems that kept her out of the field for a time. During this forced hiatus, she began work on updating and preserving the Smithsonian's vast termite collection. The materials on both neotropical and North American termites that she had collected over the years are included in the Collins Collection at the National Museum of Natural History (Smithsonian Institution). Another set comprises a reference collection of Guyana termites, housed at the Georgetown Museum in Guyana. In 1988, she moved on to taxonomic studies in collaboration with David Nickle of the Systematic Entomology Laboratory of U.S. Department of Agriculture and the Smithsonian.[76] Throughout much of this period, Collins's work on the termite collection was funded by the Women's Committee, a group of women associates at the Smithsonian.[77] Although Collins recognized the importance of the museum work, she repeatedly expressed her longing to be back in the field.[78]

In 1992, the Smithsonian's Biological Diversity Program formalized its relationship with the University of Guyana and opened the Center for the Study of Biological Diversity in the Guyanese capital.[79] In 1994, Vicki Funk, director of the Natural History Museum, suggested that Collins collaborate with Matthew Kane, a new addition to the Smithsonian staff who specialized in termite gut microbes.[80] Sufficiently recovered from the aftereffects of dengue fever, the 72-year-old Collins returned to Guyana in December 1994, accompanied by Kane. For three weeks, she collected the specimens she needed to describe a new termite species, and he collected live specimens to carry back to the bio-diversity center for extended research on the role of the methane produced by termite gut microbes in global climate change.[81] The trip marked her return to field work.

Collins's research spanned five decades and encompassed nearly the entire field of termite zoology: the evolution of desiccation resistance in termites; various termite species' tolerance of high temperatures; defensive behavior in South American termites, including chemical defenses; termite ecology; species abundance in virgin and disturbed tropical rain forests; and behavioral ecology, taxonomy, and entomology. At the time of her death in April 1996 in the Cayman Islands, Margaret Collins was perhaps the world's oldest active field scientist. She died, as she had said she wished, doing termite research in the field.[82]

NOTES

1. Margaret S. Collins, personal interview (see note 2).

2. Margaret S. Collins, personal interviews with author, 11–12 September 1995, Smithsonian Support Center, Silver Hill; 21 October 1995, Smithsonian Support Center; 16–19 March 1996, Washington, D.C., Daniel's Mountain, Va., and Institute, W.Va.; telephone conversations with author, 9, 15, and 21 August 1995; 28 September 1995; 1 January 1996; 11 February 1996; 1 March 1996; 5 March 1996; 24 March 1996. Margaret S. Collins, curriculum vitae, assorted letters, personal essays, published papers, and author's questionnaire, 12–13 September 1995. See also *American Men of Science*, 11th ed. (New York: Bowker, 1968), p. 942. *National Faculty Directory* (Detroit: Gale Research, 1988), p. 713. *Who's Who of American Women*, 1st, 2nd eds. (Chicago: Marquis

Who's Who, 1958, 1962). *Who's Who in the Frontier of Science and Technology*, 1st ed. (Chicago: Marquis Who's Who in America, 1984–1985). *Who's Who in the World*, 7th ed. (Chicago: Marquis Who's Who, 1984–1985). *People*, *Jet*, 5 February 1959, p. 25. Vivian Ovelton Sammons, *Blacks in Science and Medicine* (New York: Hemisphere, 1990), pp. 59–60.

3. Collins, personal interviews. Collins had two sisters, one older, one younger, and two older brothers. West Virginia State was then a land grant college, which had been founded for Blacks under the Morrill Act. As such, it had to have a laboratory-type high school and a laboratory-type elementary school for teacher training. The college gave up its land grant status in the 1960s, and today Blacks represent less than 20% of the student body, although the faculty is still predominantly Black (Dr. David Telfer, Dean, personal interview, 18 March 1995, West Virginia State College).

4. Collins, personal interviews. According to Collins, her father was "reared by a stern, ponderously dignified patriarch with a self-effacing wife. They were ardent churchgoers, keenly aware of their place in the society of a fairly large city—Huntington, West Virginia. My grandfather, who died before I was born, was a contractor who laid many of the beautiful brick streets in Huntington. When I was growing up we could [walk around town and] see some of his signed bricks."

5. Collins, personal interviews. For a time, Collins's mother was interested in millinery and took courses in that at West Virginia State, but afterward "never made another hat." Collins also recalled that her mother was "slightly darker than her siblings. The favorite [in her family] was her younger brother; the jealousy between them was evident throughout her life."

6. Naturalist Roy Chapman Andrews (1884–1960) was a popular author for more than four decades. Among the many books he wrote were *Under a Lucky Star: A Lifetime of Adventure* (New York: Viking Press, 1913); *Across Mongolian Plains: A Naturalist's Account of China's "Great Northwest"* (New York, London: D. Appleton, 1921); *Ends of the Earth* (New York: Putnam, 1929); *Meet Your Ancestors: A Biography of Primitive Man* (New York: Viking Press, 1945); and *Beyond Adventure: The Lives of Three Explorers* (New York: Duell, Sloan, and Pearce, 1951).

7. Collins, author's questionnaire. This was also mentioned in Collins's "The Termite Alternative," an unpublished manuscript that she was still working on at the time of her death.

8. Collins, personal interviews.

9. Ernest Thompson Seton, *Two Little Savages* and *Rolf in the Woods* (Garden City, New York: Doubleday, 1911).

10. Collins, personal interviews.

11. Collins, personal interviews.

12. Collins, personal interviews, 11–12 September 1995.

13. Collins, personal interviews.

14. Collins, personal interviews.

15. Collins, personal interviews. I was not able to determine the title of this story, but according to Collins it appeared in the pulp magazine *Amazing Stories* sometime in the late 1920s or early 1930s.

16. Collins, personal interviews.

17. Collins, personal interview, 1 March 1996. See also Collins, "Termite Alternative."

18. Collins, personal interviews. See also Collins, "Termite Alternative." (See my entry for Angie Turner King in this volume.) With regard to her lack of success in chemistry, Collins said: "I suspect the deathly fear I had of chemistry came partly from early injunctions against opening bottles with raised glass labels, necessary to a toddler meddling

around a parent's laboratory. The other early experience involved seeing the results of an explosion in the chemistry lab that disfigured a young woman and burned the high school building rather badly" (Collins, "Termite Alternative").

19. Collins, "Termite Alternative."

20. Collins, personal interviews, 11–12 September 1995.

21. Ibid.

22. Ibid. The biology professor was also the department chair, which exacerbated her problems. (She declined to mention his name.) Collins said she changed her major to music, but was so lacking in any musical ability that she drifted into ceramics. (See also Collins, "Termite Alternative.")

23. Collins, personal interviews.

24. Toye George Davis: M.S. Pennsylvania University, 1932; M.A., Ph.D. (zoology), Harvard University, 1939, 1940, respectively; M.D., Howard University, 1947 (Sammons, *Blacks in Science and Medicine*, p. 71). According to Collins, Davis had studied at Harvard under L. R. Cleveland, who specialized in protozoans (Collins, personal interviews).

25. Henry Baldwin Ward and George Chandler Whipple, *Fresh-Water Biology* (New York: John Wiley & Sons, 1918).

26. Collins, personal interviews.

27. Ibid.

28. Ibid.

29. Ibid.

30. Collins and Strickland were divorced in 1949 (ibid.).

31. Ibid.

32. Ibid.

33. Ibid. Collins worked 8-hour night shifts at the Timken Ball Bearing factory in Canton, Ohio. She recalled that at one point the workers went on strike, but that she crossed the picket lines because "I couldn't afford to strike."

34. Ibid.

35. The Alfred Emerson Termite Collection is housed at the Smithsonian Institution.

36. Collins, personal interviews. See also Collins, "Termite Alternative."

37. Alfred Edwards Emerson and Eleanor Fish, with a foreword by William Beebe, *Termite City* (New York: Rand McNally, 1937).

38. Collins, personal interviews, 17–19 March 1996. Collins did say, however, that in later years, Emerson's wife (apparently his second wife) "intervened and he would travel with her and the whole family, sometimes including female scientists." The Alfred E. Emerson Field Research Station is located at Kartabo Point, Guyana, at the confluence of the Mazaruni and Cuyuni rivers. The station was originally established by William Beebe as the New York Zoological Society's Tropical Research Station in the 1920s, when it was visited by many well-known scientists, including Emerson, who described many new termite species from the Kartabo area and later amassed one of the finest worldwide termite collections. (Information from "The Alfred E. Emerson Field Research Station," brochure provided by Margaret Collins; it appears to have been distributed sometime in the late 1970s or 1980s.)

39. The first two were Roger Arliner Young and Lillian Burwell Lewis.

40. Collins, personal interviews, 11–12 September 1995.

41. Collins, personal interviews, 17–19 March 1996; telephone conversation, 15 August 1995. The head of the department at that time was Dr. Eiliger, whom Collins described as a "white man from Denmark who had once conned Hitler into believing he could identify Jews by photographs. . . . he had wormed his way into becoming the head

of the zoology department . . . and had a penchant for throwing away valuable specimens" (Collins, personal interviews, 11–12 September 1995; 17–19 March 1996).

42. Ibid.

43. Henry Goethals, "Research and Researchers Change with Times," *Quest* 4 (spring 1995), pp. 10–11.

44. Collins, curriculum vitae.

45. Collins, personal interviews, 11–12 September 1995.

46. Herbert Louis Collins Jr. was born in 1951, and James Joseph Collins was born in 1957. She divorced Herbert L. Collins in 1963 (ibid.).

47. Ibid. See also the acknowledgments section of Margaret S. Collins, "Studies on Water Relations in Florida Termites," *Journal of the Florida Academy of Sciences* 21 (1959): 351.

48. This was most likely Florida State University, which is also in Tallahassee.

49. According to Collins, the caller said something to the effect that if "that [expletive deleted] dared to speak at the white university the science building would be blown up." Oddly, it was Collins herself who searched the building—no bomb was found (Collins, personal interviews, 11–12 September 1995).

50. The Tallahassee boycott took place around the same time as the more well-known bus boycott in Montgomery, Alabama.

51. It is useful to recall that in those days lynch mobs were sometimes peopled by local police.

52. Collins, personal interviews, 11–12 September 1995.

53. Ibid. See also Henry Goethals, "Research and Researchers Change with Times," *Quest* 4 (spring 1995), pp. 10–11.

54. Collins, personal interviews, 11–12 September 1995.

55. Ibid. I have been unable to find any reference to the title of this book in Collins's various curricula vitae. It is possible that she meant a chapter in a book, or even a paper.

56. Cold Spring Harbor "was delightful—Richard Leuontin and Barbara McClintock were there" (Collins, personal interviews, 11–12 September 1995).

57. Ibid. George W. Gore had a Ph.D. in journalism from Harvard University, according to Collins.

58. See the appendix.

59. Marc Epstein, telephone interview, 10 September 1996. Epstein said that as Collins recounted to him, an acid had been released into the air in the lab (then located at Coffee Hall), and the lining of her lungs had been burned. As a result, Collins could not tolerate being among smokers, as it aggravated her condition. Epstein also said that Collins's co-worker Richards, a physiologist, was a "pioneer in morphology and had studied under W.T.M. Forbes at Cornell University."

60. Collins, personal interviews, 11–12 September 1985.

61. Ibid. Lynette Padmore is now a geneticist at Florida A & M University, and Theodore Bremmer is a biologist at Howard University.

62. Goethals, "Research and Researchers," p. 11. See also Collins, telephone interview, 21 August 1995.

63. Goethals, "Research and Researchers."

64. Collins, curriculum vitae. See also Collins, personal interviews. Collins made a second field trip to western Mexico in 1984 with David Nickle (USDA/U.S. National Museum of Natural History, Smithsonian Institution), which was sponsored by the Smithsonian and Howard.

65. See the appendix.

66. Collins, telephone interview, 1 March 1996.

67. Collins, curriculum vitae; personal interviews, 11–12 September 1995.

68. Goethals, "Research and Researchers." The station has living quarters and table space for up to 15 investigators. Collins secured approval from the Guyana government for visiting scientists to bring in scientific equipment duty-free, and set up the means of facilitating travel in the country by foreign investigators. The station continues to operate under the auspices of the government of Guyana (Collins, curriculum vitae).

69. See the appendix.

70. Collins, "Termite Alternative."

71. The types of information regarding termites that Collins supplied to the Guyana Army ranged from building methods to avoid termite damage to the uses of termite waste products to strengthen building materials such as cement. The Voice of America broadcast an interview and account of her field work there in 1983 (Collins, personal interviews).

72. Collins, curriculum vitae. Earthwatch is a privately funded nonprofit organization begun in 1971. By 1988 it was the third largest private source of field research funding in the United States. In its first sixteen years, the organization provided $9 million in funds for 950 expeditions in 79 countries. See Aaron M. Ellison, *Expedition Briefing: Life among the Mangroves* (Earthwatch, 1988).

73. Margaret S. Collins, Letter to Earthwatch volunteers, 20 February 1984, Takutu Species Survey, Guyana.

74. Ibid.

75. According to Marc Epstein, research associates at the Smithsonian are generally not salaried, but Collins received some funding for her work, and was provided with office space at the Museum of Natural History and lab space at the Museum Support Complex in Silver Hill (Marc Epstein, telephone interview, 10 September 1996).

76. See the appendix.

77. Collins, telephone interviews; personal interview 18 September 1995.

78. Ibid.

79. Goethals, "Research and Researchers," p. 11.

80. Ibid.

81. While a graduate student, Kane and several collaborators had shown that soil-feeding termites typically generate 10 times more methane than wood feeders (ibid.).

82. Herbert Collins, telephone interview with author, 26 April 1996.

CAROL BLANCHE COTTON*

Born in 1904, psychologist Carol Blanche Cotton attended Oberlin College, a predominantly white institution in Ohio, and earned the baccalaureate degree in 1926.[1] Oberlin, then known as Oberlin Collegiate Institute, was founded in 1833 and was the first co-educational college in the United States. In 1835, the school's trustees declared that "the education of people of color is a matter of great interest and should be encouraged and sustained at this institution."[2] Cotton then obtained her master's degree at Columbia University in 1927. She finished her Ph.D. in psychology at the University of Chicago in 1939, and was elected to Sigma Xi, the scientific honor society. Her dissertation was "A Study of the Reactions of Spastic Children to Certain Test Situations."[3]

NOTES

1. Verified by Jennifer Cline, Registrar's Record Office, Oberlin College, via e-mail, 30 April 1996. Cotton is listed in the 1960 Register as Carol B. Cotton Bowie.

2. William Bigglestone, "Oberlin College and the Negro Student, 1865–1940," *Journal of Negro History* 56 (July 1971): 198–219. See also Marlene Deahl Merrill, "Oberlin College," in *Black Women in America: An Historical Encyclopedia*, ed. Darlene Clark Hine, Elsa Barkley Brown, and Rosalyn Terborg-Penn, vol. 2 (New York: Carlson, 1993; Bloomington: Indiana University Press, 1994), pp. 897–899.

3. Robert V. Guthrie, *Even the Rat Was White: An Historical View of Psychology* (New York: Harper & Row, 1976), p. 126. See also Harry Washington Greene, *Holders of Doctorates among American Negroes: An Educational and Social Study of Negroes Who Have Earned Doctoral Degrees in Course, 1876–1943* (Boston: Meador, 1946), pp. 204–205. The date of Cotton's Ph.D. degree was verified by a telephone communication with Denise Sewell, Registrar's Records Office, University of Chicago, 29 April 1996. Sewell also confirmed the date of Cotton's M.S. "for informational purposes only, not for official verification" (Sewell, ibid.).

PATRICIA SUZANNE COWINGS
Developing Wellness Strategies for Space and on Earth

From the very beginnings of America's space program, when the lists of candidates for astronaut positions were winnowed down to the famous Mercury Seven, psychology has played as crucial a role as rocket science or astrophysics. Today, research psychologist Patricia Suzanne Cowings is an important member of that tradition as the current director of the National Aeronautics and Space Administration (NASA) Psychophysiological Research Facility, a part of the Space Life Sciences Division at NASA's Ames Research Center.[1]

Speculation on how humans might deal with biomedical and psychological problems during manned space missions has long been a staple of science fiction. How would human astronauts adapt to the strong gravity forces of acceleration that are needed to escape Earth's gravity? How would they fare in the zero-gravity environment of space? How would they readapt to Earth's normal gravity after extended exposure to the microgravity environment of space? How would they deal with the adrenaline rush of traveling to new worlds, as well as the tedious time spent actually getting there? Cowings does more than speculate on such problems; she works to solve them by developing methods that enable astronauts to voluntarily control several physiological responses to a variety of environmental stressors.

Cowings entered the State University of New York at Stony Brook in 1966. As an undergraduate, she was awarded three relatively rare research assistantships in psychology: two with Professor L. Fehmi (1968–1970) and one with the psychology department chair C. Kalish (1970). After earning the baccalaureate degree in 1970 (cum laude, with honors in psychology), Cowings entered the University of California at Davis. She received a Department of Psychology scholarship (teaching assistantship, 1970–1971), followed the next year by a University of California Distinguished Scholarship Award. During her first three years at the University of California, Cowings earned a surprising amount of recognition: she spent time at the Rockefeller University of New York as a guest investigator under the supervision of N. E. Miller (1971–1972); engaged in summer research at NASA, as a research assistant (1971) and as a research psychologist (1972); and was named principal investigator at the NASA Interchange for Joint Research (CD301) with the University of California at Davis (1972–1973). Having earned the master's degree, Cowings completed her Ph.D. in psychology in only three years (1973). She had developed her specialties in physiological psychology, human learning theory, and the physiology of the autonomic nervous system.[2]

Awarded a prestigious National Research Council Post-Doctoral Associateship, Cowings conducted research at NASA's Ames Research Center (1973–1975), while continuing her work at the University of California at Davis (1972–1975).[3] Focusing on biofeedback techniques and psychophysiological methods to control space motion sickness, Cowings presented her research findings to the Space Medicine Branch at the 47th Annual Scientific Meeting of the Aerospace Medical Association in 1976, where her paper received the Best Aerospace Medicine Research Paper by a Young Investigator award.[4]

Cowings officially began her career with NASA in 1976, and in so doing became the first American woman to receive scientist astronaut training (1976–1977). In a joint program by NASA's Ames Research Center and Johnson Space Center, she trained as a payload specialist for Spacelab Mission Development (SMD II), a simulation of a mission dedicated to life sciences. She also served as the principal investigator for a Spacelab Simulation III experiment on a preventive method for zero-gravity sickness syndrome.[5] In 1977, she received both a NASA Group Achievement Award and a NASA Individual Achievement Award for her work on Spacelab Mission Development III (SMD III), as principal investigator of the project support team and as a scientist-astronaut, respectively.[6]

As the principal investigator on the planning of a series of Spacelab missions, Cowings worked steadily on designing and improving autogenic-feedback training, to ensure that the astronauts would not experience debilitating space motion sickness. Space motion sickness, which is similar to the motion sickness one might experience in a car or boat, affects approximately one-half of the astronauts who have flown in space. It involves not only individual discomforts, but is intimately associated with changes in behavior and performance such as decreased motivation, decreased muscular coordination, general feelings of malaise, inattention to detail, and a host of other problems. Because of these serious consequences, NASA has considered space motion sickness one of the most serious problems in the manned space program.[7]

Cowings's research has resulted in the Ames Autogenic Feedback Training System (AAFTS), a method for astronauts to control space sickness, for which she and NASA have a U.S. patent.[8] According to Cowings, there are three components to AAFTS. The first component is the Autogenic Feedback Training Exercise (AFTE), a six-hour training program designed to enable individuals to voluntarily control several of their own physiological responses to a variety of environmental stressors. The second component is the Autogenic Feedback System 2 (AFS-2), an ambulatory physiological monitoring and feedback system that includes a garment, transducers, signal-conditioning amplifiers, a microcontroller, a wrist-worn feedback display, and a cassette tape recorder. The third component is the Autogenic Clinical Lab System (ACLS), a physiological monitoring and training system based on a personal computer with a multiple video display adapter and Windows-environment software; the user-interactive software can directly measure and provide real-time displays of physiologic responses such as cardiac output, blood pressure, vagal tone, and total peripheral resistance.[9] Cowings's work was recognized with both NASA Group Achievement awards and NASA Individual Achievement awards in 1985 and 1993. She also received an Ames Honors Award for Excellence as a Scientist in 1985, and the Maryland Space Grant Consortium's Black Engineer of the Year Award in 1994 for her contributions to aerospace technology and science.[10]

Another space travel problem Cowings has researched is orthostatic hypotension — the low blood pressure experienced by astronauts after they return to Earth-normal gravity following extended exposure to the microgravity environment of space. Orthostatic hypotension, a symptom of cardiovascular deconditioning, has been recognized by both the American and Russian space programs as having a potentially serious impact on the health of crews aboard the Russian Mir space station and the planned international space station, as well as the crews of future planned interplanetary missions. Cowings has expanded her autogenic-feedback training to exercises that could enable individuals to voluntarily increase their blood pressure. As of this writing, her training methods were scheduled to be used on the Russian Mir space station as part of the Mir-23 mission.[11]

Cowings also explored how the techniques she was developing to control space motion sickness might also be applicable in psychosomatic disease, heart rate control, and even in promoting psychosomatic health. Her research has paid off handsomely. As NASA is supported by tax dollars, administrators in recent decades have increasingly sought to ensure that the Agency has some measure of self-support through commercial applications on its patentable breakthroughs. Cowings's system and the autogenic-feedback training exercises she has developed have commercial applications in the treatment of a range of medical problems such as hypertension, dysautonomia, autonomic neuropathy, chemotherapy-related nausea, and air and sea sickness. According to Cowings, the training exercises are also useful for alleviating low blood pressure in patients with diabetes, reducing spinal cord lesions or generalized somatic paralysis, and modifying central nervous system activity in the treatment of neuropathologic disorders such as epilepsy, attention deficit disorder, and mild head trauma. Military, commercial, and private pilots could use AFTE to help to control the physiologic arousal associated with emergency flying conditions, thus reducing the risk of human error accidents and improving crew performance and safety.[12] The system's ambulatory monitoring can allow users to determine the impact of motion sickness, fatigue, and sleep deprivation in both military and civilian operators of land, air, and sea vehicles, and help evaluate and pinpoint corrective countermeasures. Neurofeedback training has been used to alter brain activity, resulting in the ability to facilitate sleep and modify the effects of sleep deprivation on cognitive performance, thereby reducing disturbances in circadian rhythms. Individuals subject to fatigue, jet lag, insomnia, high-stress work environments, and motion sickness can use AFTE to reduce the physiologic effects of a panoply of environmental stressors.[13]

Cowings continues her work. In 1989, she won the Federal Women's Program Award, presented by the U.S. Army Tank Automotive Command, as well as the Candace Award for Science and Technology, presented by the National Coalition of 100 Black Women. In 1990, she received the Innovative Research Award, presented by the Biofeedback Society of California. Since 1991 she has also served as an Adjunct Assistant Professor of Psychiatry at the Neuropsychiatric Institute of the University of California at Los Angeles Medical School. In 1995, she was named Acting Assistant Chief of the Gravitational Research Branch, Life Sciences Division, and she is presently Director of the Psychophysiological Research Facility at the Space Life Sciences Division of NASA's Ames Research Center. Her goal is to provide basic research information on the mechanisms of behavioral or higher central nervous system influences on various space-related syndromes. The control or elimination of these mechanisms will also have earthbound applications. Her work on psychophysiologic methods is expected to figure greatly in the enhancement of crew efficiency, performance, and wellness during planned future forays into space, both on international space station missions and during extended missions in space in the twenty-first century.[14]

NOTES

1. Patricia Suzanne Cowings, fax communications to author, including curriculum vitae and current job description, 18 March 1996; e-mail messages to author, 22 March 1996 and 31 March 1996.

2. Cowings, current job description.

3. Cowings, curriculum vitae.

4. See the appendix for list of Cowings's published papers.

5. Cowings, curriculum vitae.

6. Ibid.

7. Cowings, job description.

8. "Autogenic-Feedback Training Exercise (AFTE) Method and System," United States Patent 5,694,939; 9 December 1997.

9. Cowings, job description. Cowings was the principal investigator on the 1984–1985 test procedures to validate AFT hardware as a treatment for space motion sickness. Her method was launched in Department of Defense mission STS51-C in January 1985 (Cowings, job description). Between 1985 and 1988, she served as principal investigator on a joint project between NASA's Ames Research Center and the United States Air Force, and was contracted to conduct space shuttle flight experiments on the effectiveness of AFT as a treatment for engineers suffering from space adaptation syndrome during DOD shuttle missions. She was also principal investigator for NASA Ames and the USAF AMD Brooks Air Force Base on ground-based physiologic data related to motion

sickness tests of all DOD payload specialist candidates (1983–1988). As NASA's investigator, she also collaborated with the Japanese Space Agency on a project on human health monitoring in space (1988–1992), and was principal investigator for Spacelab-J experiments on AFT as a preventative method for space motion sickness (Cowings, curriculum vitae).

10. Ibid.
11. Cowings, curriculum vitae.
12. Cowings, job description.
13. Ibid.
14. Ibid.

MARIE MAYNARD DALY

Applying Biochemistry to Medical Research

Biochemist Marie Maynard Daly's career can be viewed as realizing her father's dream. Born in Corona, New York, in 1921, into a family which placed great emphasis on education, Daly showed evidence of giftedness quite early.[1] Her father, who had been born in the West Indies, came to the United States as a young adult in hopes of receiving a scholarship to study chemistry at Cornell University. Although he received a tuition scholarship at Cornell, he was unable to meet his other living expenses and was forced to drop out after only one semester. He took a job as a postal clerk in New York City, and married Daly's mother, Helen Page, who was a native of Washington, D.C. Shortly after Marie Daly's birth, her mother's parents moved from Washington to New York.[2] Like so many of the women in this study, Daly's family encouraged reading, and from early childhood she had access to the books and other reading materials her grandfather maintained in his home library. According to Kessler, her grandfather expressed pleasure when she showed an interest in his books, and along with her parents, strongly encouraged her schoolwork.[3]

Perhaps as a reflection of her father's earlier interest in chemistry, Daly especially enjoyed reading about scientists and their work—Paul DeKruif's *The Microbe Hunters* was an early source of inspiration.[4] Daly attended Hunter College High School, an all-girls school affiliated with the City College of New York, where her science teachers encouraged her to study chemistry at the college level. After graduation, she received a scholarship to Queens College. According to Kessler, because Queens was then a new college, classes were small—a distinct advantage in terms of students receiving high-quality instruction in science courses, particularly with regard to laboratory work.[5] Daly earned the a bachelor's degree in chemistry, with honors, in 1942.

As it had earlier been with her father, money was a problem for Daly, but the faculty in the chemistry department at Queens College offered her a part-time job as a laboratory assistant and a fellowship to study for the master's degree in chemistry at New York University. She was awarded her M.S. in 1943, and she remained on the staff at Queens though 1944, when she enrolled in the doctoral program at Columbia University.[6]

With the heightened need for scientists during World War II, opportunities opened up for women. Daly received a university fellowship at Columbia, where she studied with Mary L. Caldwell, a nationally recognized specialist in nutritional chemistry.[7] During her dissertation year (1947–1948), Daly began her professional career as an instructor of physical science at Howard University, where she was hired by the noted Black physicist Herman R. Branson.[8] Daly focused on the ways in which chemicals produced in the body aid digestion. Daly received her doctorate in 1948, becoming the first Black woman to earn this degree in chemistry, and she was elected to the Sigma Xi honor society. Her dissertation was titled "A Study of the Products Formed by the Action of Pancreatic Amylase on Corn Starch."[9] Although she wanted to apprentice with A. E. Mirsky at the Rockefeller Institute of Medicine in New York City, he said she would need independent funding to do so. Daly applied for a grant from the American Cancer Society (ACS), and took a temporary job at Howard University, under Herman Branson, while she waited for a decision from the ACS.

Daly received the ACS grant in 1948, left Howard, and began work with Mirsky at the Rockefeller Institute, where she was the only Black scientist.[10] She remained there for seven years, conducting cancer research focusing on the ways proteins are constructed within cells, particularly in cell nuclei.[11] This was a particularly fruitful time in biochemistry, particularly for the study of the chemistry of cell nuclei—the function of organic acids in cell nuclei had recently been discovered, and in 1952, Watson and Crick described the structure of DNA. Two other Rockefeller Institute scientists—Lenor Michaels and Francis Peyton Rous —apparently helped Daly with discussions about the new DNA breakthroughs.[12]

In 1955, Daly became a research assistant at the College of Physicians and Surgeons at Columbia University, where she began a series of studies with Quentin B. Deming on the underlying causes of heart attacks.[13] Their research eventually focused on the role of cholesterol and other fatty substances in blocking the supply of oxygen to the heart.[14] In 1960, Daly and Deming moved their research projects to the Albert Einstein College of Medicine at Yeshiva University, where she was advanced to the rank of assistant professor of biochemistry; by 1971, she had advanced to the rank of associate professor of both biochemistry and medicine.[15] Although Daly continued to be involved in cancer research and she also did research on nucleic acids and other genetic materials, her research at Albert Einstein College generally focused on heart disease—specifically on the various causes of circulatory disease. Her work on arteriosclerosis was funded by the American Heart Association from 1958 to 1963, and she was a research scientist at the Health Research Council of New York from 1962 to 1972.[16] At various times in her career, she also did studies on the effects of sugar on arteries, and performed some interesting work on the effects of cigarette smoke on lung function.[17]

Daly was active in the National Organization for the Professional Advancement of Black Chemists and Chemical Engineers, and worked both there and in other venues to increase the participation of Black women in the chemical fields. To this end, she created a scholarship fund for minority students to study physics and chemistry at Queens College.[18] A fellow of the American Association for the Advancement of Science and the New York Academy of Sciences, Daly retired from Albert Einstein College of Medicine in 1986.

NOTES

1. James M. Jay, *Negroes in Science: Natural Science Doctorates, 1876–1969* (Detroit: Belamp, 1971), p. 60. See also *American Men and Women of Science: The Physical and Biological Sciences*, 15th ed. (New York: Bowker, 1971–), p. 488. Julius H. Taylor, Clyde R. Dillard, and Nathaniel K. Proctor, eds., *The Negro in Science* (Baltimore: Morgan State College Press, 1955), p. 182. *Who's Who among Black Americans* (Northbrook, Ill.: Who's Who among Black Americans Publishing, 1985), p. 199. "Biochemists," in *Contributions of Black Women to America*, ed. Marianna W. Davis, vol. 2 (Columbia, S.C.: Kenday Press, 1982), pp. 439–440. "From Here and There during the Month: Young Negro Scientist Engaged in Cancer Research," *Interracial Review* 24 (June 1951), p. 110.

2. James H. Kessler, J. S. Kidd, Renee A. Kidd, and Katherine A. Morin, *Distinguished African American Scientists of the Twentieth Century* (Phoenix, Ariz.: Oryx Press, 1996), pp. 57–58.

3. Ibid., p. 58.

4. Paul DeKruif, *The Microbe Hunters* (New York: Harcourt, Brace, 1926).

5. Kessler, *Distinguished African American Scientists*, p. 58.

6. Ibid., pp. 58–59. See also "Biochemists," p. 439; and Vivian Ovelton Sammons, *Blacks in Science and Medicine* (New York: Hemisphere, 1990), p. 67.

7. Kessler, *Distinguished African American Scientists*, p. 59. Mary L. Caldwell had earned her doctorate in chemistry at Columbia University in 1921.

8. Joan Rather, "Programs Provide a History Lesson—Blacks and Women in Science," *Science* 23 (April 1983): 186–187.

9. Sammons, *Blacks in Science and Medicine*, pp. 66–67.

10. "From Here and There," p. 110.

11. Sammons, *Blacks in Science and Medicine*, pp. 66–67.

12. Ibid.

13. Kessler, *Distinguished African American Scientists*, p. 60. Quentin Deming had earlier done work on how various chemicals influence the workings of the heart muscle.

14. Kessler, *Distinguished African American Scientists*, p. 60. Daly and Deming "addressed basic questions about the production of cholesterol and fatty products to clog the arteries." See also "From Here and There," p. 110.

15. Kessler, *Distinguished African American Scientists*, p. 60. See also *Who's Who among Black Americans*, p. 199.

16. Kessler, *Distinguished African American Scientists*, p. 60.

17. Ibid.

18. Collection of the National Organization for the Professional Advancement of Black Chemists and Chemical Engineers, Special Collection Archives, Robert W. Woodruff Library, Atlanta University Center. Daly was the moderator on the panel "Black Women in Chemistry, Biochemistry and Chemical Engineering: Confronting the Professional Challenges," 3 May 1980. Her name appears on various membership lists, beginning in the 1970s.

CHRISTINE VONCILE MANN DARDEN
NASA Aerospace Engineer, Sonic Boom Expert

Christine Voncile Mann Darden
Photo courtesy of NASA

People are always forming expectations of me because I am Black and because I am a woman, which means that every time I meet someone new or am thrust into a new situation, I have to "prove myself" again. White males are generally expected to be competent until they prove themselves otherwise. For us it is the exact opposite.

—*Christine Darden*[1]

Christine Voncile Mann Darden's career provides the perfect illustration of how a determined woman with degrees in mathematics and engineering can succeed in the predominantly white male world of aerospace engineering research. As an aerospace engineer at the Langley Research Center of the National Aeronautics and Space Administration (NASA), Darden has performed vital research in sonic boom prediction, sonic boom minimization, and supersonic wing and flap design. She has explored computational and analytical methods in supersonic aerodynamics, including wind-tunnel and flight test programs to validate or discard methods and designs. Since joining NASA in 1967, Darden has seen many changes: "I have stood on the shoulders of some women who were at NASA ahead of me, and a lot of women and blacks who are here now are standing on my shoulders. . . ."[2]

Darden was born on September 10, 1942, in Monroe, North Carolina, and grew up in a family atmosphere where spirituality and education were paramount.[3] Her mother, who had come from Pennington, Alabama, was the youngest of eleven children and had gone away to boarding school at Annie Manie, Alabama, with the help of her mother and some of her older sisters. Her mother had then gone on to complete high school at Knoxville College, a Presbyterian school with missionary-type teachers. According to Darden, "Most of her older sisters and brothers were not educated, so Mother was very lucky in this respect."[4] After high school, her mother attended college at Knoxville. Darden's father hailed from Curryville, Georgia, and he was the youngest of nine children. His father, a farmer, had been able to send his youngest two sons away to finish high school and college at Knoxville College. Darden's parents met at Knoxville, and they married after he graduated from college in 1927; her mother had only completed her sophomore year by then. The young couple then started out as teachers in Georgia, but after their fourth child was born it became too difficult to make ends meet, especially during the summers when teachers were not paid. As a result, her father decided to change careers, and moved the family to Charlotte, North Carolina, where he took a job with the North Carolina Mutual Life Insurance Company.[5]

Darden recalls the value that her parents placed on education:

> Both my parents felt very strongly about the value of education. I don't remember it being discussed a lot, but everybody in the house grew up knowing that they were going to college. Being much younger, I saw all my sisters and brothers go off to college (three were in school at the same time on two different occasions).[6]

Two of Darden's siblings graduated from Johnson C. Smith College, one graduated from Fisk University, and another graduated from Tennessee State. Both her sisters are teachers; one brother completed the seminary at Johnson C. Smith and became an ordained Presbyterian minister, as well as completing a degree in dentistry at Meharry Medical College and serving on the Meharry faculty.[7] Her other brother was the second Black in the medical school at the University of North Carolina; after completing his degree, he practiced in Denver for a number of years and is now vice-president for quality control at a Denver hospital. Darden firmly believes that the accomplishments of her siblings illustrate "the ideas of education, honesty, hard work, and character that permeated our household."

> People have often asked me if we were poor—I don't remember thinking about it as a child. We were warm, fed, and Dad always had a car. We certainly had a lot less than the neighborhood doctor, but we were comfortable. I remember my Dad asking me every night what I had learned that day. He also spent a lot of time playing cards with me and taking me along on his insurance routes. My parents were also very strong in church attendance, putting on church programs, and so on. My spiritual life has always been very important. . . .
>
> My parents were always very active in the church. Dad was an elder who took me to Sunday school every Sunday. Mom often presented programs and plays

for the children at church. . . . I grew up in the church. I saw my father on his knees every night before he went to bed.[8]

Darden does not recall exactly how old she was when she started to read, but she recalls starting school at age three, and remembers people commenting on her being so small and being able to read. At that time, her mother taught grades one through four in a two-room schoolhouse in Union County, North Carolina. Being a strong cotton growing area, the school year started in July and went on hiatus in September so children could help with the harvest. The July before her fourth birthday, Darden recalls, her mother "took me to school with her mainly as a way of babysitting me. She allowed me to go out and play, but since I had no one to play with, I usually stayed in the classroom and did the work. After that first year, she sent me to the [segregated] public school in Monroe, [where I wound up] in the second grade [at age four]."[9]

> There were all kinds of books and magazines around my house, but I don't remember exactly what I read. There were four children older than I — from eight to thirteen years older — and they probably helped me in reading. I also learned to play all sorts of card games from them at a very early age.[10]

Darden also remembers going to the library up the street from her house and checking out books, especially in the summertime.

As she neared her teenage years, Darden was sometimes home alone for a couple of hours in the afternoon after school, and she also recalls that she was "beginning to put a lot of pressure on [my mother] about dating and going to dances like my classmates (who were two years older) were doing. I think maybe she also wanted to get me away from some of the influence of Monroe."[11] Darden's mother decided that a boarding school would be the best choice, and sent her to Allen High School, a segregated parochial school that had been established by the Methodist Episcopal Church for the specific purpose of educating poor Black girls from the Appalachian Mountain area. A majority of the girls in the school came from single-parent homes, making Darden one of a very few who came from a two-parent home. Every girl at the school, regardless of income level, had to do duty work each day — cleaning a bathroom, a classroom, a hall, or waiting on tables in the dining hall. Girls who could not afford the tuition did scholarship work, in the form of ironing or washing clothes, washing dishes, or cooking meals.[12] Darden recalls that almost all of the teachers at Allen were white and that they behaved "almost like missionaries" — all students were required to take Bible study classes, attend weekly Methodist Youth Fellowship meetings, and learn to teach Sunday School.[13] "We had study hall every night from 7:00 to 8:30 and lights were out at 9:30," Darden said. "I believe this routine put a lot of discipline into my study habits."[14]

After graduation from Allen, Darden entered college at Virginia's Hampton Institute, an institution that had an educational philosophy similar to Allen's.[15] Darden arrived at Hampton in the late 1950s, and she recalls that several of her classmates there had been affected by various public school systems in Virginia closing their doors to avoid integration. Civil rights demonstrators staged sit-ins at other schools around the South, but at Hampton, life went on pretty much as it always had—study and scholarship were the watchwords.

Although she steadily enjoyed support from family, several teachers along the way provided Darden with needed encouragement. "My love for mathematics [had been] cemented during my junior year in high school with my geometry teacher," she said. "My scores were almost perfect in her class and my love for math really blossomed." At Hampton, her freshman physics teacher related so many everyday things to physics principles that her love for that specialty developed.[16] She earned the baccalaureate degree from Hampton in 1962, with a major in mathematics and a minor in physics. After graduation, Darden taught math for a year (1962–1963) at Russell High School in Lawrenceville, Virginia, then married and bore her first child.[17] Darden continued to teach along with her child care duties, at this point at Norcom High School in Portsmouth (1964–1965). During this time Darden also began taking in-service classes at Virginia State College, in Petersburg.[18] Again she experienced some crucial support, as she recalled:

> Dr. Reuben McDaniels was the head of the math department. I did very well in his classes, and ultimately, he arranged an assistantship for me which enabled me to get my master's. He also often gave me advice on dealing with people and on making it in a real world.[19]

Darden was awarded a research assistantship in aerosol physics (1965–1966) at Virginia State by the Department of Health Education and Welfare, and she obtained continued funding the following year (1966–1967) as a graduate mathematics instructor. She completed the master's degree in applied mathematics in 1967.[20]

Immediately after completing her master's program, Darden took a position as a data analyst (1967–1973) with NASA at its Langley Research Center in Virginia. Three years later, she gave birth to her second child and, as before, continued working.[21] Meanwhile, she had begun taking engineering courses at Langley and discovered she did as well or better than the non-minority males in the classes who would be managing NASA in a few years.[22] Her success in the engineering classes helped Darden make the decision to pursue a doctorate.

In 1973, Darden received promotion to the position of aerospace engineer, and she enrolled in the doctoral program at the School of Engineering and Applied Science at George Washington University in Washington, D.C, the first integrated school she had ever attended. As with a number of other Black women scientists, it actually took longer for Darden to acquire the degree than she had expected:

> I think in many instances I have not always had the confidence in my own abilities that I should have had. I know my career progression has been slow because of race and gender but also, to some degree, because of the energy I put into school and my family.
>
> It took ten years going to school part time, working part time, and juggling the duties of Girl Scout mom, Sunday School teacher, trips to music lessons, and homemaker before I was able to complete the doctorate. And even with that

schedule, I was one of the few [of those I started out with] who was able to complete the degree. . . .

There have also been a couple of supervisors at NASA who have taken an interest in me and given me advice. I really consider all of these people mentors who have affected my life at some point. Once while working on my doctorate, when I had a particularly hard semester because of lots of family problems, I was really discouraged and ready to quit; it was one of my NASA supervisors who encouraged me to keep going.[23]

Although Darden describes herself as a part-time worker, her professional accomplishments at NASA during this period indicate far more than a part-time commitment. Between 1970 and 1983 she developed SEEB, a computer code that defines the equivalent area distributor needed for an airplane to generate a minimized sonic boom signature at given flight conditions. Darden also developed MMOC, a quasi three-dimensional higher order sonic boom prediction method, and she designed and tested the wing-body concepts to validate her theory. All sonic boom minimization work since the late 1970s has been based on Darden's SEEB code.[24]

Darden completed the D.Sc. degree in mechanical engineering, with a specialty in fluid mechanics, at George Washington University in 1983.[25] Then between 1983 and 1987, she and a co-author, Harry W. Carlson, developed AERO2S, a subsonic analysis code, and WINGDES, a subsonic/supersonic design code. Both codes are based on modified linear theory and include the effects of leading-edge thrust and vortex flow, and both are highly used in preliminary design work throughout the aerospace industry.[26]

In June 1989, Darden was named technical leader of NASA's Sonic Boom Group of the Vehicle Integration Branch of the High Speed Research Program. Responsible for devising, advocating, implementing, and coordinating the sonic boom research program within NASA and in industry and academia, Darden and her team conducted research on analytical methods for predicting aircraft aerodynamics in supersonic flow, and devised experimental programs required to support or supplement the theoretical research. Their research has both strategic and economic implications, as currently more than 50 countries have laws prohibiting commercial supersonic flight over their territory because of sonic booms. If the sonic boom could be reduced to acceptable levels, then these laws might be rescinded and the economic outlook for a supersonic aircraft—as well as for the production of supersonic aircraft—would be heightened significantly. Right now, according to Darden, the prospects are for over-water flight only.[27]

Darden has produced nearly 60 technical publications and presentations in the area of highlift wing design in supersonic flow, flap design, and sonic boom prediction, and sonic boom minimization.[28] As a group leader, Darden reviewed technical presentations and papers, planned future work with the members of the Sonic Boom Group, and authored or was principal author of thirty-four formal publications, twenty-three unrefereed technical papers, and two nontechnical papers. She has given fourteen technical briefing presentations, and received five NASA technical brief awards for her computer codes and twelve technical brief awards for her published research since 1990.

In October 1994, NASA named Darden a deputy program manager of its high speed research program involving the TU-144, a Russian-made supersonic aircraft.[29] Darden works with the flight management team—whose members come from Boeing, Douglas, NASA Dryden, NASA Lewis, Rockwell, GE, Pratt & Whitney, and NASA headquarters—in planning and managing the aircraft's modification with Black Jack engines to transform it into a supersonic flying laboratory. As part of the program she also works on the planning and coordination of the six flight experiments and two ground engine tests of the TU-144 in Russian facilities.[30] "Tremendous progress has been made in the design and testing of airplane concepts which display significant reductions in the sonic boom area," Darden says.[31]

Not only has Darden received the NASA medals for equal opportunity and for achievement in leading the sonic boom program, but also received the 1987 Candace Award for Science and Technology from the National Coalition of 100 Black Women, and the 1988 Black Engineer of the Year Award for Outstanding Achievement in Government from the Mobil Oil Corporation, the Council of Engineering Deans of Historically Black Colleges and Universities, and Career Communications, the publishers of *Black Engineer* magazine. Ordained an elder of the Presbyterian Church in 1980, she has been an active church participant on both the local and national levels. Along with her NASA duties and the other commitments that come with wide recognition as a role model, she presently struggles with another "continuing conflict"—finding more time to spend with her three grandchildren.[32]

NOTES

1. Christine Darden, e-mail questionnaire from author, 24 April 1996.

2. Ibid.

3. Christine Darden, personal communications from Langley, Va., including telephone conversation, 5 March 1996; curriculum vitae received by fax 5 March 1996; e-mail communication, 10 April 1996; e-mail questionnaire, 24 April 1996. See also *Who's Who among Black Americans* (Northbrook, Ill.: Who's Who among Black Americans Publishing, 1980). *Who's Who in Aviation and Aerospace* (Boston: National Aeronautical Institute, 1983). Speaking of People, *Ebony*, July 1977. Speaking of People, *Ebony*, January 1985. *Black Enterprises*, February 1983. "Christine Darden," *Management Magazine* 6, no. 4 (1987), inside cover. People section, *Essence*, September 1988, p. 38. Patricia Foote, *Girls Can Be Anything They Want* (New York: Julian Messner Publishers, 1980), pp. 34–39. "Engineers and Space Technologists," in *Contributions of Black Women to America*, ed. Marianna W. Davis, vol. 2 (Columbia, S.C.: Kenday Press, 1982), pp. 460–461.

4. Darden, e-mail questionnaire.

5. Ibid.

6. Ibid.

7. According to Darden, her brother retired from the Meharry faculty a couple of years ago, probably in the early 1990s (ibid.).

8. Ibid.

9. Ibid.

10. Ibid.

11. Ibid.

12. Ibid.

13. Ibid.
14. Ibid.
15. For more information on Hampton Institute, see the entry on Flemmie Kittrell.
16. Darden, e-mail questionnaire.
17. Darden's husband already had a daughter by a previous marriage.
18. Virginia State was then an historically Black college and hence segregated.
19. Darden, e-mail questionnaire.
20. Darden, curriculum vitae.
21. Ibid.
22. Darden, e-mail questionnaire. See also "Engineers and Space Technologists," p. 460.
23. Darden, e-mail questionnaire.
24. The computer code was named SEEB (which is not an acronym) because the theory used in the code was developed by a man named Seebass. Darden, e-mail questionnaire.
25. The D.Sc. at George Washington University is equivalent to the Ph.D. — different schools call their degrees by different titles.
26. Darden, e-mail communication, 10 April 1996; Darden, curriculum vitae.
27. Darden, e-mail communication.
28. Darden, curriculum vitae. NASA publishes technical briefs, which are distributed to companies and businesses that might be interested in using a technology, process, or computer code. Research items written up in *Tech Brief Magazine* receive cash awards and a certificate. NASA also publishes *Research Highlights* each year. Significant accomplishments are generally written up and put into this document for a given year, generally receiving a quicker release than a formal paper. Papers, technical briefs, and research highlights have typically been very important at NASA and are used to justify promotions.
29. Darden, e-mail communication.
30. Ibid.
31. Ibid.
32. Ibid.

GERALDINE CLAUDETTE DARDEN

Molding a New Generation of Mathematicians at Hampton College

Once a student of Marjorie Lee Browne, mathematician Geraldine Claudette Darden is currently chair of the Mathematics Department at Hampton College.[1] Born in Nansemound County, Virginia, in 1936, Darden attended segregated elementary and secondary schools before enrolling at Hampton Institute, where she received the baccalaureate degree with honors in mathematics in 1957.[2]

Although, by her own account, Darden was a good student, no one had ever suggested that she might consider graduate school. Instead, she intended to teach mathematics at the secondary school level, because people in those jobs provided the only role models she was aware of. After her graduation from Hampton Institute in 1957, Darden taught at S. H. Clarke Junior High School in Portsmouth, Virginia, for two years. By the end of the first six months, she said, she realized that secondary teaching was "a mistake" for her.[3]

The successful launch of the Russian satellite Sputnik in 1957 created a window of opportunity for Darden and other aspiring scientists and mathematicians in the United States. To close the sudden space gap, the federal government became generous with funding in the sciences and related fields. In the summer of 1958, Darden received a National Science Foundation grant to attend a summer institute in mathematics held at North Carolina Central University.[4] The institute, directed by mathematician Marjorie Lee Browne, was a turning point for Darden: when she initially reported to Browne's office, the first thing Browne asked her was "Why aren't you in graduate school?"[5] From that time on, Darden said,

> [Browne] encouraged me . . . and during that summer I had one of the most intense studies in linear algebra that anyone has ever had in any institution. She was a very good teacher; she knew her algebra and she was very demanding. As a result of that I went to graduate school at the University of Illinois, in September 1959, on an Academic Year Institute [Fellowship].[6]

Despite receiving Browne's support and the fellowship, Darden was still unsure of her abilities: "I was still leery about this mystical Ph.D.—it frightened me to death. The institute at Illinois was good, and although I did very well, I was still not sure that I could go to a regular graduate mathematics class and compete."[7]

Darden received the master's degree in mathematics education in 1960, and accepted a position as an associate professor of mathematics at Hampton Institute, where she discovered that she enjoyed teaching at the college level.[8] "I decided that college teaching was what I wanted. But in order to stay at the college level, I would have to return to graduate school, and there was that graduate education in mathematics again."[9]

Intending to begin "gradually," Darden asked for a leave to study for a second master's degree, but Hampton's president would only grant a leave if she studied for the doctorate. Thus, Darden enrolled at Syracuse University, which accepted some of her credits from the University of Illinois. As a result, Darden said,

> I didn't start out in the first year algebra class, but in the advanced course. I put in many hours and at the end of the year I went to pick up my final exam. I told my professor [Arthur Sagle] I had come to see how I'd done.
> He said, "You did pretty well. You came in third."
> I said, "Are you sure you know who I am?"
> I was the only female and the only black in the class.
> He laughed, and said, "No, you did well!"
> There were some people in the class who had high reputations in the department, and having come in third, I knew I had beaten somebody out. . . . From that point on, even though there were many times that I doubted that I would do it, I believed that I *could* do it.[10]

Sagle encouraged her to continue on for the doctorate, assuring her that she was well able to complete the work. Without the encouragement of Sagle, Marjorie Browne, and an earlier professor, Walter Talbot, Darden said she would not have earned the doctorate.[11] She completed the Ph.D. at Syracuse University in 1967. Her dissertation, "On the Direct Sums of Cyclic Groups," was done under the direction of James Reid.[12]

NOTES

1. Geraldine Darden, interview by author, GWSE Conference, National Academy of Science, Washington, D.C., October 1995. See also Vivian Ovelton Sammons, *Blacks in Science and Medicine* (New York: Hemisphere, 1990), p. 68.

2. Ibid. See also Patricia Clark Kenschaft, "Black Women in Mathematics in the United States," *Journal of African Civilizations* 4 (April 1982): 63–83. Kenschaft, "Black Women in Mathematics," *Association for Women in Mathematics Newsletter* 8 (September 1988): 5–7. Kenschaft, "Marjorie Lee Browne: In Memoriam," *Association for Women in Mathematics Newsletter* 10 (September/October 1980): 8–11. Geraldine Darden, personal account quoted in "Excerpts from the Association for Women in Mathematics Panel on Black Women in Mathematics," ed. Patricia Clark Kenschaft, Atlanta, 7 January 1978. My thanks to Dr. Kenschaft for providing me with a copy of this paper.

3. Darden, personal account.

4. Ibid.

5. Ibid. As with many other women at that time, Darden was so unsure of her intellectual abilities that at first she wasn't sure Browne was addressing her. Darden replied that she had had to go to work as soon as she had finished college. She reports that Browne quipped, "Oh, so you could earn that big car you are driving out there?" Darden's reply —that she needed the car because she came from a big family and that she had younger brothers and sisters at home who needed some financial help—seemed to mollify Browne.

6. Ibid.

7. Ibid.

8. Sammons, *Blacks in Science and Medicine*, p. 68. Kenschaft, "Black Women in Mathematics in the United States," p. 73.

9. Sammons, *Blacks in Science and Medicine*, p. 68. Kenschaft, "Black Women in Mathematics in the United States," p. 73.

10. Ibid.

11. Ibid.

12. Kenschaft, "Black Women in Mathematics in the United States," p. 74.

MARY LOVINIA DECONGE-WATSON

Religious Vows and Mathematics

Mathematician Mary Lovinia Deconge-Watson, also known as Sister Mary Sylvester DeConge, was born in Wycliff, Louisiana in 1933. She attended elementary and secondary schools in Louisiana, where she was "inspired by a high school mathematics teacher."[1] She entered the convent of the Order of the Sisters of the Holy Family in 1949, at age sixteen, and later took religious vows.[2] From 1952 to 1955 she taught in parochial elementary schools in the Baton Rouge and Lafayette Dioceses.[3]

In 1959, she earned the baccalaureate degree in mathematics and science at Seaton Hill College, in Greensburg, Pennsylvania, and returned to Louisiana to teach mathematics and science at Holy Ghost High School in Opelousas. During the 1961–1962 academic year, she was awarded a National Science Foundation grant to study for the master's degree at Louisiana State University, where she would earn her M.A. in 1962. She then returned to Holy Ghost High School, where she remained until 1964. Between 1962 and 1964, she also taught mathematics at Delisle Junior College.[4]

In 1964, Deconge entered St. Louis University, and received the Ph.D. in mathematics with a minor in French in 1968. Her dissertation, "2-Normed Lattices and 2-Metric Spaces," was written under the direction of Raymond Freese.[5] She taught for three years at Loyola University (1968–1971) before joining the faculty of Southern University in Baton Rouge, where she spent the remainder of her career, directing theses for the master of teaching in mathematics degree.[6] Despite the fact that her career was heavily oriented toward teaching, Sister Deconge also produced a number of research papers, particularly on Cauchy problems, during the early 1970s.[7] She is currently a professor of mathematics and the executive director of the Center for Mathematics in Science, Engineering, and Technology at Southern University.[8]

NOTES

1. Patricia C. Kenschaft, "Black Women in Mathematics in the United States," *Journal of African Civilizations* 4 (April 1982): 74. Although none of the sources specify that she attended segregated schools, it is clear that they were, as even Catholic schools in Louisiana (and in other southern states) were segregated in the 1940s.

2. Ibid.

3. Apparently, Sister Deconge taught at the elementary level in Catholic schools without a college degree.

4. Vivian Ovelton Sammons, *Blacks in Science and Medicine* (New York: Hemisphere, 1990), p. 72. See also Virginia K. Newell et al., eds., *Black Mathematicians and Their Works* (Ardmore, Penn.: Dorrance, 1980), p. 283.

5. Kenschaft, "Black Women in Mathematics," p. 75. See also Sammons, *Blacks in Science and Medicine*, p. 72; Newell, *Black Mathematicians*, p. 283.

6. Kenschaft, "Black Women in Mathematics," p. 75.

7. Sammons, *Blacks in Science and Medicine*, p. 72. See Mary DeConge, "D2 Lattices," *Notices of the American Mathematical Society* 18 (January 1971): 94 (abstract).

8. Department of Mathematics and Computer Science at St. Louis University [web page]: http://euler.slu.edu/Dept/events.html [accessed 20 December 1989].

GIOVONNAE ANDERSON DENNIS

Electrical engineer Giovonnae Anderson Dennis recalled that she had always had an interest in the field, but during her undergraduate years her classmates challenged her interest, suggesting that the proper female role would be to emotionally support, not compete with, her male classmates. "So I tried to hide by changing my major to physics," she said.[1] As the physics courses were offered in the same building as the science education courses (which were acceptable for women), everyone assumed she was majoring in education. "The irony," Dennis said, "is that by trying to blend in with the crowd, I ended up with what turned out to be a better undergraduate major for me."[2]

Like many of the women in this volume, Dennis stressed the importance of early parental support in her education:

> Every Saturday morning my brother and I would curl up in bed with our parents, and we would all talk about the things we were learning, the things there were to do in life, and all the possible careers we might follow. These sessions started when we were as young as four, and I will never forget them. In our minds, our work would be an extension of our play.[3]

That support extended itself to Dennis's penchant for nontraditional pursuits. As a child, she was something of a tinkerer. When she was eight years old she found a broken electrical extension cord, took it apart, and discovered the plug was worn out. As she recalled in a 1981 interview, she went to a hardware store to buy a new one:

> Of course the people at the hardware store assumed my parents had sent me. I went home, put the cord back together the same way I'd taken it apart, and it worked. I felt high about it for a whole week. My father loved it, my mother worried that it wasn't lady-like, but the result was that I gradually became the household fix-it person.[4]

Dennis earned her bachelor's degree in physics and her master's degree in electrical engineering from the University of California at Davis. In 1979 she completed the Ph.D. in electrical engineering, with research on the effect of non-uniform channel doping at high frequencies.[5]

Dennis began her career at the Hewlett-Packard Technology Center in Santa Rosa, California, working on the types of instrumentation used by sound and radar engineers and scientists. She was involved in the design, fabrication, and testing of solid-state devices used in test equipment for high-frequency communications and radar systems. In the early 1980s she collaborated on the development of a mixer to convert high frequency radio signals to lower frequency to facilitate more reliable signal analysis and re-propagation.

As a minority woman in a technical field, Dennis has faced a subtle form of discrimination—companies are so eager to list such individuals on their rosters they do not make enough demands on their special skills. "It's like being there is just enough," she said.[6] As a result, minorities, both men and women, often wind up languishing in entry-level jobs that do not require them to remain abreast of current trends in their fields. Dennis has a response for discrimination because she is Black and a woman:

> The best way to prepare for [discrimination] is to get as much exposure as you can to people from different cultures. . . . With those who don't know how to relate to me, I work to find a common denominator. . . . But if people just don't want to work with me because I am Black or a woman, I tell them that it's their own problem to deal with.[7]

NOTES

1. Katie Nutter, "Giovonnae Anderson," *The Black Collegian* 12, no. 1 (August–September 1981), p. 187.

2. Ibid.

3. Ibid.
4. Ibid.
5. She titled her dissertation "The Effect of Nonuniform Channel Doping in High Frequency JFET and MESFET." See Vivian Ovelton Sammons, *Blacks in Science and Medicine* (New York: Hemisphere, 1990), p. 12.
6. Ibid.
7. Ibid.

EUGENIA V. DUNN*

Biologist Eugenia V. Dunn, also known for a time as Eugenia D. Christian, spent the bulk of her career teaching science to women at Spelman College in Atlanta.[1] Dunn was born in Henderson, Kentucky, around 1916, and she received her baccalaureate degree from Louisville Municipal College (now known as the University of Louisville) in 1938. She apparently embarked on study for the master's degree immediately after graduation and received the M.S. from Atlanta University in 1940.

In February 1940, Dunn secured a teaching position in the biology department at Spelman College, most probably at the instructor's level, where she remained until 1943. Then, in February 1943, she moved on to Bethune-Cookman College, in Florida, and stayed until the end of the 1944 spring semester. She returned to Spelman for the academic year 1944–1945, before returning to Bethune-Cookman, where she was named head of the biology department.

Dunn remained at Bethune-Cookman until 1948, when she returned again to Spelman College as the head of its biology department.[2] Eleanor Ison Franklin, one of her students, is discussed in an entry later in this volume.

NOTES
1. I have no substantive information on the name change, which was fairly brief. I assume that Dunn was married briefly and then divorced. If she had been widowed, she would most likely have retained her husband's name.
2. See *Catalogue of Spelman College* for the years 1941–1943, 1944–1945, 1949–1950, 1952–1953, and 1953–1954. See also Vivian Ovelton Sammons, *Blacks in Science and Medicine* (New York: Hemisphere, 1990), p. 54; James M. Jay, *Negroes in Science: Natural Science Doctorates, 1876–1969* (Detroit: Belamp, 1971), p. 176.

GEORGIA M. DUNSTON
Medical Researcher

Microbiologist Georgia M. Dunston has been a member of the graduate faculty at Howard University in the departments of genetics and human genetics, microbiology, surgery, and oncology.[1] Dunston earned a B.S. degree in biology at Norfolk State College in 1961. She was awarded a Carver Research fellowship, and earned the M.S. degree, also in biology, at Tuskegee Institute in 1965. She then attended the University of Michigan at Ann Arbor, where she made the decision to direct her career path toward working at a Black institution:

> I made a commitment that I would give the best that I could to my studies with the desire that at some point I would be *able* to give my best . . . through research and teaching to make a significant contribution to the Black community.[2]

She earned the Ph.D. in human genetics in 1972. She declined a postdoctoral fellowship in Salt Lake City, and instead took a position as an assistant professor at Howard University School of Medicine.[3]

In 1975–1976, Dunston accepted a postdoctoral fellowship at the National Cancer Institute, where she specialized in tumor immunology. She also completed additional study, supported by a United States Public Health Service traineeship grant. Dunston moved from the study of cancer to research in a broader range of diseases with possible genetic links. She consulted or was a member of various projects for the Job Corps Sickle Cell Anemia Program of the United States Department of Labor, the Cancer Coordinating Council for Metropolitan Washington, the National Institute of Environmental Health Sciences, and the Genetic Basis of Disease Review Committee of the National Institute of General Medical Sciences. In 1982–1983 she served as a scientist with the National Cancer Institute Laboratory of Immunodiagnosis, investigating the immunogenetic properties of human natural killer cells, and received grants from several agencies of the National Institutes of Health.

After Dunston was appointed director of the Human Immunogenetics Laboratory at Howard University in 1985, she began to focus her research in two areas. She is interested in diseases particularly prevalent in Black communities, such as diabetes, vitiligo, Grave's disease, hypertension, and cancer. And she also investigates human major histocompatibility complex (MHC) genes and antigens that are unique within the American Black population. Not only would better identification and definition of these MHC genes and antigens improve matching techniques for kidney and bone marrow transplants, but this might also serve as a unique anthropological probe to study the organization of MHC, providing new information on the fundamental role of individual genes and antigens in immunological processes. In 1988–1989, Dunston became the co-principal investigator on a grant from NIAID for a five-year program to screen, characterize, and acquire reagents for histocompatibility testing of transplantation antigens in Blacks and Native Americans.[4]

Dunston is a recipient of a Howard University Excellence in Teaching Award, and has published more than forty articles and abstracts, including research on the role of specific genes in insulin-dependent diabetes mellitus among the Black population.[5]

NOTES

1. Rosalyn Mitchell Patterson, "Black Women in the Biological Sciences," *Sage* 6 (fall 1989): 11.

2. Dunston, quoted in Patterson, "Black Women."

3. Ibid.

4. Ibid.

5. Patterson, "Black Women." G. M. Dunston, C. K. Hurley, R. J. Hartzman, and A. H. Johnson, "Unique HLA-D Region Heterogeneity in American Blacks," *Transplantation Proceedings* 19 (1987): 870–871. She has also been named to the Eminent Scholars Research Program at Norfolk State University.

BARBARA JEANNE DYCE*

Biochemist Barbara Jeanne Dyce was born in Chicago, Illinois, in 1928. Various references do not indicate where she earned the baccalaureate degree, other than that she attended a number of colleges. She earned the M.S. degree from the University of Southern California School of Medicine in 1971, but did not advance to the doctoral level.[1] She apparently began her career as an assistant professor of pharmacology and medicine at the USC School of Medicine in Los Angeles, where she became laboratory director of research in biochemistry. Her research focused on methods of cancer detection and treatment, as well as the treatment of cystic fibrosis. She was also the technical director of the Radioimmunoassay Laboratory of Southern California, and published articles in a variety of scientific journals.[2]

NOTES

1. Among the schools Dyce attended are Loyola University, the University of Illinois, Evansville College, the University of Chicago, and the University of Illinois Medical School. Vivian Ovelton Sammons, *Blacks in Science and Medicine* (New York: Hemisphere, 1990), p. 80. See also *Ebony Success Library*, vol. 1 (Chicago: Johnson Publishing, 1973), p. 101. *Who's Who among Black Americans*, 4th ed. (Northbrook, Ill.: Who's Who among Black Americans Publishing, 1985), p. 243. "Biochemists," in *Contributions of Black Women to America*, ed. Marianna W. Davis, vol. 2 (Columbia, S.C.: Kenday Press, 1982), p. 440.

2. The Radioimmunoassay Laboratory of Southern California is located at 10510 Chapman Avenue, Garden Grove, California 92640.

ANNIE EASLEY*

Annie Easley was included among the twenty-four biographies in the government's *Black Contributors to Science and Energy Technology*.[1] Described as a native of Birmingham, Alabama, where she was born in 1932, Easley joined the National Aeronautics and Space Administration's predecessor agency, the National Advisory Committee on Aeronautics (NACA), in 1955. She worked at NASA's Lewis Research Center in Cleveland, Ohio, developing and implementing computer codes used in solar, wind, and other energy projects. She worked on "studies to determine the life of storage batteries (such as those used in electric vehicles) and to identify energy conservation systems that offer the greatest improvement over commercially available technology."[2] While working at NASA, she earned an advanced degree in mathematics from Cleveland State University in 1977, and continued to attend specialized courses over the years at NASA.[3]

NOTES

1. *Black Contributors to Science and Energy Technology* (Washington, D.C.: U.S. Department of Energy, Office of Public Affairs, 1979), p. 19. Crystal Wade, NASA Lewis Research Center, Office of Human Resources, had no data on Easley (telephone interview, 12 April 1996).

2. *Black Contributors*, p. 19.

3. Ibid. Although neither contain additional information, see also "Engineers and Space Technologists," in *Contributions of Black Women to America*, ed. Marianna W. Davis, vol. 2 (Columbia, S.C.: Kenday Press, 1982), p. 461; and Mary Mace Spradling, ed., *In Black and White: A Guide to Magazine Articles, Newspaper Articles, and Books Concerning More Than 15,000 Black Individuals and Groups*, 3rd ed., 2 vols. (Detroit: Gale Research, 1980), p. 289.

CECILE HOOVER EDWARDS

Howard Nutritionist

Even though nutritionist Cecile Hoover Edwards has more than 140 publications to her credit, little has been written about her.[1] Born in East St. Louis, Illinois, in 1926, Edwards attended de facto segregated schools there, and graduated from high school with honors. She earned the baccalaureate in nutritional chemistry at Tuskegee Institute in 1946, and was then awarded a Carver Foundation research fellowship sponsored by the Swift Meat Packing Company to study for the master's degree.[2] Tuskegee awarded her the M.S. in organic chemistry in 1947. Immediately after obtaining her master's degree, Edwards received a two-year General Education Board fellowship to enter the doctoral program at Iowa State University; during her final year of study she held an Iowa State University Department of Foods and Nutrition research assistantship. Edwards earned the doctorate in nutrition in 1950 and was elected to the Sigma Xi honor society; her dissertation was "Utilization of Nitrogen by the Animal Organism: Influence of Calorie Intake and Methionine Supplementation on the Protein Metabolism of Albino Rats Fed Rations Low in Nitrogen and Containing Varying Proportions of Fat."[3]

Edwards began her career as an assistant professor of foods and nutrition at Tuskegee Institute in 1950, and became head of the department in 1952. Concurrently, she was also a research associate at the Carver Foundation.[4] During this period, Edwards obtained a number of research grants: she was the principal investigator on a Carnegie research project (1951–1952); the director of the Amino Acid Analysis Project, contracted by the Human Nutrition Research Division of the U.S. Agricultural Research Service (1952–1955); the principal investigator on a National Institutes of Health project (1952–1956); and the principal investigator on a Tuskegee Foundation nutrition project (1953–1954). All of her research projects were conducted either at the Tuskegee Institute or the Carver Foundation laboratories.[5] Either alone or with collaborators, she produced nine research papers between 1948 and 1956.[6] At some point during this period, she met and married Gerald Alonzo Edwards.[7]

Edwards left Tuskegee in 1956 to become a professor of nutrition and research at North Carolina State Agricultural and Technical University; she was named chair of the Department of Home Economics in 1968, and remained there until 1971.[8] Physical chemist Gerald Edwards apparently accompanied his wife to North Carolina, where the pair began a long research collaboration. Their first co-authored paper was published in 1959 in the *International Journal of Applied Radiation and Isotypes*.[9] While at North Carolina State, Edwards continued to attract grant money: she was principal investigator on a National Institutes of Health project from 1956 to 1968, was director of the Undergraduate Research Participation Program in Nutrition (1960–1966), and was director of the Vegetable-Protein Research Project (1964).[10] She was chair of the Department of Home Economics from 1968 to 1970.[11]

Between 1957 and 1971, Edwards authored or co-authored more than 35 research papers. Her research focused extensively on the amino acid composition and availability in various foods, the utilization of protein from vegetable sources, the utilization of methionine for nutritional needs as well as its use as a potential protective agent against irradiation, and the use of timatine and digitonin as precipitating agents in the elimination of cholesterol. She was also interested in the nutritional needs of women and children, particularly poor women and children, both in the United States and in India. Finally, Edwards was a pioneer in the study of pica—a then little-known dietary aberration that expresses itself as an abnormal craving for non-food substances such as dirt, paint chips, cornstarch, or clay.[12]

Although a majority of the Black women scientists working during the 1940s and 1950s did not know one another, the women in the nutrition and home economics field appear to have been an exception. Nutrition and home economics were largely female-dominated fields during that period, and the field was well organized in terms of networking.[13] The letters in the Flemmie Kittrell Collection, for example, indicate a great deal of correspondence among the various home economists and nutritionists, not only on a national but also an international scale. Additionally, the high level of government funding in these fields was also an important factor, as were the number and vigor of the fields' professional journals.

Edwards was well placed at Tuskegee Institute and the Carver Foundation in the early to mid-1950s. Both the Institute and the Foundation sponsored symposia and conferences that were well attended, particularly by Black scientists working in the chemistry, nutrition, and home economics fields. Flemmie Kittrell visited Tuskegee in 1952, and developed a close working and personal relationship with Edwards. As early as 1952, Kittrell had begun a correspondence with Edwards, and her letters over the years clearly indicate a high awareness of Edwards's work. As a result, when Kittrell began to approach retirement, she appears to have favored

Edwards as the best candidate to take over her position as head of the Department of Home Economics at Howard University.[14]

Although Edwards was still officially listed as chair of the Department of Home Economics at North Carolina State A&T University for the 1970–1971 academic year, she apparently took a leave of absence and accepted a position on the Howard University home economics faculty during that period. At the end of the 1971 spring semester, she officially left North Carolina; in August 1971, she took up a one-year post as chair of the Howard University Home Economics Department, replacing Kittrell who was on leave in Africa.[15] Kittrell wrote to another friend and colleague, Marguerite Burk, that "we are fortunate in having Dr. Cecile Edwards as our new Chairman. She was the choice of our Home Economics faculty."[16] Kittrell returned briefly as chair in 1972; when she retired in 1973, Edwards became permanent chair. The friendship between Kittrell and Edwards was indicated by Edwards hosting Kittrell's retirement party in her own home.[17] Edwards apparently shared Kittrell's interest in India, and it is likely that she also spent time teaching nutrition and home economics there, as some of her publications indicate.[18]

Edwards and her husband continued their research after the move to Howard. She also served as chair of the White House Conference on Nutrition (1969) and was a member of the Advisory Committee to the Director of the National Institutes of Health (1972–1975). In 1978, Edwards received a Ford Foundation fellowship to serve as consultant in nutrition at the University of Khartoum in the Sudan.[19] In 1974, Flemmie Kittrell's long cherished dream of elevating the Department of Home Economics into the School of Human Ecology finally occurred. Cecile Edwards was named Dean of the School of Human Ecology at Howard University in 1974, and held the position until her retirement in 1990.[20]

NOTES

1. Most of the information for this study of Cecile H. Edwards has been pieced together from data contained in the Flemmie P. Kittrell Collection housed in the Moorland-Spingarn Research Center at Howard University. I tried for more than two years to contact Edwards through letters and by requests sent by mutual acquaintances, all to no avail.

2. Cecile H. Edwards, curriculum vitae, quoted in "Proposal for a Program Leading to the Doctor of Philosophy Degree in Nutrition in the Department of Home Economics, Howard University, prepared by Flemmie P. Kittrell, Chair, May 6, 1971" (Kittrell Collection, Moorland-Spingarn Research Center, Howard University). See also Vivian Ovelton Sammons, *Blacks in Science and Medicine* (New York: Hemisphere, 1990), p. 82. "Food Chemist," in *Contributions of Black Women to America*, ed. Marianna Davis, vol. 2 (Columbia, S.C.: Kenday Press, 1982), pp. 440–441. *American Men and Women of Science: The Physical and Biological Sciences*, 14th ed. (New York: Bowker, 1971–), p. 804. *Who's Who among Black Americans*, 4th ed. (Northbrook, Ill.: Who's Who among Black Americans Publishing, 1985), p. 248. Julius H. Taylor, Clyde R. Dillard, and Nathaniel K. Proctor, eds., *The Negro in Science* (Baltimore: Morgan State College Press, 1955), p. 177.

3. Edwards, curriculum vitae. See also Sammons, *Blacks in Science and Medicine*.

4. Edwards, curriculum vitae.

5. Ibid.

6. See the appendix.

7. *Who's Who among Black Americans* lists Gerald Alonzo Edwards as her husband. In addition, a letter from Flemmie Kittrell to Cecile Edwards, dated February 21, 1952, closes with this remark: "It was a pleasure to meet your husband and to know about the fine work that he is doing in physical chemistry" (Kittrell Collection).

8. Edwards, curriculum vitae.

9. Ibid. Between 1959 and 1971 they co-authored, often with other collaborators, 23 papers. See the appendix for a list of publications.

10. Ibid.

11. Ibid. It appears that Edwards retained her position as chair of the Department of Home Economics at North Carolina State through 1971, although she was apparently on leave for the final year.

12. This assessment was arrived at from the content of the various publications.

13. Both the Association of Administrators of Home Economists and the American Association of Home Economists were very active. In addition, there was a great deal of activity involving nutritionists, home economists, and various businesses, as is indicated in the Kittrell Collection. Food companies regularly conducted tests or asked academic nutritionists to conduct food tests for them. There were also organizations such as Meals for Millions that apparently made and tested various diets for Third World consumption—they, too, requested testing from Kittrell, Edwards, and others in the field (Kittrell Collection).

14. See correspondence in the Kittrell Collection, Moorland-Spingarn Research Center, Howard University.

15. Letter to Cecile H. Edwards from Robert L. Owens, III, Dean, Howard University, dated August 11, 1971 (on file in the Kittrell Collection). The letter reads, "The President of the University has authorized the following recommendation, pending approval of the Board of Trustees: That Cecile Edwards be appointed Chairman of the Department of Home Economics for the period beginning August 1, 1971 and terminating automatically June 30, 1972."

16. From a letter written by Kittrell to Dr. Marguerite C. Burk, Springfield, Virginia; dated September 3, 1971 (Kittrell Collection).

17. Among those attending were Burk (see note 16), Anna B. Camp (Home Economics Department at Howard), Effie B. Crockett (who was at one time in the Home Economics Department at Howard), Dr. Lichu Hsu, Dr. Ruth B. Jefferson, Adele B. McQueen, Emma J. Mettam, Mary D. Morris (Home Economics at Howard), Dr. Barbara K. Nordquist, Josephine Price, Dr. Marianna B. Sewell (Home Economics Department at Howard), Jewel Terrell, and Dr. Clarence F. Winchester (Home Economics Department at Howard). This information is from a "thank you" list of the attendees (Kittrell Collection).

18. Definitive documentation of Edwards's interest in India is not available, but see, for example, Cecile H. Edwards, "Contribution and Obligations of Pre-primary Education in a Democratic Society," *College Annual* (Institute of Education), Mysore, India, 1968; and Cecile H. Edwards, "The Importance of Teaching Nutrition in Secondary Schools in India," *Indian Journal of Home Science* 2 (1968): 74.

19. *Who's Who among Black Americans*, p. 248.

20. The exact date of Edwards's retirement is sketchy, but it occurred sometime between 1989 and 1991—at which time the School of Human Ecology at Howard University was closed down, thus marking the end of the heyday of home economics.

ANNA CHERRIE EPPS

Microbiologist Anna Cherrie Epps was denied admission to medical school because she was female, but has nevertheless moved to positions of power as a medical school administrator. Born in New Orleans, Louisiana, in 1930, she was the daughter of a former schoolteacher, Anna Cherrie, and a physician, Ernest Cherrie Sr. From an early age, she was programmed for a career in medicine, and routinely accompanied her father on house calls. "The kind of exposure I had as a youngster was the most outstanding a young person could have. . . . The kind of admiration for my Dad and the life that he lived, the compassion he showed . . . [provided me with] a role model for demonstrating all the things that I admired."[1]

She attended Catholic schools—Corpus Christi Elementary School and Xavier University Preparatory High School—in New Orleans, before entering Howard University at age sixteen, where she studied under Margaret Strickland Collins, an early mentor, and Louis A. Hansborough.[2] She earned the bachelor of science degree in zoology in 1951 when she was only nineteen. Epps applied to medical school both at Howard and Meharry Medical College, but despite her high grades she was denied admittance to both, ostensibly because she was too young and because she "might take the place of a young man who possibly would then have to go into the military" and thus lose the chance of becoming a physician.[3] Upset by these colleges' decisions, she held one-year positions as a technologist at the clinical laboratories of Our Lady of Mercy Hospital in Cincinnati, Ohio (1953–1954), and Flint-Goodbridge Hospital in New Orleans, Louisiana (1954–1955).

Epps next obtained a position as an instructor and served as acting head of the medical technology department at Xavier University (1954–1960). Between 1957 and 1959 she was in study for the master's degree, and also worked summers as a blood bank technology specialist in the clinical laboratory blood bank department at Freedman Hospital, in Washington, D.C. She earned her M.S. in microbiology at Loyola University in 1959.

From 1959 to 1960, Epps was a part-time research technologist in the Department of Medicine at Louisiana State University, where she assisted in research on protein, mucopolysaccharides, cytology, and radioactive carbon-14. During the summer of 1960 she did radioisotope studies at the University of California at Berkeley.[4] In 1961 she became an assistant professor of microbiology at the Howard University College of Medicine.

At that point, Epps's interests had turned toward immunology, with a focus on the study of tropical and infectious diseases. And, in addition to her teaching, she embarked upon doctoral studies at Howard. Margaret Grigsby, who was also a faculty member at Howard at that time, was the resident expert in those fields.[5] Epps was awarded the Ph.D. in 1966, and was elected to Sigma Xi.[6] Her dissertation, "Immunological Responses in the Chick Embryo to Chick and Mouse Limb Bud Grafts," was done under the supervision of Louis A. Hansborough.[7] Epps's subsequent research interests have focused on immunological studies, specifically autoimmune diseases, immunological embryology, transplantation immunology, and hepatitis immunology.[8]

Epps remained at Howard until 1969, when she received a United States Public Health Service Faculty Research Fellowship in the Department of Medicine at the Johns Hopkins University School of Medicine, where she became an assistant professor. A second USPHS research fellowship allowed her to do research at Tulane University from 1969 to 1971. She then took a position as an associate professor of medicine at Tulane (1971 to 1975), and was named Director of Medicine at the Tulane Medical Center (1975–1980).[9] She became the assistant dean of student services at Tulane Medical Center in 1980, then became associate dean for student services in 1986. In 1995, she was named interim vice-president for academic affairs at Tulane, and was also named dean of the School of Medicine at Meharry Medical College, both positions that she currently holds.[10]

Epps is on the Board of Trustees of Children's Hospital in New Orleans and the Board of Regents of Georgetown University in Washington, D.C. She has served on the Minority Enhancement Awards Special Emphasis Panel of the National Cancer Institute, and the Minority Health Advisory Committee for the Centers for Disease Control (CDC), as well as the National Institutes of Health Recombinant DNA Advisory Committee. She has also been a consultant at the NIH Office of Health Resources Opportunities. She is a member of the National Research Council of the National Academy of Sciences, the National Science Foundation, and the Howard Hughes Medical Institute Doctoral Fellowship Program.[11] She is a past editor of the *Medical Education Journal*.[12]

Epps has long been an advocate of efforts to recruit and retain minority students in the medical sciences, as she noted in a 1996 interview:

> The bulk of my life has been devoted to assisting minority students in pursuit of medical careers. . . . A few years back, I had a great deal of fear. . . . Just when schools were more attentive to minorities, the bottom fell out: tuition skyrocketed, other professions became more attractive, and the cream of the student crop were being skimmed off. . . . But in the last three to five years, there's been a turnaround. Twenty-five years ago the attitudes were more socially responsible. What I felt then needs to occur now . . . [but due to recent negative attitudes toward affirmative action] I'm not so sure we'll be able to revisit that.[13]

Shortly after her arrival at Tulane, Epps designed, developed, and implemented the Medical Education Reinforcement and Enrichment Program (MEdREP) to identify and recruit minority students capable of pursuing careers in health fields, particularly medicine. The initial eight-week MEdREP summer program is designed to give those college students who are "on the fence" the tools they need to become more competitive, should they decide to pursue medicine:

> [The program] gives [students] an idea of what it's like in medical school and the worth of their undergraduate education and its application to service and research. . . . The primary purpose is to address atypical students (for example,

older or students with interruptions in their education) and give them a chance to get into the mainstream so they don't have to play catch up.[14]

Once students have opted to enter medical studies at Tulane, the MEdREP program monitors their performance and progress, and provides strong tutoring, testing assistance, and even preparation for certification and licensing exams after graduation. In addition, the program frequently fosters partnerships with other institutions and has developed an outside mentoring program with physicians in the field to help bolster the success of its participants. According to Epps, 70 percent of the program's students have opted to pursue medical careers, a success rate that is twice the national average.[15]

NOTES

1. Epps, quoted in Mike Dejoie, "Doctor Anna Cherrie Epps, Ph.D.: Medical Educator and Miracle Worker," *The New Orleans Tribune*, June 1996, pp. 14–15. (My thanks to Dr. Epps for providing me with a copy of this article.)

2. Anna Cherrie Epps, personal communication, 21 June 1996; Anna Cherrie Epps, curriculum vitae, brief biographical sketch, 24 June 1996. See also Vivian Ovelton Sammons, *Blacks in Science and Medicine* (New York: Hemisphere, 1990), p. 85. *Who's Who among Black Americans*, 4th ed. (Northbrook, Ill.: Who's Who among Black Americans Publishing, 1985), p. 256. *American Men and Women of Science: The Physical and Biological Sciences*, 15th ed. (New York: Bowker, 1971–), p. 880. *Who's Who of American Women* (Chicago: Marquis Who's Who, 1979, 1981). "Immunologist," in *Contributions of Black Women to America*, ed. Marianna W. Davis, vol. 2 (Columbia, S.C.: Kenday Press, 1982), pp. 429–430. Yolanda Scott George, "The Status of Black Women in Science," *The Black Collegian*, May/June 1979, p. 70. Also see my entry for Margaret Strickland Collins (and comments on Louis A. Hansborough therein). Epps's brother, Ernest Cherrie Jr. is a physician. Epps married a physician, Joseph Epps Sr., who had two children from a previous marriage. The children also went into the medical fields: daughter Grace Epps (deceased) was a physical therapist and medical educator, and son, Joseph Epps Jr. is a neurosurgeon (Epps, personal communication). Joseph Epps Sr. died in 1984 (Epps, curriculum vitae).

3. Dejoie, "Doctor Anna Cherrie Epps," p. 14.

4. Epps, curriculum vitae.

5. See the entry on Margaret E. Grigsby in this volume.

6. Sammons, *Blacks in Science and Medicine*, p. 85; "Immunologist," pp. 429–430.

7. Epps, personal communication.

8. See the appendix.

9. Sammons, *Blacks in Science and Medicine*, p. 85.

10. Epps, curriculum vitae. Verified with Yolanda Chaisson (executive assistant) via telephone, 21 June 1996. See also Sammons, *Blacks in Science and Medicine*, p. 85; *Who's Who among Black Americans*, p. 256.

11. Epps, curriculum vitae.

12. "Immunologist," p. 439; Sammons, *Blacks in Science and Medicine*, p. 85; *Who's Who among Black Americans*, p. 256; George, "Status of Black Women," p. 70.

13. Epps, curriculum vitae.

14. Epps, quoted in Dejoie, "Doctor Anna Cherrie Epps," p. 14.

15. Ibid. Also Epps, personal communication.

PRINCILLA VIOLET SMART EVANS

The Power of Inspiration

Fisk University biochemist Princilla Violet Smart Evans grew up in more home towns than most people visit in a lifetime. Evans was born in San Antonio, Texas, the daughter of a United States Air Force career enlisted man. The family soon moved to Amarillo, Texas, and from there to Bangor, Maine, before her father was stationed in England. Eventually they returned to the United States and lived in Idaho, before her father was again assigned to England, where Evans attended high school.[1]

Evans recalls that her parents were very supportive of both her and her sister's education and studies:

> There was plenty of reading material in our home, sets of encyclopedias and so on. My sister started reading when she was three, she [eventually] majored in math. I always had an inquisitive, creative mind. My parents didn't go to college, but they encouraged us to do so.[2]

As early as junior high school, every time Evans entered a science fair at school, she would "try very hard to win."[3] She was still in junior high when she made an artificial kidney for a science fair; although she says it didn't actually work on blood, she did devise a way for it to purify the water sent through it.

Evans attended Lakenheath High School, in England, which she believes was, at that time, the only American high school in that country. At Lakenheath, she encountered a female science teacher, Miss Kaplan, who was a microbiologist by training. Evans found Kaplan's biology class both interesting and fun:

> As part of our biology class we grew germs in petri dishes. We coughed into the dishes, smeared raw hamburger onto them, scraped our skin and the inside of our cheeks into the various dishes. It was a fun, interesting project and all the "bugs" grew. I was already interested in science, but this captured me. By the tenth grade I had decided to major in science.[4]

After her graduation from Lakenheath in 1968, Evans attended Fisk University. "I went to Fisk because my father was from Nashville," she said. "He [had] told me it was a hard school to get into, and that, of course, challenged me."[5] She majored in biology because Kaplan, her role model, was a microbiologist. At Fisk she encountered a biochemistry teacher, Henry Moses, whose class she loved. Biochemistry, she said, "kind of bridged the gap between my growing boredom with biology—which just wasn't doing anything."[6]

Evans had a number of mentors at Fisk, both male and female, including the aforementioned Moses (who was adjunct faculty at Fisk and at Meharry Medical College), Fisk faculty geneticist Mary McKelvey, botanist Marian Williams, and chemist Samuel von Winbush. Prince Rivers, who has mentored generations of Black science students, also influenced Evans, although she did not actually take any of his courses.

They were all my role models, very stimulating, very demanding but very supportive. I think that's the most striking thing about the best mentors/ teachers—they are incredibly demanding. It's a shame that generally only those who teach at the large universities get recognition. Those who teach at the smaller colleges really are very influential on others' lives. Dr. [Marian] Williams had a tremendous influence on me. . . . She would come into class— she was brilliant, by the way—with never a note. She walked down the steps to the blackboard with a piece of chalk in her hand—a piece of chalk and what was in her head—and she would begin to write—plants, phyla, everything that was interesting about botany.[7]

Evans completed her bachelor's degree in 1972, and began graduate study at Howard University. Her research at Howard focused on biochemistry, and she earned her Ph.D. in chemistry in 1977. In recalling the special gift that Marian Williams had for inspiring students, Evans says, "I made up my mind that some-day I'd be able to do that, and I've succeeded."[8]

NOTES

1. Princilla Violet Smart Evans, telephone conversation with author, 29 April 1996.
2. Ibid.
3. Ibid.
4. Ibid.
5. Ibid.
6. Ibid.
7. Ibid. Mary McKelvey is chair of the Biology Department at Fisk. Marian Williams is retired. Samuel von Winbush retired from the State University of New York, College at Old Westbury in 1996. As of 1996, Prince Rivers was at the Atlanta University complex.
8. Ibid.

ETTA ZUBER FALCONER

Changing the Topography of Mathematics Education

Etta Zuber Falconer
Photo courtesy of Spelman College Archives, Atlanta, Georgia

Born in Tupelo, Mississippi, in 1933, mathematician Etta Zuber Falconer entered Fisk University when she was only sixteen years old.[1] Discovered and nurtured by mathematics professors Lee Lorch and Evelyn Boyd Granville, Falconer earned the baccalaureate degree in 1953. Strongly encouraged by Lorch, she entered the University of Wisconsin, where she earned the M.S. degree in mathematics in 1954.[2]

Falconer was an instructor of mathematics at Okolona Junior College from 1954 to 1963, and taught high school in Chattanooga for a year (1964–1964), before she became an assistant professor of mathematics at Spelman College in 1965.[3] While on the faculty at Spelman, Falconer entered the doctoral program at Emory University, and earned the Ph.D. in mathematics in 1969.[4] Her dissertation, "Quasigroup Identities Invariant under Isotopy," was directed by Trevor Evans.[5]

During the late 1960s, along with mathematician/chemist Shirley Mathis McBay, Falconer was awarded a National Science Foundation Faculty Fellowship (1967–1969) and then left Spelman in 1971 to take a position as an assistant professor of mathematics at Norfolk State College. During this period, Falconer published at least two research papers that focused on interests she had developed during her dissertation work.[6] She returned to Spelman in 1972 as an associate professor, and when McBay was named chair of the school's Division of Natural Sciences in late 1972, Falconer became chair of the mathematics department.[7] In 1975, when McBay left Spelman to take a position at the National Science Foundation, Falconer took over as chair of the Division of Natural Sciences.

Falconer continues today as the division chair.[8] During her tenure, Spelman created a chemistry department (1976), and program majors in biochemistry (1978) and computer science (1984). In less than fifteen years, first under McBay's and then under Falconer's guidance, Spelman had upgraded its science program to include strong departments of biology, mathematics, chemistry (including biochemistry), and computer science.[9] Between 1987 and 1992, more than 37 percent of Spelman's students majored in science and mathematics[10]—as Falconer noted, this is "convincing evidence of the vitality of the science program at Spelman."[11] By 1991, approximately half the students in the College Honors Program were majoring in science or engineering (one of the program's requirements is a science research project). One legacy of the combined leadership of Falconer and McBay was that by 1993 the all-female Spelman had acquired a reputation as having one of the best undergraduate science and mathematics education programs in the country.

Funding for the various science and engineering programs at Spelman is provided by the National Academy of Science's Women in Science and Engineering (WISE) Scholars Program, providing financial assistance for approximately 50 science and engineering majors per year; the National Institutes of Health's Undergraduate Minority Access to Research Careers Training Program; and the Alcohol, Drug Abuse and Mental Health Administration's Research Training Program. In 1993, the W. K. Kellogg Foundation, one of the nation's largest private philanthropic organizations, identified Spelman as one of the ten premier historically Black colleges and awarded the school a $3 million grant to establish a Center for Scientific Applications of Mathematics. The effort to secure the grant and initiate the center was driven by Falconer, by then Associate Provost for Science Programs and Policy.[12]

NOTES

1. I have deduced Falconer's age at her entry into college by piecing together the available information. Vivian Ovelton Sammons, *Blacks in Science and Medicine* (New York: Hemisphere, 1990), p. 88. See also *American Men and Women of Science: The Physical and Biological Sciences*, 15th ed. (New York: Bowker, 1971–), p. 934. Virginia K. Newell et al., eds., *Black Mathematicians and Their Works* (Ardmore, Penn.: Dorrance, 1980), p. 286. *National Faculty Directory* (Detroit: Gale Research, 1988), p. 1109. Patricia Clark Kenschaft, "Black Women in Mathematics in the United States," *American Mathematical Monthly* 88 (October 1981): 599–600. Kenschaft, "Black Women in Mathematics in the United States," *Journal of African Civilizations* 4 (April 1982): 63–83. Etta Z. Falconer, "A Story of Success: The Sciences at Spelman College," *Sage* 6 (fall 1989): 36–38.

2. Kenschaft, "Black Women in Mathematics in the United States," pp. 75–76.

3. Shirley Mathis McBay, who also taught mathematics (and chemistry) at Spelman, was on leave at that time, studying for the doctorate. (See entry for McBay in this volume.)

4. McBay returned to Spelman in 1966, and although the records are not clear as to whether Falconer took a leave to do her doctoral studies, it is probable that she would have been free to do so after McBay's return to full-time teaching. Sammons, *Blacks in Science and Medicine*. See also Kenschaft, "Black Women in Mathematics in the United States," and Newell, *Black Mathematicians*.

5. Sammons, *Blacks in Science and Medicine*; Kenschaft, "Black Women in Mathematics in the United States."

6. According to Kenschaft, "Black Women in Mathematics in the United States," pp. 599–600. The articles are Etta Zuber Falconer, "Isotopy Invariants in Quasigroups," *Transactions of the American Mathematical Society* 151 (1970): 511–526; and Falconer, "Isotopes of Some Special Quasigroup Varieties," *Acta Mathematica of the Academy of Sciences of Hungary* 22 (1971): 73–79. Also see Sammons, *Blacks in Science and Medicine*.

7. Sammons, *Blacks in Science and Medicine*; Kenschaft, "Black Women in Mathematics in the United States."

8. Although I tried repeatedly to contact Falconer by mail, and by telephone when I was in Atlanta in November 1995, to date she has not replied to any of my queries.

9. Falconer, "A Story of Success," p. 37.

10. Based on a report on the top 100 degree producers conducted by the journal *Black Issues in Higher Education* (May 1993); see "$3 Million Grant Establishes Math Center," *Inside Spelman*, 1993.

11. Falconer, "A Story of Success," p. 37.

12. "$3 Million Grant."

ELEANOR ISON FRANKLIN
Working to Increase the Number of Black Women in Science

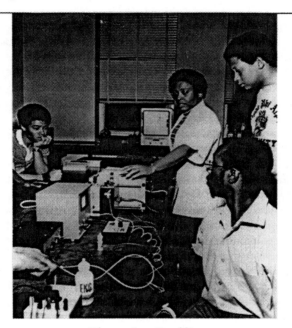

Eleanor Ison Franklin
Photo courtesy of Spelman College Archives, Atlanta, Georgia

During her career, endocrinologist and medical physiologist Eleanor Ison Franklin has been active in efforts to ease the entry of Black women into the scientific and medical professions—lately moving from research to policy making to facilitate that goal. Born in Doublin, Georgia, in 1929, she attended segregated schools, graduating as valedictorian from Carver High School in Monroe. She then attended the all-female Spelman College in Atlanta, where she studied under Eugenia V. Dunn, and received the baccalaureate in biology, magna cum laude, in 1948.[1]

Franklin remained at Spelman for the next year as a biology instructor, and then entered a master's program in zoology at the University of Wisconsin in 1949. Awarded the M.S. in 1951, Franklin returned to Spelman, where she remained until 1953.[2] She then obtained a General Education Board Fellowship from the Rockefeller Foundation for graduate study. She entered the graduate program in endocrinology and medical physiology at the University of Wisconsin, at Madison, was awarded the doctorate in 1957, and was elected to the Sigma Xi honor society.[3]

Franklin has had an active career as a research scientist and educator.[4] Immediately after completing the doctorate, she became an assistant professor in the

Department of Physiology and Pharmacology at the Tuskegee Institute's School of Veterinary Medicine.[5] In 1963, she declined an offer for postdoctoral study at Sloan Kettering Memorial Institute in New York City, and accepted a position as an assistant professor of physiology at Howard University's College of Medicine, where she became involved in cardiovascular research.[6] She developed an interest in studying the relationship between hypertension and the autonomic nervous system, and began a series of animal model investigations into the behavior of the heart and circulation in response to hypertension and chronically imposed interventions, particularly with regard to various mechanisms associated with modulation of left ventricular mass.[7] At times, she collaborated with Edward W. Hawthorne, who was also at the College of Medicine.[8]

Franklin was appointed associate dean for administration of the College of Medicine in 1970—the first woman to hold such a post—and advanced to a full professorship by 1971. She received research grants from the National Aeronautics and Space Administration's Ames Research Center, the National Institutes of Health, and the Washington Heart Association.[9] Active in the American Physiological Society (APS), she served on several of its committees, including the Porter Development Committee, which she co-chaired. Due to her involvement with the Porter Development Committee, which donates modern electronic instruments to colleges in need, she played an important role in upgrading laboratory equipment in various departments at Howard University.[10] With a Porter grant she produced a video teaching demonstration module for physiology courses, "Cardiovascular Responses to Selected Drugs and Pacing: A Student Participation Demonstration."[11]

In 1972, Franklin was named Associate Dean for Academic Affairs, a position she held until 1980, when she returned to teaching and research duties in the restructured and renamed Department of Physiology and Biophysics. At that time, she also assumed the directorship of the Cardiovascular Research Laboratory.[12] In addition to her scientific and teaching responsibilities, Franklin served on several panels of the National Science Foundation, the National Academy of Sciences, and the MARC Program of the National Institutes of Health. From 1972 to 1976, Franklin was president of the National Alumnae Association of Spelman College and was the Northeast Regional Coordinator (1979–1981). In 1973 she was appointed to the Board of Trustees of Spelman College and was chair of the Committee on Educational Policy from 1974 to 1979.[13]

From her position at Howard University, Franklin was instrumental in assembling a profile of the typical Black woman in science and medicine. She found that such women tend to be assertive, competitive, and highly motivated, resulting from their having to confront societal and professional pressures. According to Franklin, the Black woman who seeks a place in science and medicine must be "one whose identity of self is strong, whose coping mechanisms have been nurtured within a supportive ethnic environment, whose career choice is incidental to the more important need to achieve academically, and who entered an institution which traditionally (though within limits) accepted the fact that women have a role in the medical profession."[14]

In 1989, Franklin was named a graduate professor at the Howard University School of Arts and Sciences. She has served two terms as Graduate Faculty Trustee on the Howard University Board of Trustees.[15]

NOTES

1. "Alumnae News: Spelman Alumnae Elect New President," *Spelman Messenger*, August 1972, pp. 25–26. See also Vivian Ovelton Sammons, *Blacks in Science and Medicine* (New York: Hemisphere, 1990), p. 94. Rosalyn Mitchell Patterson, "Black Women in the Biological Sciences," *Sage* 6 (fall 1989): 8–9. "Toward Preparation of Black Women as Doctors," in *Contributions of Black Women to America*, ed. Marianna W. Davis, vol. 2 (Columbia, S.C.: Kenday Press, 1982), pp. 376–393. *National Medical Association Journal* 62 (September 1970): 369.
2. "Alumnae News."
3. "Alumnae News." See also Sammons, *Blacks in Science and Medicine*, and Patterson, "Black Women." Patterson incorrectly lists Franklin's graduate degrees as being in zoology.
4. "Alumnae News."
5. "Alumnae News"; Patterson, "Black Women."
6. Patterson, "Black Women."
7. Ibid.
8. See the appendix.
9. Patterson, "Black Women."
10. "The Mammalian Physiology in Action Program," *Spelman Messenger*, May 1971, p. 37. A picture accompanying the article shows Franklin at Howard, instructing students on the use of a new electrocardiogram machine. The Porter Development Committee of the American Physiological Society worked in conjunction with the Harvard University Apparatus Foundation.
11. Ibid.
12. "Black Women."
13. Ibid.
14. Quoted in Eleanor Ison Franklin, "Cultural Conflicts in Professional Training of Minority Women," in *The Minority Woman in America: Professionalism at What Cost?* (San Francisco: University of California Press, 1979), p. 29. See also "Preparation of Black Women as Doctors," p. 382.
15. Patterson, "Black Women."

SADIE CATHERINE GASSAWAY*

Mathematician Sadie Catherine Gassaway was born in Memphis, Tennessee, in 1916. She earned her bachelor's degree at LeMoyne College in 1941; her master's degree from the University of Illinois in 1945; and her Ph.D. from Cornell University in 1961 with her dissertation "The Effectiveness of Continued Testing."[1] Gassaway spent her entire career at Tennessee State University, where she moved from instructor to assistant professor of mathematics between 1949 and 1962. In 1962 she achieved the rank of full professor of mathematics, and was named chair of the Department of Physics and Mathematics in 1968.[2] She died in 1976.[3]

NOTES

1. Vivian Ovelton Sammons, *Blacks in Science and Medicine* (New York: Hemisphere, 1990), p. 98. See also *Who's Who in America*, 20th ed. (Chicago: A. N. Marquis, 1976). *Who's Who among Black Americans* (Northbrook, Ill.: Who's Who among Black Americans Publishing, 1975), p. 226. People, *Jet*, 29 July 1976, p. 14.

2. Sammons, *Blacks in Science and Medicine*, p. 98.

3. People, p. 14.

EVELYN BOYD COLLINS GRANVILLE

Research Mathematician, Teacher

The employment and funding opportunities resulting from World War II benefited mathematician Evelyn Boyd Collins Granville, as did the Cold War atmosphere and the resulting "space race" of the 1950s and 1960s. Unlike Katherine Coleman Johnson, her immediate predecessor in space science, Granville entered the field with a doctorate at the height of the country's commitment to space exploration; as a result, her career began at a higher level than Johnson's had.[1]

The second of two daughters of William Boyd and Julia Walker Boyd, Granville was born in Washington, D.C., in 1924. Her parents separated during the Depression, and Granville was raised by her mother and her mother's twin sister, Louise Walker. She attended Washington's segregated Dunbar High School, where she displayed an early gift for mathematics and graduated as one of five valedictorians in her class.[2] Although Granville was aware that segregation placed "many limitations on Negroes," she has always tended to focus on the more positive aspects of that time:

> Daily one came in contact with Negroes who had made a place for themselves in society; we heard about and read about individuals whose achievements were contributing to the good of all people. These individuals, men and women, served as our role models; we looked up to them and we set our goals to be like them. We accepted education as the means to rise above the limitations that a prejudiced society endeavored to place upon us.[3]

One of Granville's role models was the famous Black educator Mary McLeod Bethune, but she also viewed the teachers she came in contact with as strong role models. Granville noted that "the 'colored' school system in Washington attracted outstanding administrators and teachers," many of whom were exceptionally well educated, and perhaps most importantly, dedicated to uplifting the race through education.[4] In particular, Dunbar High School had a national reputation and had developed a tradition of nurturing students; it was a training ground for many future Black leaders.

Despite the constraints of segregation, Granville said she had "never heard the theory that females aren't equipped mentally to succeed in mathematics":

> And my generation did not hear terms such as "permanent underclass," "disadvantaged," or "underprivileged." Our parents and teachers preached over and over again that education is the vehicle to a productive life, and through diligent study and application we could succeed at whatever we attempted to do.[5]

Encouraged by her high school mathematics teacher Mary Cromwell, among others, Granville applied to and was accepted at both Mount Holyoke and Smith Colleges in Massachusetts, although neither school offered funding.[6] She chose Smith, even though she was unsure about whether she would actually be able to attend. At that point in Granville's life, the support of other women proved crucial. Granville's aunt, Louise Walker, was determined that she would attend a northern college. To this end, her aunt offered to pay half of her college tuition for a year. Granville also received a partial scholarship from Phi Delta Kappa, the national sorority of Black women teachers. Her mother would pay the remainder of her school fees.[7]

Granville entered Smith in the fall of 1941, and worked as a waitress at the Faculty Club to help meet her other expenses. Being Black was not a problem at what was then the largest women's college in the country—students were valued, and valued one another, on the basis of their academic achievements. By the end of her freshman year, Granville had performed so well that Smith awarded her a full tuition scholarship. Three months after Granville enrolled at Smith, the United States entered World War II; for Granville and many other Blacks of her generation, the war opened up new opportunities—she worked for three summers at the National Bureau of Standards in Washington, D.C., first as a technical aide, then as a computist, and finally as a mathematician.[8] Her summer employment at the National Bureau of Standards enabled her to acquire valuable professional experience.

By her own account, Granville's time at Smith was both happy and fruitful. In high school she had been bothered by her shyness: "Academically I was a success, but socially I was a flop—shy and quite unsophisticated."[9] In the all-women's atmosphere at Smith, however, she thrived and excelled in mathematics, physics, and astronomy. Granville studied mathematics with Neil McCoy and Susan Rambo, and astronomy with Marjorie Williams. For a time she even considered becoming an astronomer, but did not like the idea of spending long hours isolated in an observatory.[10] She was elected to Phi Beta Kappa and Sigma Xi in her senior year, and she earned the baccalaureate, summa cum laude, in 1945 with honors in mathematics.[11]

The Smith College Student Aid Society awarded Granville a scholarship to attend graduate school, and she chose Yale University over the University of Michigan, because it offered additional financial aid. In 1946, she earned the master's degree in mathematics and theoretical physics. In addition to the Smith scholarship, she was awarded two Julius Rosenwald Fellowships (1946–1948) and an Atomic Energy Commission Pre-doctoral Fellowship (1947–1949).[12] At Yale, Granville was mentored by Einar Hille, a former president of the American Mathematical Society and a specialist in functional analysis who also served as her dissertation advisor. When Yale awarded her the doctorate in 1949, Granville and Marjorie Lee Browne tied as the first Black women awarded a doctorate in mathematics; no other Black women earned the doctorate in mathematics during the next ten years.[13] (As detailed in a previous entry, Browne was ten years older than Granville and her pursuit of the doctorate had progressed at a slower pace.) Granville's dissertation was titled "On Laguerre Series in the Complex Domain."[14]

Beginning in 1949, Granville worked for a year as a part-time research assistant at New York University's Institute of Mathematics while searching for a full-
time teaching position. Patricia Clark Kenschaft, who has done a great deal of
research on Black mathematicians, says that "[Granville] interviewed for one job
in the New York area, [but] they merely laughed at her application" (it is not clear
whether it was racial or sexual discrimination).[15] The following year, Granville
took a full-time position as an associate professor of mathematics at Fisk University in Nashville, Tennessee. At that time, Lee Lorch, a white mathematics professor, had been hired to head the mathematics department at Fisk.[16]

It may seem odd to feature a white man, however briefly, in a work on Black
women scientists and mathematicians. But white mathematician Lee Lorch
had an uncommon effect in helping springboard Black women into the higher
reaches of mathematics. In the decade between the late 1940s and the late 1950s,
Lorch lost four teaching jobs at colleges in the United States because of his involvement in Black educational and civil rights issues.[17] Patricia Kenschaft recounts how in the late 1940s he lost a teaching position at a New York college
as a result of his efforts to integrate the Stuyvesant Town housing development
where he lived. He then secured a post at Pennsylvania State University, but was
fired after only a year because he and his wife, Grace Lonergan Lorch, had permitted a Black family to live in their apartment while they were out of town. Lorch
moved on to Fisk University in 1950. A month after the 1954 *Brown v. Board
of Education of Topeka* ruling, the Lorches attempted to remove their daughter
from the whites-only school across town where she was being bused and register
her in the Black neighborhood school across the street from their home. This
caused a furor, and eventually resulted in Lorch's being subpoenaed by the U.S.
House of Representatives Committee on Un-American Activities. He was subsequently fired (without formal charges) by a split vote of Fisk University's predominantly white Board of Trustees.[18] Lorch and Granville were together on the
Fisk faculty from 1950 to 1952. Between them, in that short period of time, they
mentored Vivienne Lucille Malone Mayes and Etta Zuber Falconer; Lorch mentored his last student at Fisk, Gloria Conyers Hewitt, after Granville left.[19]

Granville apparently felt the marked contrast between small, segregated Nashville and the large, cosmopolitan bustle of New York City and Washington, D.C.[20]
In July 1952 she left Fisk and took a position as a mathematician at the National
Bureau of Standards in Washington, D.C. For the next sixteen years, Granville
worked as a research mathematician in government and industry. At the National
Bureau of Standards, Granville's group worked on mathematical analyses of
ballistic missile fuses — research fueled by the Cold War. The research group was
later transferred to the Department of the Army, and renamed the Diamond
Ordnance Fuze Laboratory. During her time at DOFL, development of electronic computers was just beginning, and Granville became interested in research on the application of computers to scientific studies.[21]

In January 1956, Granville joined the International Business Machines Corporation (IBM) and developed programs for the IBM 650 computer, which was
then the state-of-the-art machine.[22] Two years later, in 1958, she became part of
the IBM team at the Vanguard Computing Center, in Washington, D.C., which

had been awarded a National Aeronautics and Space Administration (NASA) contract to plan, write, and maintain computer programs for Project Vanguard, and later for Project Mercury.[23] Granville described this period when she was a member of a group responsible for writing computer programs to track the paths of vehicles in space as "the most interesting" in her life.[24]

While working on Project Vanguard in 1958, Granville visited California, where she met the Reverend Gamaliel Mansfield Collins. They were married in 1960, and she left the IBM/NASA program in November to move to California, where she joined the staff of the Computation and Data Reduction Center of Space Technology Laboratories. There, she continued to do research on new methods of computing orbits of manned and unmanned space vehicles. According to Granville, "there was a great demand for mathematicians and scientists to work in private industry [in the early 1960s] as companies increased their staffs to perform contract work for NASA and defense agencies. It was not unusual in that era for a person to switch jobs often as more interesting (and more lucrative) positions opened up."[25] In 1962 she moved to North American Aviation (NAA), which had been awarded a NASA contract to work on Project Apollo; she worked as a research specialist with the Space and Information Systems Division of NAA in a group providing technical support to the engineering departments on celestial mechanics, trajectory and orbital computations, numerical analysis, and digital computer techniques. She returned to the Federal Systems Division of IBM in 1963, where she did research in trajectory analysis and orbital computations for NASA's Apollo Program.[26]

In 1967, Congress instituted the first of many wide-ranging cuts to NASA's funding, leading to a reduction in the number of research contracts and staff in the space industry. At the same time, Granville's marriage to Gamaliel Collins ended. Although she was offered another position with IBM in Washington, D.C., Granville decided to remain in California and accepted a post as an assistant professor of mathematics at California State University at Los Angeles. She taught classes in computer programming and numerical analysis, and was also assigned to teach a required mathematics course for prospective elementary school teachers.

Granville's course for elementary school teachers led to a new research interest: developing teaching programs for what was then called the "new math." In 1968, under the State of California's Miller Mathematics Improvement Program, college faculty were recruited to teach at the elementary level; Granville spent a year teaching part-time in a second and a fifth grade class at a Los Angeles elementary school. In 1969, she directed an after-school mathematics enrichment program for students in kindergarten through fifth grade. She taught the upper four grades herself, while continuing with her college course load.

Granville's experience in learning how to teach mathematics effectively to elementary school youngsters led to an association with another mathematician, Jason Frand. Their research evolved into a college mathematics textbook for elementary school teachers. Granville also taught in a number of National Science Foundation summer institutes for elementary mathematics teachers.[27] After ten years of involvement in mathematics education, Granville once again turned her interests toward computer science and the mathematical aspects of computer languages.[28]

Granville married Edward V. Granville, a Los Angeles real estate broker, in 1970. In 1984, she retired with the rank of full professor from the California State University at Los Angeles, and moved to Texas. In 1985 she joined the faculty of Texas College, a small, predominantly Black four-year college in Tyler, where she taught mathematics and computer science until 1988. In 1990, she became the Sam A. Lindsey Professor of Mathematics at the Tyler campus of the University of Texas, a position she currently holds.[29]

In a 1983 interview, Granville said, "I always smile when I hear that women cannot do mathematics."[30] And, despite the fact that her first marriage ended in divorce, she has said she "never encountered any problems in combining a career and private life. Black women have always had to work."[31] Granville felt that, because Black women have had a long history of work outside the home, Black females tend to be better educated than their male counterparts — and "hence the more aggressive professionally."[32]

During her long career, Granville served as a member of the U.S. Civil Service Commission Panel of Examiners of the Department of Commerce (1954–1956); was appointed to a three-year term by then Governor Edmund Brown to the Psychology Examining Committee of the Board of Medical Examiners of the State of California (1963–1970); was a member of the Advisory Committee of the National Defense Education Act Title IV Graduate Fellowship Program of the Office of Education (1966–1969); was a member of the Board of Trustees of the Center for the Improvement of Mathematics Education of the National Council of Teachers of Mathematics and the California Mathematics Council; and served as president of the Beverly Hills Chapter of the American Association of University Women (AAUW) and a member of the AAUW State Committee on "Bridging the Gap between the Scientist and the Layman" (1968–1970).[33]

NOTES

1. See entry on Katherine Coleman Johnson in this volume.

2. Evelyn Boyd Granville, "My Life as a Mathematician," *Sage* 6 (fall 1989): 44–46. Sylvia G. L. Dannett, *Profiles of Negro Womanhood*, vol. 2: *Twentieth Century* (Negro Heritage Library. Yonkers, N.Y.: Educational Heritage, [1964–1966]), pp. 82–86. "Space Computing Mathematician," *Ebony*, August 1960, p. 7. Marianna W. Davis, "Black Women Achievers as Mathematicians, Engineers, and Space Technologists," Chapter 4 in *Contributions of Black Women to America*, vol. 2 (Columbia, S.C.: Kenday Press, 1982), pp. 453–473. Patricia Clark Kenschaft, "Blacks and Women in Mathematics," paper presented at the Science and Technology Seminar, The City University of New York, 3 March 1983, in *Educational Policy Seminar Papers* (New York: New City University of New York, Office of Special Programs, 1986), pp. 22–23. Kenschaft, "Evelyn Boyd Granville," in *Women of Mathematics: A Bibliographic Sourcebook*, ed. Louise S. Grinstein and Paul J. Campbell (New York: Greenwood Press, 1987), pp. 57–60. James H. Kessler, J. S. Kidd, Renee A. Kidd, and Katherine A. Morin, *Distinguished African American Scientists of the Twentieth Century* (Phoenix, Ariz.: Oryx Press, 1996), pp. 127–131.

3. Granville, "My Life as a Mathematician." Granville uses the term "negro," saying, "we were not referred to as Blacks in those days."

4. Ibid. According to Kenschaft, at Dunbar High School Granville studied with "people educated in mathematics at Yale University and the University of Pennsylvania" ("Blacks and Women in Mathematics," p. 23).

5. Granville, "My Life as a Mathematician." See also Davis, "Black Women Achievers," and Kenschaft, "Blacks and Women in Mathematics."

6. Cromwell had a mathematics degree from the University of Pennsylvania; another of Granville's mathematics teachers was Ulysses Bassett, a graduate of Yale University. The faculty at Dunbar High School had a tradition of encouraging students to apply to Ivy League colleges; according to Kenschaft, Cromwell's sister Otelia had earned a doctorate in English at Yale University around the turn of the century and had fostered this tradition (Kenschaft, "Evelyn Boyd Granville," 1987, pp. 57–58).

7. Granville, "My Life as a Mathematician." See also Davis, "Black Women Achievers," and Kenschaft, "Blacks and Women in Mathematics." Apparently, Granville's aunt was willing to continue with this arrangement for her entire college career, but when Granville received scholarship money her aunt no longer had to help out. According to Kenschaft, Granville's mother, Julia Walker Boyd, and her aunt, Louise Walker, were both high school graduates; in addition, Walker had a degree in teacher training from Minor Normal Teachers College (now part of the University of the District of Columbia). Both women were examiners at the U.S. Bureau of Engraving and Printing. Walker never married and apparently devoted her life to helping educate Granville and her older sister, Doris. Doris Boyd left college in 1942 and began a career as a statistical assistant at the U.S. Census Bureau ("Evelyn Boyd Granville," 1987, pp. 57–60).

8. Dannett, *Profiles of Negro Womanhood*, p. 84.

9. Ibid., p. 82.

10. Granville, "My Life as a Mathematician," and Kenschaft, "Evelyn Boyd Granville," 1987, p. 58. Granville recalled, "I was fascinated by the study of astronomy and at one point I toyed with the idea of switching my major to this subject. If I had known then that in the not too distant future the United States would launch its space program, and astronomers would be in great demand in the planning of space missions, I might have become an astronomer instead of a mathematician."

11. Granville, "My Life as a Mathematician." See also Kessler, *Distinguished African American Scientists*; Davis, "Black Women Achievers"; and Kenschaft, "Evelyn Boyd Granville," 1987.

12. Vivian Ovelton Sammons, *Blacks in Science and Medicine* (New York: Hemisphere, 1990), p. 104. See also Granville, "My Life as a Mathematician"; Kessler, *Distinguished African American Scientists*; and Davis, "Black Women Achievers."

13. James M. Jay, *Negroes in Science: Natural Science Doctorates, 1876–1969* (Detroit: Belamp, 1971). See also Patricia Clark Kenschaft, "Black Women in Mathematics in the United States," *American Mathematical Monthly* 88 (October 1981): 599–600; Granville, "My Life as a Mathematician"; and Kessler, *Distinguished African American Scientists*, p. 129.

14. Sammons, *Blacks in Science and Medicine*, p. 104.

15. Kenschaft, "Blacks and Women in Mathematics," pp. 23–24. In a later article, Kenschaft stated, "A woman adjunct on that faculty told this author that when the interviewing committee discovered [Granville] was black, 'they just laughed.' Later a male mathematician familiar with the situation said, 'It was not her race that was the problem. The mathematicians wanted to hire her, but the dean said they would "have to change the plumbing."' Whether it was due to racial or sexual discrimination, the refusal to hire her was totally inconsistent with her superb academic credentials" (Kenschaft, "Evelyn Boyd Granville,"1987, pp. 58–59).

16. Vivienne Malone Mayes, "Lee Lorch at Fisk: A Tribute," *American Mathematical Monthly* 83 (November 1976): 708–711. Lee Lorch is discussed more fully in the entry on Vivienne Malone Mayes.

17. Kenschaft, "Evelyn Boyd Granville," 1987; Mayes, "Lee Lorch at Fisk."

18. Patricia Clark Kenschaft, "Black Women in Mathematics in the United States," *Journal of African Civilizations* 4 (April 1982): 63–83. Also see Kenschaft, "Blacks and Women in Mathematics." The Nashville School Board's members — 4 whites and 1 Black — voted 4 to 1 against the move. The Fisk University Board of Trustees, like those of most historically Black colleges and universities of that time, was dominated by white members. After being fired from Fisk, Lorch next fetched up at Philander Smith College in Little Rock, Arkansas, where he taught from 1955 to 1959. During the Little Rock school integration crisis in 1957–1958, the Lorches continued their active support of integration. Grace Lorch became involved in the rescue of Elizabeth Eckford, a fifteen-year-old Black student attempting to enter Little Rock High School, and wound up facing an angry mob herself. The following account of Grace Lorch's involvement in the Little Rock High School event is from a *Time* magazine article ("The South: Making of a Crisis in Arkansas," September 16, 1957):

> Fewer than 100 people (not counting reporters, pupils and militiamen) were outside Central High when the test came. Most of the Negro children came in a group, accompanied by adults, and left quietly when told by a National Guardsman that "Governor Faubus has placed this school off limits to Negroes." But little Elizabeth Eckford, 15, stepped alone from a bus at the corner of 14th and Park Streets. In a neat cotton dress, bobby-sox and ballet slippers, she walked straight to the National Guard line on the sidewalk. The Guardsmen raised their rifles, keeping her out.
>
> Elizabeth, clutching tight at her notebook, began a long, slow walk down the two blocks fronting the school. She turned once to try the line again — and again the rifles came up. A militia major shielded her from the crowds, escorted her to a bus-stop bench, left her. "Go home, you burr head," rasped an adult voice. Elizabeth sat dazed as the crowd moved in. Then Mrs. Grace Lorch, wife of a Little Rock schoolteacher, sat down on the bench and slipped her arm around the child's shoulders. "This is just a little girl," she cried at the crowd. "Next week you'll all be ashamed of yourselves."
>
> After 35 minutes a bus finally pulled up. Mrs. Lorch took Elizabeth's arm and shoved through the crowd. "I'm just waiting for one of you to touch me," said she. "I'm just aching to punch somebody in the nose." The crowd gave way before the white-haired woman and the little girl — and that was about as close as Little Rock came all week to Orval Faubus' manufactured "violence." (pp. 24–25)

In the ensuing furor, Mississippi Senator James Eastland summoned Grace Lorch before his United States Senate committee, and Lee Lorch was fired once again. At that point, he was unable to find another academic post in the United States, and the family moved to Canada where he continued to teach. He was elected to the Royal Society of Canada, twice elected to the Council of the American Mathematical Society, and eventually chaired a (U.S.) National Science Foundation research grant committee (Kenschaft, "Blacks and Women in Mathematics"). See also Kenschaft, "Black Women in Mathematics in the United States," p. 68.

19. See my entries on Vivienne Malone Mayes, Etta Zuber Falconer, and Gloria Conyers Hewitt.

20. Granville, "My Life as a Mathematician."

21. Ibid. See also Kessler, *Distinguished African American Scientists*.

22. At a later period in her employment with IBM, she also worked on the IBM 704 computer (Kenschaft, "Evelyn Boyd Granville,"1987, p. 59).

23. Granville, "My Life as a Mathematician."

24. Ibid.

25. Ibid.

26. Ibid. See also Kessler, *Distinguished African American Scientists,* p. 130.

27. Jason L. Frand and Evelyn B. Granville, *Theory and Applications of Mathematics for Teachers* (Belmont, Calif.: Wadsworth Publishing, [1975]); the second edition was published in 1978.

28. Granville, "My Life as a Mathematician."

29. Kessler, *Distinguished African American Scientists,* pp. 127, 131. See also Patricia Clark Kenschaft, "Evelyn Boyd Granville," in *Black Women in America: An Historical Encyclopedia,* ed., Darlene Clark Hine, Elsa Barkley Brown, and Rosalyn Terborg-Penn, vol. 1 (Brooklyn, N.Y.: Carlson, 1993; Bloomington: Indiana University Press, 1994), p. 499.

30. Kenschaft, "Blacks and Women in Mathematics," p. 24.

31. Kenschaft, "Evelyn Boyd Granville," 1987, p. 59. Although Granville's first husband had three children and her second husband had two, she had no children of her own.

32. Dannett, *Profiles of Negro Womanhood,* p. 86.

33. Kenschaft, "Evelyn Boyd Granville," 1993, p. 499.

BETTYE WASHINGTON GREENE

Industrial Research Chemist

Research chemist Bettye Washington Greene was born in Fort Worth, Texas, in 1935.[1] She attended segregated public schools and graduated from I. M. Terrell High School around 1952. She then entered Tuskegee Institute in Alabama, where she earned a B.S. degree in chemistry in 1955. She married William M. Greene in July of 1955, and bore two children—a daughter in 1957, and a son in 1958.

Sometime after the birth of her son, Greene began graduate studies in chemistry at Wayne State University in Detroit, Michigan. During her time in graduate school, she bore a third child, a daughter, in 1964. She was awarded the Ph.D. in physical chemistry in 1965, and elected to Sigma Xi. Her dissertation, "Determination of Particle Size Distributions in Emulsions by Light Scattering," was published later that year under the title "Determination of Size Distributions in Emulsions Using Scattering Spectra."[2] It is not clear whether Greene received funding for graduate school, but she did teach undergraduate chemistry at Wayne State during this period.

In 1965, Greene joined the Dow Chemical Company as a research chemist at the company's E. C. Britton Research Laboratory in Midland, Michigan, where she conducted fundamental research in colloid and latex chemistry, particularly in the areas of stabilization and characterization.[3] In 1970, she was promoted to the level of senior research chemist at Dow. She joined the company's Designed Polymers Research Division in 1973, where her research focused on polymers that could be used to improve the various properties of latex. Her work included studies on the electrokinetics and rheological properties of pigment dispersions used in paper coatings; an exploration of factors that would impart and control the stability of re-dispersible latexes and the re-dispersibility of latex powers; and the use of latexes as beater additives in the manufacture of paper products.[4]

Greene published a number of scientific papers between 1965 and 1970, the last year for which information is available on her career. She left Dow Chemical in 1985.[5]

NOTES

1. Bettye W. Greene, biographical data sheet, Dow Chemical Company Public Relations Department, 17 August 1970. Dow Chemical Company Public Relations Department, news release, 25 January 1971. "B. W. (Bettye) Greene," *Blue Flyer* 2, no. 61 (Dow Chemical Midland Division, News and Information Services, 25 July 1975). "Dr. Bettye W. Greene—Biographical Information," Dow Chemical Company, 25 August 1970. (I am indebted to Doug Draper and archivist Kathy Thomas of the Dow Chemical Company's Post Street Archives in Midland, Michigan, for providing me with copies of this information.) See also Vivian Ovelton Sammons, *Blacks in Science and Medicine* (New York: Hemisphere, 1990), p. 105. Speaking of People, *Ebony*, July 1973, p. 6. *American Men and Women of Science: The Physical and Biological Sciences*, 15th ed. (New York: Bowker, 1971–), p. 293. Note that Sammons lists Greene's birthplace as Palestine, Texas, but Greene herself lists it as Forth Worth, as does the *Ebony* article.

2. Greene, biographical data sheet. See also Sammons, *Blacks in Science and Medicine*. See the appendix for publication information.

3. "B. W. (Bettye) Greene."

4. Ibid.

5. I have been unable to determine whether Greene retired or moved on to another company. None of the sources provided any information in this regard.

MARGARET E. GRIGSBY

Tropical Medicine and Infectious Disease Specialist

A specialist in tropical medicines and infectious diseases, physician Margaret E. Grigsby was born in Prairie View, Texas, in 1923. She earned the baccalaureate at Prairie View College in 1943, and was awarded the M.D. degree from the University of Michigan in 1948. She completed an internship and residency at Homer G. Phillips Hospital in St. Louis, Missouri (1948–1950), and completed her second residency at Freedman's Hospital in Washington, D.C. (1950–1951). Grigsby was then awarded a postdoctoral Rockefeller Foundation research fellowship at Harvard Medical School (1951–1952).[1]

Grigsby began her career as an instructor at Howard University Medical College in 1952. In 1956, she was awarded a China Medical Board Fellowship to study at the School of Tropical Diseases at the University of Puerto Rico. Upon her return to Howard, she was advanced to the rank of assistant professor. In 1960, she toured the Iron Curtain countries and presented papers in Moscow, Warsaw, and Prague.[2]

At the end of the 1961 academic year, Grigsby took a leave of absence to study abroad. She received an advanced degree (Doctor of Tropical Medicine and Health) from the University of London in 1963, and was promoted to a full professorship at Howard in 1966.[3] During this period, she also served as an expert advisor to the Peace Corps. In 1967, she took another leave from Howard to serve as a Distinguished Professor of Medicine at the University of Ibadan in Nigeria,

West Africa.[4] Grigsby's research has focused on tropical medicine and infectious diseases, antibiotics, and electrophoresis of proteins.

NOTES

1. Vivian Ovelton Sammons, *Blacks in Science and Medicine* (New York: Hemisphere, 1990), pp. 107–108. See also *American Men and Women of Science: The Physical and Biological Sciences*, 15th ed. (New York: Bowker, 1971–), p. 321. "Outstanding Women Doctors: They Make Their Mark in Medicine," *Ebony*, May 1964, p. 72. *Who's Who among Black Americans* (Northbrook, Ill.: Who's Who among Black Americans Publishing, 1985), p. 336.

2. "Outstanding Women Doctors," p. 72.

3. Sammons, *Blacks in Science and Medicine*.

4. Marianna W. Davis, ed., *Contributions of Black Women to America*, vol. 2 (Columbia, S.C.: Kenday Press, 1982), p. 388.

GLORIA CONYERS HEWITT

Chair of the Department of Mathematical Science, University of Montana

Gloria Conyers Hewitt

Born in Sumter, South Carolina, in 1935, mathematician Gloria Conyers Hewitt spent her first two years as an undergraduate at Fisk University studying with

Lee Lorch.[1] Although her period of study under Lorch was fairly brief, he nevertheless exerted a lasting influence on her career. Lorch continued to act as Hewitt's mentor even after he had lost his position at Fisk, and without her knowledge recommended her for graduate study at two universities. According to historian Patricia Clark Kenschaft, Hewitt was "amazed" in her senior year when she received invitations from two graduate schools on the basis of his recommendations. Until that time, she had not considered graduate school, nor did she realize that Lorch considered her worthy of performing research:[2] "As [Lorch] had taught me for such a short time, it amazes me to this day that he felt I could handle graduate school in mathematics. . . . the thought of entering graduate school never crossed my mind; I never knew it crossed his mind until I heard of his recommendations."[3]

Hewitt completed the baccalaureate degree in mathematics at Fisk in 1956, and accepted a fellowship at the University of Washington, in Seattle.[4] Due to Lorch's departure before the start of her junior year at Fisk, Hewitt's preparation in mathematics was not what it ought to have been; although she was reluctant to ask for help, the other graduate students—all white males—were unusually supportive, both intellectually and socially.[5] Hewitt's study group consisted of eight or ten other students, three of whom—Carl Stromberg, Kenneth Ross, and Robert Phelps—were particularly helpful. Her study group and the continued support of her parents, Crenella and Emmett Conyers, proved crucial to her success: "It was almost as if they had gotten together and decided that I should get this degree."[6] Hewitt also had a mentor at the University of Washington, Richard Pierce, "whose faith in me remained even when my own faltered," Hewitt recalled.[7] She earned the master's degree in 1960; and she completed her doctorate in 1962 with her dissertation, "Direct and Inverse Limits of Abstract Algebras."[8]

Hewitt began her professional career as a predoctoral associate at the University of Montana in 1961. After she had completed her doctoral program, she was moved to the rank of assistant professor; she was promoted to the rank of associate professor in 1966. During academic year 1965–1966, she received a National Science Foundation postdoctoral science faculty fellowship for research at the University of Oregon. She achieved the rank of full professor of mathematics at the University of Montana in 1972. She spent her 1980–1981 year of sabbatical leave at the Case Institute of Case Western Reserve University doing research on algebraic topology and homologic algebra.[9]

During a 1991 interview with Kenschaft, Hewitt said that despite her success at the University of Montana, her early experiences with her white male colleagues in the 1970s required some adjustment: "They were so busy loving me that they didn't notice that they didn't respect me—so, of course, neither did I." According to Kenschaft, a colleague had recommended Hewitt for a merit award and the department had supported her. When the dean denied the award, Hewitt decided to appeal—and she was asked if she could prove that her sex and race had nothing to do with the widespread recognition she had received.[10] As Hewitt told Kenschaft,

> Nobody around here knew more math than me. . . . They weren't doing that
> much research either. One day I got mad. "I don't care what you think," I said,

and I meant it. "I'm tired of trying to please you. What you think of me doesn't matter to me any more."

Then they had to begin respecting me. I believe now that my colleagues both respect and like me.[11]

Hewitt was named associate chair of the Department of Mathematical Science in 1993, and took over the chairmanship in 1995, a position she currently holds.[12]

During her tenure at Montana, Hewitt has been a visiting lecturer for the Mathematical Association of America (1964–1972); Associate Director of the National Science Foundation's Academic Year Institute for College Teachers (1969–1971); and Associate Director and Director of the Office of Education Summer Institute for College Teachers (1972 and 1973, respectively). She was elected to the National Organization of Women's Board of Directors (1974–1977); and the Executive Council of Pi Mu Epsilon, the national mathematics honorary society (1972–1975). She also served as a member of the Educational Testing Service's Graduate Record Examination Committee (1976–1986; committee chair, 1984–1986); on the College Board's Advanced Placement Calculus Development Committee (1987–1991); and was elected to a three-year term on the Board of Governors of the Mathematical Association of America in 1995.[13]

Hewitt's research has focused on abstract algebras, including algebraic topology and homologic algebra, group theory, and Noetherian Rings.[14] In 1992, she conducted a survey under the auspices of the National Association of Mathematicians on the status of Black graduate students in mathematics,[15] and she is currently involved in conducting a national survey of minority graduate students in mathematics for the Mathematical Society of America (through its SUMMA Program) and the National Association of Mathematics, with funding provided by the Alfred E. Sloan Foundation.[16]

NOTES

1. Gloria Conyers Hewitt, curriculum vitae, 6 January 1996. See also Patricia Clark Kenschaft, "Gloria Conyers Hewitt," in *Black Women in America: An Historical Encyclopedia,* ed. Darlene Clark Hine, Elsa Barkley Brown, and Rosalyn Terborg-Penn, vol. 1 (Brooklyn, N.Y.: Carlson, 1993; Bloomington: Indiana University Press, 1994), pp. 557–558. Kenschaft, "Black Women in Mathematics in the United States," *Journal of African Civilizations* 4 (April 1982): 63–83. Kenschaft, "Blacks and Women in Mathematics," paper presented at the Science and Technology Seminar, The City University of New York, 3 March 1983, in *Educational Policy Seminar Papers* (New York: New City University of New York, Office of Special Programs, 1986), pp. 21–31. Vivian Ovelton Sammons, *Blacks in Science and Medicine* (New York: Hemisphere, 1990), p. 118. *National Faculty Directory* (Detroit: Gale Research, 1988), p. 1655. *Winning Women into Mathematics* [brochure] (American Mathematical Society, 1991). Hewitt is also included in the Chicago Museum of Science and Industry exhibit "Black Achievers in Science." See the section on Lee Lorch in the entry on Evelyn Boyd Collins Granville in this volume.

2. Kenschaft, "Gloria Conyers Hewitt," p. 557. Kenschaft, "Blacks and Women in Mathematics," p. 25.

3. Kenschaft, "Black Women in Mathematics in the United States," p. 68.

4. Hewitt, curriculum vitae.

5. As Kenschaft relates, "The other graduate students . . . would come to her office and volunteer to explain 'interesting' problems. Once one said, 'The trouble with you is

that you don't understand trigonometry. I am going to teach it to you.' [Another said,] 'Isn't this an interesting problem? Let me show you how interesting it is!'" (Kenschaft, "Black Women in Mathematics in the United States"). "Eventually, she felt free to ask them questions whenever she was puzzled. She felt at home socially in the group. When they went to a bar in the evening she went too. If they went to the best restaurant in Seattle, she came along" (Kenschaft, "Blacks and Women in Mathematics," pp. 25–26).

6. Kenschaft, "Black Women in Mathematics," pp. 68–69.
7. Ibid., p. 69.
8. Hewitt, curriculum vitae.
9. Hewitt, curriculum vitae.
10. Kenschaft, "Gloria Conyers Hewitt."
11. Ibid. She received the award eleven years later.
12. Hewitt, curriculum vitae.
13. Ibid.
14. Noetherian rings are named after woman mathematician Emmy Noether.
15. Gloria C. Hewitt, "1992 Survey of Black Graduate Students in Mathematics," a report for the National Association of Mathematicians, 1993 (Hewitt, curriculum vitae).
16. Ibid. According to Hewitt, this is the first such national survey ever conducted.

MARY ELLIOTT HILL

Part of a Husband and Wife Research Team

Chemist Mary Elliott Hill, who preferred to be known as Mrs. Carl McClellan Hill, was born in South Mills, North Carolina in 1907. She attended Virginia State College and graduated with a bachelor of science degree in chemistry in 1929. She began her career as a teacher of chemistry in the Laboratory High School at Virginia State in 1930. In 1932, she moved on to a position at Hampton Institute, where she remained until 1936.[1]

At some time during that four-year period, Hill's career began roughly to parallel that of her husband, chemist Carl McClellan Hill.[2] He was first a chemistry teacher at the Laboratory High School of Hampton Institute (1931–1939), then became an assistant professor of chemistry at the college (1939–1940), and returned to the Laboratory School as principal (1940–1941). During the latter part of this period, from 1938 to 1942, Mary Hill returned to Virginia State College, where she taught chemistry at the college level. During this period, she also began work on her master's degree in analytic chemistry at the University of Pennsylvania, studying during the summers. She was awarded the M.S. in 1941, and moved on to Bennett College for the 1942–1943 academic year.

Meanwhile, in 1941, Carl Hill had moved to an associate professorship at the Agricultural and Technical College of North Carolina; in 1944, he moved again and became head of the chemistry department at Kentucky State University (1944–1951), during which time he also worked on chemical research projects for the Tennessee Valley Authority (1948–1952). In 1944, Mary Hill also moved to Kentucky State University, where she advanced from assistant professor to associate professor, and served as acting head of the department during the 1951–

1952 academic year when her husband's duties with the TVA took up too much of his time.

Carl Hill was named chair of the chemistry department at KSU in 1953, and served in that capacity until 1958, when he moved on to become dean of the College of Arts and Sciences. He served as dean until 1962, when he returned to the chemistry department. He later served as president of Hampton Institute from 1976 to 1978.[3]

Mary Hill did research in ultraviolet spectrophotometry; Carl Hill did research in aliphatic, alicyclic, and arylunsaturated ethers and Grignard reagents. Together, they continued the research he had begun with his dissertation on ketenes, particularly on the synthesis and properties of their monomeric and dimetric forms.[4]

NOTES

1. See Vivian Ovelton Sammons, *Blacks in Science and Medicine* (New York: Hemisphere, 1990), p. 120. See also *American Men of Science*, 11th ed. (New York: Bowker, 1968), p. 2295. Julius H. Taylor, Clyde R. Dillard, and Nathaniel K. Proctor, eds., *The Negro in Science* (Morgan State College Press, 1955), p. 183. "Food Chemist," in *Contributions of Black Women to America*, ed. Marianna W. Davis, vol. 2 (Columbia, S.C.: Kenday Press, 1982), p. 441.

2. Carl McClellan Hill (B.S., Hampton Institute, 1931; M.S., 1935; Ph.D., Cornell University, 1941). See Sammons, *Blacks in Science and Medicine*, p. 119. See also James M. Jay, *Negroes in Science: Natural Science Doctorates, 1876–1969* (Detroit: Belamp, 1971), p.150. *Ebony Success Library*, vol. 1 (Chicago: Johnson Publishing, 1973), p. 152. *American Men and Women of Science: The Physical and Biological Sciences*,16th ed. (New York: Bowker, 1971–), p. 710. *Who's Who among Black Americans* (Northbrook, Ill.: Who's Who among Black Americans Publishing, 1977), p. 417.

3. Sammons, *Blacks in Science and Medicine*, p. 119.

4. Ibid.

ESTHER A. H. HOPKINS

Industrial Research Chemist, Lawyer

Research chemist Esther A. H. Hopkins spent the bulk of her career in industrial research. Born in Stamford, Connecticut, in 1926, Hopkins earned the baccalaureate degree at Boston University in 1947 and the master's degree at Howard University in 1949. She began her career as a chemistry instructor at Virginia State College, but soon moved on to industry, where she was employed as a control chemist at a Stamford, Connecticut, corporation.

During the 1950s a number of corporations with strong research laboratories were expanding their facilities in Connecticut, and in 1955 Hopkins was offered a position as an assistant researcher in biophysics at the New England Institute for Medical Research in Ridgefield. She then moved into a position as a research chemist at American Cyanamid's Stamford Research Laboratory, where she remained for a number of years. During this period, she returned to college and earned a second master's degree, in chemistry, at Yale University in 1962. She completed her doctorate at Yale in 1967, and was elected to both the Phi Beta Kappa and Sigma Xi honor societies. The dedication page of her dissertation reads, "To the three generations of my family whose dreams and sacrifices are inextricably woven into this degree and all that it means."[1]

Immediately after receiving her doctorate, Hopkins obtained a position as a supervisory research chemist with the Polaroid Corporation in Cambridge, Massachusetts, where she has conducted analyses on the emulsion coatings used on films and supervised experiments on the development of new types of film. While at Polaroid, she completed a law degree, with a special focus on patent law.[2] Hopkins is now the deputy general counsel for the Massachusetts Department of Environmental Protection.[3]

NOTES

1. "Industrial Chemist," in *Contributions of Black Women to America*, ed. Marianna W. Davis, vol. 2 (Columbia, S.C.: Kenday Press, 1982), pp. 442–443.

2. Ibid.

3. Susan A. Ambrose, Kristin L. Dunkle, Barbara B. Lazarus, Indira Nair, and Deborah A. Harkus, *Journeys of Women in Science and Engineering: No Universal Constants* (Philadelphia: Temple University Press, 1997), p. 213.

RUTH WINIFRED HOWARD

Minister's Daughter, Social Worker, Psychologist

As the first Black woman to earn a doctorate in psychology, Ruth Winifred Howard (Beckham) made a unique contribution to the field, not only with her research on triplets, but also with regard to the ways different ethnic groups view and work with one another.

Howard had originally intended to be a librarian, but after three years of study at Howard University she decided to give up her scholarship and change her major to social work. A 1920 scholarship from the National Urban League allowed her to transfer from Howard University to Simmons College in Boston. In a personal account published in 1983, Howard recalled that for many of her classmates at Simmons fieldwork meant a tremendous adjustment: it represented their first contact with members of other ethnic groups as well as the poor. Howard claimed this was not so in her case because she was "a Negro," and also because during her childhood her father, the Reverend William T. Howard, had always counseled members of his congregation at home.[1]

In Boston, Howard came in contact with Dr. Lucille Eaves, who directed an agency in which the idea of women doing important work was promoted. Eaves introduced Howard to a new concept regarding the scope and role of women's work:

Today this sounds naive, but in the first quarter of the century the pattern was different. Then women [working in high level positions] were unusual. Those working in "white collar" levels were teachers or clerks. Below them were domestic workers.[2]

Eaves's influence caused Howard to examine her own future role in social work. As a result, from that time, she seemed to want to do more than just follow established patterns, both in her work experience and in her life.

During a 1921 National Urban League summer internship, directed by League sociologist Dr. Charles Johnson, Howard studied a small Long Island community. The young people there appeared to have no purpose—they were not attending school, had no jobs, and had few recreational outlets other than loitering on street corners. The experience convinced Howard that there was a tremendous "need for community planning for young people during the summer in educational, recreational, and counseling programs."[3] Later that year, she completed her baccalaureate in social work.

In 1922, Howard began her career in Cleveland, Ohio, as a social worker with the Cleveland branch of the National Urban League. In that position she became aware of the dichotomy between the needs of the individuals she counseled and the broader needs of the communities in which they lived:

> [The] position involved a new dimension of human relationships. As a case worker and counselor I focused on one person or one family group for the solution of a problem. Work with a community group involved a composite of attitudes and objectives which were to be welded together for a common goal. Neighbors may share a common concern, but they could have widely different perceptions of the problem's solution. Here I had an education in social psychology.[4]

Howard then moved on to a position with the Cleveland Child Welfare Agency, where she counseled children "living with their own inadequate families or in foster homes."[5] Her work territory included different ethnic neighborhoods, and it soon became clear to Howard that the agency's predominantly white staff "didn't know about and, more important, didn't understand or sympathize with cultural groups other than their own":

> This was markedly true about Negroes for whom they had firmly fixed preconceived ideas. They did not bother to learn about the social milieu of these people, [they] did not explore the abilities, attitudes, frustrations, and ambitions of the children and their guardians. Since the staff did not know about the cultural climate in which their clients lived and how it influenced their behavior, their clients' visits to the mental health clinic were of little value.[6]

Howard quickly realized that she needed to know more about the "dynamics of how a person thinks, feels, and behaves."[7] That meant her training in social work was not enough—she would have to return to school and study psychology. Funding was the primary obstacle, but sometimes such things just seem to work out, and that is what happened to Howard.

In 1929, she visited her sister in New York, and while she was there she decided to "try her luck" at the Rockefeller Foundation. One Saturday afternoon, without an appointment, Howard just dropped in for a visit at the Foundation's office. She chatted with an official, explained why she wanted to study psychology, and in a few weeks received a letter awarding her a Laura Spelman Rockefeller Fellowship.[8] In September, she enrolled at Columbia University, where she was mentored by Lois Meek Stolz. At the end of the 1930 spring term, she secured an additional Rockefeller grant, and transferred to the Institute of Child Development at the University of Minnesota.

Even in the highly feminized field of psychology, the Institute of Child Development was fairly unusual: the majority of the faculty was female, as were all the students (with a single exception), and the number of graduate students was small. Among the women Howard studied under was Florence Goodenough, who functioned as her chief advisor and mentor while she was validating research data on the "Draw-A-Man" test. Howard also studied under Mary Shirley, who was just finishing up research on a study of twenty-five babies from birth to the age of two; Edna Heidbreder, who among other courses taught the history of psychology; and Edith Boyd, an anatomist doing laboratory research on bone growth.[9] All were heavily engaged in research, which Howard later said set the tone for their students.[10]

Howard already had strong interests related to child development, and she decided to concentrate on a study of the nature-nurture dichotomy. Studies had already been made of twins, but she believed a study of triplets might result in more interesting data because triplets offered "three biological possibilities: three persons from different eggs; three persons from a single egg; or two persons from one egg and one person from another egg."[11] Her research was based on data obtained on 229 sets of triplets, and on a more detailed study of 29 sets with whom she had personal contact. A nurse, Ellen Church, collected the biological data, and Howard focused on the psychological. In 1934, Howard was awarded the Ph.D. in psychology from the University of Minnesota. Her dissertation, "A Study of the Development of Triplets," was the first published study to examine a sizeable group of triplets of varying ages and from several different ethnic groups.[12]

Within a few months of her graduation, Howard married Albert Sidney Beckham, also a psychologist, and moved to Chicago. She turned down an offer of a staff appointment at a teacher college in Washington, D.C., because she "was marrying and joining [her] husband," who already had established himself in Chicago. She then obtained an internship at the Institute of Juvenile Research (1935–1936), which was affiliated with the University of Illinois Medical School.

At that time, the Institute was moving into the types of approaches to the problems of youth that Howard had earlier visualized during her undergraduate research on Long Island. Saul Alinsky was working on a pioneering technique in youth therapy with street-corner groups which was aimed at rehabilitating neighborhood gangs, and Rayser Lane, of the Chicago Urban League, was involved in developing a community self-help program called the Block Club. The idea of the Block Club was that neighbors joined together to improve their immediate block and then spread out to improve the larger community. For Howard, the Institute provided a "two-way reach"—a problem beginning with a child-parent conflict, for example, could be treated on the individual level, then the Institute

staff could suggest additional neighborhood resources to ameliorate the parent-
child problem.[13]

When Howard completed her internship there were few jobs in the private
sector because of the Depression. In 1937, she was recruited for a supervisory position with the National Youth Administration, a federal effort aimed at providing youth with marketable work skills that could later be used as the economy
improved. As the Administration's Director of Mental Health and Training Program in the Chicago area, Howard assessed the suitability of each young person's work assignment, and determined the progress each was making within the
program. In addition, she supervised the adults who were employed to train the
young clients, monitored the attitudes of both the youth workers and their supervisors, and provided counseling where needed. Howard found that "anti-ethnic
feelings influenced negative attitudes of some supervisors toward trainees. [For
some] this was their closest contact with young people not of their own ethnic
grouping."[14]

All participants in the program—the young people, the trainees, and the adult
supervisors—received a modest stipend, which was welcome as other employment was almost nonexistent:

> I learned that both youth trainees and adult supervisors felt that the NYA wage
> was a means of at least partial support. It was not a dole which could defeat self-
> worth. The work program for all participants afforded an opportunity to learn
> and practice new skills.[15]

Some clients, however, found the experience degrading, according to Howard.
For her, observing the personalities of those involved in the National Youth
Administration project provided case studies, which she and other psychologists
might later use in other types of situations: Did attitudes and behaviors change
with time? Did resistant individuals learn to accept their situations, and could a
psychologist do something to trigger such an improvement? Might a psychologist
introduce techniques for changing attitudes and thus create constructive behaviors?[16]

Following her stint with the National Youth Administration, Howard became
more selective with regard to the type of professional psychological experience
she wanted. In 1940, she and her husband set up a private practice, the Center for
Psychological Services, which focused on developmental and clinical psychology. At the same time, both continued outside careers. Howard functioned as the
Center's co-director until her retirement in 1964.

Howard's professional interests were divided between the physical health
fields, where she acted primarily as a psychologist and a lecturer, and reading and
play therapy, where she continued to do developmental research on children.
From 1940 to 1964 she was a psychologist for Chicago's Provident Hospital
School of Nursing, which had originally been set up as a training school for Black
nurses.[17] She also consulted at historically Black schools of nursing in Missouri
and Florida, and conducted clinics at Kentucky State College and Edward Walters College, in Jacksonville, Florida. She was a lecturer and a psychological consultant to adolescents for the Evanston, Illinois public schools (1953–1955), was

a reading therapist at the University of Chicago's Reading Clinic (1955–1956), served as a staff member at Worthington and Hurst Psychological Consultants (1966–1968), and worked as a psychologist for the McKinley Center for Retarded Children (1964–1966) and the Mental Health Division of the Chicago Board of Health (1968–1972).[18] She also served for a time as a staff psychologist at Abraham Lincoln Centre, an agency set up to meet the needs of Chicago's thickly populated and ethnically and economically diverse community.

In the early 1950s, Howard had enrolled as a postdoctoral student at the University of Chicago.[19] After a period of a few years, she ended up in classes with Helen Robinson, a reading therapist, and Virginia Axline, a play therapist. After working with Robinson, Howard joined the staff at the Reading Clinic, and later published an article based on her experiences with play therapy. Howard believed that women psychologists had played a substantial role in the growth of their discipline and had contributed to its value for human progress; she became involved with a number of professional organizations, such as the American Psychology Association, the International Psychology Association, the International Council of Women Psychologists, the International Reading Association, and the American Association of University Women. In the 1940s, she also helped organize the National Association of College Women, a Black-based group aimed at raising the academic standards at historically Black colleges and universities.[20]

NOTES

1. Ruth Winifred Howard, personal account, in *Models of Achievement: Reflections of Eminent Women in Psychology*, ed. Agnes N. O'Connell and Nancy Felipe Russo (New York: Columbia University Press, 1983), p. 56. See also Robert V. Guthrie, *Even the Rat Was White: An Historical View of Psychology* (New York: Harper & Row, 1976).

2. Howard, personal account, p. 57.

3. Ibid.

4. Ibid.

5. Ibid.

6. Ibid., pp. 57–58.

7. Ibid., p. 58.

8. Ibid.

9. Ibid., p. 59.

10. Ibid.

11. Ibid., p. 60.

12. Guthrie, *Even the Rat Was White*, p. 139. Howard continued to work on the triplets for a number of years. An updated version, "The Developmental History of a Group of Triplets," was published in the *Journal of Genetic Psychology* in 1947 (ibid.).

13. Howard, personal account, pp. 61–62.

14. Ibid., pp. 62–63.

15. Ibid., p. 62.

16. Ibid.

17. Ibid., p. 63. See also Guthrie, *Even the Rat Was White*.

18. Guthrie, *Even the Rat Was White*, pp. 139–140.

19. It appears that she did this in an unofficial capacity: "I could enroll in classes, practicums, and seminars on a non-fee basis" (Howard, personal account, pp. 64–65).

20. Ibid.

DEBORAH J. JACKSON
Succeeding in the Rarefied Air of Experimental Physics

Deborah J. Jackson

Deborah J. Jackson is a member of an exceptionally rare breed—she is a young Black woman with a doctorate in physics, and she is a researcher in the Spacecraft Transponder Group of the Jet Propulsion Laboratory (JPL) of the California Institute of Technology, a posting many would say represents the crème de la creme of American physics. Developing photonic flight hardware for the Cassini Space Probe has been one of her research projects at JPL.

Born in Topeka, Kansas, in 1952, Deborah Jackson's childhood was that of the quintessential "military brat"—between kindergarten and the completion of high school, she attended thirteen different schools, both in the United States and abroad. When she, her mother, and her three siblings could not travel with their career military father to his postings, their home community was either a U.S. military base or Memphis, Tennessee. She attended Catholic schools in the first, seventh, and eighth grades; at other times she and her siblings attended various Department of Defense schools in Europe. Jackson earned her high school diploma at the AFCENT International School, in Brunssum, Netherlands, in 1970.

Jackson was an early reader, and was much taken with the Nancy Drew mystery series for girls. She moved on to adult mystery stories, and then to science fiction. As with so many other budding scientists, the science fiction inspired the science:

> I studied science because I wanted to be a science fiction writer and needed the background to write well. Then I came to be challenged and to love the science. Math came naturally—I started doing well on math tests in junior high school—but I had no instinct for science; this was learned and acquired slowly and painfully.[1]

Part of a tight-knit family, Jackson consistently received parental encouragement to excel in school. As she recalled,

> Sometimes a random teacher here or there encouraged me. One told my parents I had above average IQ. Even before that my parents were consistent with encouragement and emphasis on academic excellence. Education was important to them and they sometimes rewarded high grades. They left no doubt that we were all expected to attend college.[2]

An excellent high school record and very high scores on the Scholastic Aptitude Test (SAT) and National Merit exams, coupled with the affirmative action climate of the early 1970s, prompted a number of colleges to contact her with scholarship offers.[3]

Prior to attending college, Jackson had had no science role models; journalism was her first choice as a major. Nevertheless, she chose to attend MIT, where she majored in physics, and where she said there were "many women, white and black, and also some black men who mentored me — most gave support to women and to minorities where needed.

> I was also mentored by professors, other graduate students [who] gave me pep talks on occasion, and my thesis advisor. Knowing that they had succeeded, I felt I could too. They gave me hope to deal with difficult times. . . .
>
> Professors sometimes interfered with the system on my behalf, sometimes without my knowledge, and recommended me for awards or for introduction into professional societies.[4]

She completed her B.S. in physics at MIT in 1974, and was accepted for graduate study at Stanford University. Initially, she was awarded a teaching assistantship in the Advanced Lab and Intermediate Optics courses at Stanford (1974, and 1977–1978). Jackson received a Ford Foundation Fellowship (1976), and a Bell Labs Cooperative Fellowship (1976–1980). Between 1977 and 1980 she was also a research assistant with the Schawlow/Hansch Group at Stanford University, working on infrared high resolution spectroscopy using state of the art color center lasers, which provided the research for her doctoral thesis.[5]

For Jackson, race and gender were definitely an issue in graduate school. At first, she said, fellow Black students formed support groups, then she and a group of friends formed their own support and study group. Much of the support she had received from mentors and graduate students at MIT stopped once she had "left the protective fold"[6] for Stanford. After going through several thesis advisors at Stanford, she "finally found one who was supportive."[7] She completed the M.S. and the Ph.D. degrees in physics at Stanford, in 1976 and 1980, respectively.

Jackson stressed that the continued support of mentors is needed to "learn the ropes." "After graduate school," she said, "I had only minimal support from a mentor — which meant I had to learn about the pitfalls through experience."[8] Lack of strong mentoring notwithstanding, her postings before joining JPL in 1992 provide a blueprint of upward mobility in optical physics. In the summer of 1974 she was a staff physicist at the IBM General Products Division in San Jose, California, where she worked on Schlieren optics for magnetic disk quality assurance. From 1976 to 1977 Jackson worked as a research assistant at the Stanford Synchrotron Radiation Lab in Stanford, California, where she also acted as liaison for Stanford to the Bell Laboratories x-ray effort on transition metal films. From 1981 to 1983, she was a visiting scientist at IBM in Yorktown Heights, New York, where she worked on nonlinear optics with high power lasers. Between 1983 and 1988, she was on the technical staff at Hughes Research Labs in Malibu, California, where she advanced to program manager for research on monolithic integrated opto-electronic devices. Between 1988 and 1992, she was a senior member of the technical staff for Defense Policy and Analysis at the RAND Corporation, in Santa Monica, California, where she worked on a methodological approach to developing photonic processors.[9]

In her work as a scientist, Jackson's niche of choice would be "surfing the edge in photonics."[10] Two key results have come from her research: the discovery of quantum interference effects between different optical harmonies, and a methodology for developing a photonic processor. Despite research successes, Jackson said her career development is "still in progress. . . . Relative to my cohort, males from Stanford and MIT, I feel it has taken me longer to gain access to the necessary funding to develop research programs."[11]

An elected member of the scientific honor society Sigma Xi, the American Physical Society, the New York Academy of Sciences, and the National Society of Black Physicists, Jackson has served on the National Research Council committees on women in science and engineering (1994–1995) and the use of the international space station for engineering research and technology development (1995).[12]

NOTES

1. Deborah J. Jackson, interview by author, Washington, D.C., October 1995. Jackson, author questionnaire, 1 November 1995. Jackson, résumé, 1 November 1995. I attended the October 1995 Graduate Women in Science and Engineering (GWSE) Conference in Washington, D.C., and had checked in too early; as a result, I ended up at the National Academy of Science with nothing much to do for a couple of hours. Serendipity is a wonderful thing. A chance meeting with Deborah Jackson, who was another early arrival, wound up with the two of us walking to a restaurant, where, during a lengthy lunch, I interviewed her for my dissertation. She mentioned during the interview that Shirley Ann Jackson—the Black woman physicist who just a few months earlier had assumed the chairmanship of the Nuclear Regulatory Commission (NRC)—had been both her heroine and role model. And although they were not related, having the same last name as well as similar interests had always made Shirley Ann Jackson particularly special to Deborah J. Jackson. Many of the women I've interviewed over the years have mentioned the importance of role models or heroes—sometimes it's just a convenient thing to say, other times it is heartfelt. For Deborah Jackson, it was heartfelt.

A couple of days after our lunch together, the GWSE conferees eagerly awaited the appearance of guest speaker Shirley Ann Jackson. Some waited more eagerly than others. Deborah Jackson worried that the other Jackson would not arrive at all, that she might be held up in Washington traffic or sidetracked by a problem at one of the nation's aging nuclear power plants. Then a buzz and slight stir at the back of the auditorium signaled the arrival of Shirley Jackson, and we all scrambled for seats as she strode purposefully to the front of the room.

Seated two rows behind Deborah Jackson, I observed the impact of her role model's presence. The younger Jackson perched on the edge of her seat, straining her head forward, ear cocked toward the elder Jackson so as not to miss a word. A rock star couldn't have had a more attentive fan. To my mind, Shirley Jackson's speech that day was a bit disappointing; she seemed somewhat constrained by her new position as chair of the NRC. But watching the younger Jackson's body language, I came to the conclusion that what Shirley Jackson said or didn't say wasn't what mattered. What mattered to Deborah Jackson was that here was a women — not only a physicist like herself, but the *first* Black women in the country to earn a doctorate in physics — who had made it to the top in academia, in physics research, in industry, and now in government. Clearly, Shirley Ann Jackson had become the embodiment of a dream, and I realized that if she could provide that bit of magic for Deborah Jackson, she could do it for all young women with a talent for physics and mathematics. It took me a bit longer to realize that Deborah Jackson herself also embodied the dream, and that she, too, could serve as a role model.

Someday soon, perhaps, I will attend a conference and be seated two rows behind another young women with a penchant for physics. Maybe Deborah Jackson will be the honored speaker. Perhaps I will watch this other young woman perch on the edge of her seat, straining her head forward, ear cocked so as not to miss a word as her role model recounts the experiences in becoming a physicist. Surely, Shirley Ann Jackson would be pleased to share the mantle.

2. Jackson, questionnaire. In fact, according to Jackson, all her siblings attended college. "My dad and all my siblings except one out of seven worked their way through college. My mother completed high school and college after the kids left home."

3. This is Jackson's own account.

4. Ibid. Jackson also received funding for work as an instructor of introductory physics for MIT's Project Interphase (January–August 1973).

5. Jackson, résumé.

6. Jackson, questionnaire.

7. Jackson, questionnaire and interview.

8. Ibid.

9. Jackson, résumé.

10. Jackson, questionnaire, ibid.

11. Ibid. Also Jackson, interview.

12. Jackson, résumé.

JACQUELYNE JOHNSON JACKSON

Studying African-American Social and Medical Issues

In 1968, medical sociologist Jacquelyne Johnson Jackson became the first full-time tenured Black faculty member at the Duke University Medical School. She and her fraternal twin Jeanne were born in Winston-Salem, North Carolina, in 1932 and grew up in Tuskegee, Alabama.[1] Jackson was a product of the Black elite; her parents, James A. Johnson and Beulah (Crosby) Johnson, were both well educated. James Johnson, who had both a bachelor's and a master's degree from New York University, established and directed Tuskegee Institute's School of Business. Beulah Johnson had a bachelor's degree in English and Latin from Shaw University, and later earned a master's degree. Both parents were involved in local politics and were active members of the Tuskegee Civic Association, which began a campaign in the 1930s to register Black voters. Family friends included George Washington Carver; Black sociologist Charles Goode Gomillion, the head of the Tuskegee Civic Association; Benjamin O. Davis Jr., who became the first Black general in the United States Air Force in 1954; and Portia Washington Pittman, the daughter of Booker T. Washington, who was Jackson's piano teacher.[2]

Jackson's parents placed strong emphasis on education: "Education was emphasized as part of child-rearing by my parents. . . . I was taught to think and to think independently. . . . In my early environment, I did not experience the negativity associated with being female."[3] After excelling in high school, she earned both the bachelor's and master's degrees at the University of Wisconsin at Madison in 1953 and 1955, respectively.

In 1960, Jackson became the first Black woman to earn a doctorate in sociology at Ohio State University. Immediately after earning the doctorate, Jackson took a number of temporary teaching positions—at Howard University, Jackson State College, and Southern University.

Jackson was awarded a 1966–1968 postdoctoral fellowship at Duke University's Center for the Study of Aging and Human Development, where her research focused on families headed by Black females and problems associated with aging among Blacks. She became full-time tenured faculty in 1968 after she completed her fellowship. She has contributed numerous articles to scholarly journals and is the author of two books.[4] Between 1972 and 1975, she edited the *Journal of Health and Social Behavior*, and helped found the *Journal of Minority Aging*. In the late 1970s, Jackson received a one-year John Hay Whitney Fellowship to study epidemiology at the University of North Carolina's School of Public Health in Chapel Hill, and a National Science Foundation fellowship to study cultural anthropology. During this time, she became the first woman chair of the Association of Black Sociologists, and one of the primary architects of the National Caucus on Black Aged.[5] Divorced and the mother of one child, Jackson serves on the Board of Directors of the Carver Research Foundation and the National Council on Family Relations.[6]

NOTES

1. "Jacquelyne Johnson Jackson," in *Notable Black American Women*, ed. Jessie Carney Smith, vol. 1 (Detroit: Gale Research, 1992), pp. 554–555. See also *Ebony Success Library*, vol. 1 (Chicago: Johnson Publishing, 1973), p. 167. *Who's Who among Black*

Americans (Detroit: Gale Research, 1990). According to Smith, Beulah Johnson insisted on traveling from Tuskegee to Winston-Salem to have the services of the same Black Winston-Salem physician who had earlier delivered her (p. 554).

2. Ibid.
3. Ibid.
4. See the appendix.
5. "Jacquelyne Johnson Jackson," p. 555.
6. Ibid.

MARY WINSTON JACKSON*

Scientist and mathematician Mary Winston Jackson completed the B.S. degree in physical science and mathematics at Virginia's Hampton Institute in 1942. In a 1977 interview, she said she had always liked math and had originally begun work at National Aeronautics and Space Administration's forerunner agency, the National Advisory Committee on Aeronautics (NACA) as a research mathematician in 1951.[1] Involved in a special training program, she was promoted to the position of aerospace engineer in the Theoretical Aerodynamics Branch of the Subsonic-Transonic Aerodynamics Division at the Langley Research Center, where she used model airplanes in wind tunnels to simulate flight conditions experienced by airplanes flying at subsonic speeds. The research was designed to provide a clearer understanding of air-flow dynamics, including such conditions as thrust and drag.

Jackson, who at the time of her interview was married and the mother of two children, also formed a neighborhood science club for Black youngsters, and "helped them construct a miniature wind tunnel to promote an interest in science and math outside the classroom."[2]

NOTES

1. Shawn D. Lewis, "She Lives with Wind Tunnels," Ebony, August 1977, p. 116. For more information on NACA, please see my entry on Katherine Coleman Goble Johnson in this volume.
2. Ibid.

SHIRLEY ANN JACKSON

*High Energy Physicist Becomes Power Player in
Science, Technology, and Politics*

Shirley Ann Jackson

Before theoretical physicist Shirley Ann Jackson assumed the chairmanship of
the Nuclear Regulatory Commission (NRC) in July 1995, she had, according to
Howard University physics professor Ronald E. Mickens, made important theoretical contributions to several areas of physics: the three-body scattering problem, charge density waves in layered compounds, polaronic aspects of electrons
on the surface of liquid helium films, and optical and electronic properties of
semiconductor strained layer superlattices.[1] As chair of the NRC, Jackson oversees an agency with a $500 million yearly operating budget and a mission to pro-
tect the public's basic health and safety vis-à-vis the nuclear industry. She is responsible for regulating the safety of the nation's aging nuclear power plants,
some of which have been repeatedly cited for safety violations; dealing with the
problems of storing an ever-growing stockpile of nuclear waste at a time when
nuclear plants are running out of on-site storage capacity; making politically sensitive decisions regarding re-licensing, license renewal, or license extensions on
nuclear power facilities; and overseeing the safe decommission of plants that shut
down before their license terms expire.[2]

Like so many of her other accomplishments, the chairmanship of the NRC is fraught with both perils and challenges — technological, scientific, and political. Jackson was appointed to the NRC at a time when the United States nuclear industry was under attack from a variety of fronts: industry workers pointed to hazards resulting from substandard, poorly implemented, or outdated safety measures; consumer groups charged that everything from increased cancer rates to birth defects stem from nuclear facilities; various public advocacy groups viewed the NRC's new approval system for renewing licenses of older nuclear plants as a relaxation of already weak safety standards; and politicians were all too aware of the public's growing lack of confidence in the industry. The poor perception of the nuclear industry has resulted in no new nuclear power facilities being constructed in the United States in more than two decades.[3] On the other side of the issue are the proponents of nuclear energy, such as the industry-supported Nuclear Energy Institute, which views the NRC as an overly prescriptive agency that interferes in licensees' management prerogatives.[4]

Jackson is no stranger to difficulty, controversy, or fighting the status quo, however. She was born in Washington, D.C., in 1946 and grew up in the then predominantly white Petworth area of the city. Barred from attending the all-white Barnard Elementary School near her home, she and her sister Gloria attended a Black school across town.[5] Both of Jackson's parents stressed education — her social worker mother, Beatrice Jackson, encouraged her reading and writing skills, and her postal worker father, George Jackson, spurred her interest in math and science and supported her early attempts at scientific experiments. The sisters built soapbox carts with tools their father provided, and Jackson raised bumblebees in the basement.

Jackson attended high school at a time when national and international events coincided to spur better opportunities for females and minorities with an interest in, and a talent for, science. The 1954 *Brown v. the Board of Education of Topeka* Supreme Court decision had ruled segregation illegal; a decade later, in 1964, the Johnson Administration succeeded in passing the first civil rights bill. In 1957, the Soviets successfully launched their Sputnik satellite into space, sending United States leaders into paroxysms of effort to eliminate the perceived Soviet scientific hegemony. Suddenly, the federal government became heavily involved in promoting scientific talent in students, regardless — on the surface, at least — of color or gender. Jackson has described herself as a beneficiary of both school desegregation and the nation's Sputnik-spurred interest in science and technology.[6] Interested mathematics teachers and an accelerated math and science program at Roosevelt High School provided Jackson with the types of gatekeeper courses she would need to major in science at college. A straight-A student, she wound up as valedictorian of her 1964 high school class.[7]

In a perfect world, one could simply state that Shirley Ann Jackson entered the Massachusetts Institute of Technology in 1964 and completed the bachelor of science degree in physics in 1968, but it is not a perfect world. The world Jackson entered at MIT was a lonely one — she was one of only forty-five women, and one of only ten to fifteen Blacks (mostly male) in the 900-member freshman class.[8] The predominantly white male world of her physics classes meant she experienced near total isolation—neither the men nor the handful of women physics majors would work with her. According to Jackson,

> The irony is that the white girls weren't particularly working with me, either. I had to work alone. I went through a down period, but at some level you have to decide you will persist in what you're doing and that you won't let people beat you down.[9]

To add insult to injury, Jackson said her white female classmates not only made it plain that they didn't want her in their study groups, but they even refused to sit at the same cafeteria table with her.[10] Jackson later described her isolation at MIT as lasting until "people found out what I could do and that I was as serious as they were."[11] Jackson received scholarships from the Prince Hall Grand Masons (1964–1968) and the Martin Marietta Aircraft Corporation (1964–1968). She thrived academically, working in the materials science lab at MIT, and earned the baccalaureate in physics in 1968.[12]

Jackson was accepted to graduate physics programs at Harvard University, Brown University, and the University of Chicago, but she chose to remain at MIT,[13] where—with support from the university itself and fellowships from the Martin Marietta Aircraft Company (1972–1973), the National Science Foundation (1968–1971), and the Ford Foundation (Advanced Study, 1971–1973)—she concentrated on theoretical high-energy particle physics.[14] What Gibbons has described as Jackson's "trademark combination of self-help and help for others," may account for her decision to stay at MIT, where she had already been active in urging the university to admit more minorities.[15] Given her own successful academic record, her work with the Black Student Union (the co-chair for two years as an undergraduate and an advisor as a graduate student) received greater recognition. She was concerned about the overall scarcity of Blacks in science, and she set up recruiting committees, got a commitment from MIT to make admission requirements more flexible, and worked to establish support programs for incoming minority students. Although the Black enrollment rate at the college inched slowly upward, the attrition rate for Blacks in the late 1960s and early 1970s hovered at a dismal 50 to 60 percent. As Jackson said in a 1974 interview,

> I knew it would be hard for [other Black students]. It may have been even harder for me, because there was no such program when I got there and the one I helped establish wasn't going to help me that much . . . I had made whatever adjustments I was going to make on my own. . . .
>
> In a lot of ways, [physics is] like a club or fraternity and blacks don't have entrée. High energy physics is very ethnic and black representation in small. There may be 30 blacks with Ph.D.s in physics in the country—eight in high energy physics.[16]

In 1973 she became the first Black woman to earn a Ph.D. at MIT, and the first Black woman in the nation to earn a doctorate in physics. Her research in theoretical elementary particle physics was directed by James Young, the first full-time tenured Black professor in the physics department.[17]

In a 1974 interview, Jackson said her then biggest problem was "gaining acceptance in a white male–dominated profession. . . . I hesitate to call that a real problem because I'm not really looking for acceptance, just respect, and that comes—grudgingly—but it comes."[18] As she built her reputation, others' acceptance, even respect, for her began to accrete, and Jackson's career as a theoretical physicist changed gears. She spent her first postdoctoral year as a research associate in theoretical physics at the Fermi National Accelerator Laboratory (1973–1974). She spent the next year on a Ford Foundation Individual Grant at the European Organization for Nuclear Research (CERN), in Geneva, Switzerland (1974–1975), returned to Fermi National Accelerator Laboratory (1975–1976), then spent time at the Stanford Linear Accelerator Center and the Aspen Center for Physics (1976–1977). In 1976 she joined the theoretical physics staff at the AT&T Bell Telephone Laboratories in Murray Hill, New Jersey, and in 1978 began work at the company's Scattering and Low Energy Physics Research Laboratory. Until 1991 her research at Bell Laboratories focused on gasses, films, semiconductors, and the relevant theories in solid-state, quantum, and optical physics.[19]

During her time at Bell Labs, Jackson widened her scope beyond the laboratory. She married fellow physicist Morris A. Washington, and had a son, Alan.[20] She served on the boards of directors of a New Jersey bank, an electric power company, an oil and gas company, and the MIT Corporation. Appointed to the New Jersey Commission on Science and Technology in 1985 by then Governor Thomas Kean, Jackson also served on an advisory panel to the United States Secretary of Energy, on research councils of the National Academy of Sciences, and on the Advisory Council of the Institute of Nuclear Power Operations.[21] A strong supporter of women's roles in science, Jackson has served on committees with the National Academy of Sciences, the American Association for the Advancement of Science, and the National Science Foundation to promote science and research and to discuss and examine the challenges and opportunities in science and related fields for women and minorities.[22]

At a 1992 conference on women in science, Jackson told the audience, "If you want to participate in a profession at the highest level, you have to be good and put your uniqueness to an advantage. There are few black women in physics, so people remember Shirley Jackson."[23] The price for such recognition has often been steep. In the highly collaborative world of theoretical physics research, Jackson has had to work hard to overcome the protective habits of the loner that she had acquired during her early years at MIT. "I tended to do my own thing, and that's not always the best way to do things in science. That's why when women are isolated—or blacks or any minority—it can be very destructive."[24] Yet, as Jackson acknowledged, as one of only a few women and the only Black, the isolation she felt could also act as a spur: "If you give a physics paper, it had better be good—because people will remember."[25]

In 1991, Jackson changed her working relationship with Bell Labs—she left her full-time position to become its consultant in semiconductor theory. This change allowed her the freedom to accept a position as professor of physics at Rutgers University in Piscataway, New Jersey. "I wanted to have graduate students, to build my own research groups," she said in 1993.[26] She then took a leave of absence from Rutgers to accept the position as chairman of the NRC.

When Jackson accepted the NRC chair, she was clear about her qualifications for the position. She pointed to her own laboratory work, cited the number of nuclear facilities at which she had worked, and indicated her broad experience:

> I was involved on the board of a nuclear utility, and I have been on the advisory council of INPO. And that is only the tip of the iceberg. I have had a lot of management, research management, executive managerial, executive oversight, corporate board, non-profit board, state commission experience, and on and on. I have also had a lot of financial planning experience, so I feel all of this has perfectly prepared me for this job. It all flows together.[27]

She had experience in coping with challenges beyond the realm of the scientific and technological. Jackson took political and practical leadership of an NRC operating in a changing financial, regulatory, and statutory environment—she and the agency would face challenges from any number of quarters.

In 1995, the same year Jackson took office, Congress, lobbied by interests within the nuclear power industry, entertained legislative proposals that would have placed a freeze on some NRC regulations, retroactive to November of 1994. At the time, Jackson said, "We would hope that in the end, the nature of the industries that we regulate should affect how we are treated in any regulatory reform legislation."[28] Issues related to other government agencies came into play, as well. In terms of storage requirements for spent nuclear fuel, for example, the NRC must contend with the standards and needs of other agencies such as the United States Department of Energy (DOE). The NRC has its own research organization independent of DOE. Housed within the Southwest Research Institute in San Antonio, Texas, the NRC's research group studies key areas related to the licensing of spent nuclear fuel repositories; ideally, the results of its work will play a part in creating a regulatory structure to provide effective review of license applications for spent nuclear fuel storage from the DOE, among others. According to Jackson, the research group will also be important under the integrated waste management approach to centralized interim storage, which includes NRC-certified storage and transportation casks.[29] Jackson believes that

> pushing harder in the risk area, in terms of risk-based approaches to regulation, will enable better focus on those things that are of particular safety significance. The [nuclear energy] industry is pushing on its side, particularly since it is interested in regulatory relief. We are paralleling this, with somewhat different activities going on, and I think those activities have the potential to relieve regulatory burden.[30]

The NRC, like most government entities, must also contend with the bureaucratic penchant for creating scores of acronymed agencies, committees, or groups, each of which must then zealously guard its own style of conducting business. While acknowledging that some past differences between the NRC and the EPA had to do with regulatory approach or style, Jackson said that developing a consistent regulatory approach would "revolve around some common understanding or common approach to risk. What constitutes acceptable risk? . . . At the heart of it is really coming to concurrence on what constitutes acceptable risk—the creation of ISCRS allows the opportunity to have a forum for getting at this problem."[31] The ISCRS, the Interagency Steering Committee on Radiation Standards, was created jointly by the NRC and the Environmental Protection Agency (EPA) in April 1995 to decide on radiation standards that would apply (uniformly, one presumes) to all government agencies. This was an important specific step, according to Jackson, toward developing a consistent regulatory approach between EPA and NRC. In an era of increasing demand for electrical power, Jackson's main concern is for public safety, both in the long and short terms:

> What the technology is that will be used to generate that power is a question no one can definitively answer. . . .
> It is not for the NRC to decide what kind of energy generation choices get made in this country—only that if the choice happens to be nuclear, that it is as safe as it can be.[32]

In a 1974 *Ebony* magazine article, Jackson said that she had always been concerned about the scarcity of Blacks in the sciences, and she cited the particular need for concrete scientific expertise in developing African countries. "We [Blacks] need to be watchdogs. . . . Because a number of scientists are advisers on national policy. . . . we should have a voice."[33] Now, nearly twenty-five years later, Shirley Ann Jackson has become a watchdog—one with an eloquent, reasoned voice.

NOTES

1. Ronald E. Mickens, "Shirley Ann Jackson," in *Notable Black American Women*, ed. Jessie Carney Smith, vol. 1 (Detroit: Gale Research, 1992), pp. 565–566. Jackson was nominated to the NRC by President Bill Clinton and confirmed by the U.S. Senate. Information provided by Gerry Schuetze, personal assistant to Jackson, written communications to author, February 1996; and Office of Public Affairs press/information release, 26 February 1996 (Washington, D.C.: United States Nuclear Regulatory Commission, February 1996). Jackson was sworn in as an NRC commissioner on May 2, 1995, and assumed the chairmanship two months later, on July 1, 1995 (Schuetze, written communication). In communications from Jackson's office, as well as in recent articles, her position is listed as "chairman," not chairwoman or chairperson. I use what is obviously the preferred title. See also Allen Zeyher, "A Nuclear News Interview: Jackson Takes the Reins," *Nuclear News* 38 (November 1995): 26–29. Daniel Southerland, "Equation for Success: A Life of Struggle Takes D.C.'s Shirley Jackson to NRC's Helm," *The Washington Post*, 4 May 1995, pp. B10–13. Robert Cohen, "A Positive Reaction: Rutgers Physicist Sails Through Hearing as Nuclear Nominee," (Newark, N.J.) *Star Ledger*, 17 February 1995, p. 11.

2. Southerland, "Equation for Success," pp. B11–13. See Zeyher, "Jackson Takes the Reins," pp. 26–29. With regard to decommissioning, there are trust fund set-asides that Jackson must make sure are available in the event of a premature shutdown/decommissioning of a plant (Jackson, cited in Zeyher).

3. Southerland, "Equation for Success," pp. B10–11. With the exception of the Long Island Lighting Company's nuclear power plant debacle at Shoreham, New York, no

new nuclear plants have been constructed in the United States in nearly three decades. Although the Shoreham plant was completed and tested at 25% capacity, it never went online and was decommissioned after a sustained battle involving Long Islanders, ratepayers, the utility, and New York state politicians led by then Governor Mario Cuomo.

4. See Towers Perrin, *Nuclear Regulatory Review Study* (Nuclear Energy Institute, October 1994), cited in Zeyher, "Jackson Takes the Reins." In reviewing the candidacy of Dan M. Berkovitz, President Clinton's other nominee for a seat on the NRC, Sen. Lauch Faircloth (R-N.C.), a member of the Senate Environment and Public Works Committee, charged, along with seven other Republican senators, that Berkovitz "would impose burdensome regulations" on the nuclear industry. Faircloth suggested that Berkovitz's lack of private business experience was a fatal flaw (Southerland, "Equation for Success," pp. B10–13). Berkovitz's nomination was rejected. During the Senate hearings on Jackson's candidacy, Faircloth saw her as "exactly what we need in a nominee" (quoted in Southerland, ibid.). Jackson was also strongly supported by New Jersey Democratic Senators Bill Bradley and Frank Lautenberg (ibid.).

5. According to Southerland, Jackson's sister, Gloria Jackson Joseph, is now an attorney and director of administration at the National Labor Relations Board in the District of Columbia.

6. Jackson, quoted in Southerland (ibid.).

7. Mickens, "Shirley Ann Jackson," pp. 565–566. Southerland, "Equation for Success," pp. B11–12.

8. Ann Gibbons, "Pathbreakers: Gaining Standing—By Standing Out," *Science* 260 (16 April 1993): 393. Hattie Carwell, "Dr. Shirley Jackson: Theoretical Physicist," in *Blacks in Science: Astrophysicist to Zoologist* (Hicksville, N.Y.: Exposition Press, 1977), p. 60.

9. Jackson, quoted in Gibbons, "Pathbreakers," p. 393.

10. Ibid.

11. Southerland, "Equation for Success."

12. "Physicists," in *Contributions of Black Women to America*, ed. Marianna W. Davis, vol. 2 (Columbia, S.C.: Kenday Press, 1982), pp. 447–448. See also Mickens, "Shirley Ann Jackson," p. 565.

13. "Nuclear Physicist at Fermi Lab: Young Ph.D. Holds Her Own in White Male-Dominated World," *Ebony*, November 1974, pp. 114–116, 118, 120, 122.

14. "Physicists," pp. 447–448. Mickens, "Shirley Ann Jackson," pp. 565–566.

15. Gibbons, "Pathbreakers," p. 393.

16. "Nuclear Physicist at Fermi Lab," pp. 120, 122.

17. Zeyher, "Jackson Takes the Reins," p. 26. The title of Jackson's dissertation is "The Study of a Multiperipheral Model with Continued Cross-Channel Unitarity."

18. "Nuclear Physicist at Fermi Lab," p. 115.

19. "Physicists," pp. 447–448; Mickens, "Shirley Ann Jackson," pp. 565–566. See also Zeyher, "Jackson Takes the Reins." At FermiLabs and at CERN she worked on theories of strongly interacting elementary particles (Mickens).

20. Schuetze, written communication; USNRC press/information release.

21. Southerland, "Equation for Success." Governor Kean reappointed Jackson to a five-year term on the Science and Technology Commission in 1989. Jackson has held boards of director positions at New Jersey's Public Service Enterprise Group and its subsidiary, Public Service Electric and Gas Company, Sealed Air Corporation, CoreStates Financial Corporation, CoreStates/sNew Jersey National Bank, and New Jersey Resources Corporation (Mickens, "Shirley Ann Jackson"; also Schuetze, written communication, and USNRC press/information release).

22. Mickens, "Shirley Ann Jackson."

23. Quoted in Gibbons, "Pathbreakers." The conference was held at Trinity College, Washington, D.C.

24. Quoted in Gibbons, "Pathbreakers."

25. Ibid.

26. Quoted in Gibbons, "Pathbreakers."

27. Zeyher, "Jackson Takes the Reins," pp. 27–28. INPO is the Institute of Nuclear Power Operations.

28. Ibid.

29. Quoted in Zeyher, pp. 28–29. See the proposed Upton Bill (HR 1020), 1995 Congressional Session.

30. Quoted in Zeyher, p. 29.

31. Quoted in Zeyher, p. 29.

32. Ibid.

33. "Nuclear Physicist at Fermi Lab." In Jackson's original response, she cited "biological warfare," but I have omitted this portion of her statement, as it does not actually pertain to the situation the overall quote illustrates.

CHARLENE DREW JARVIS

Medical Pioneer's Daughter Uses Her Science Training to Carve Out a Political Career

Charlene Drew Jarvis

On the one hand, Charlene Drew Jarvis was destined to become a scientist. On the other hand, she was destined to become a politician and public servant. Jarvis is the daughter of Charles Richard Drew, the physician whose pioneering research in the 1930s on blood preservation led to the founding of blood banks. As a professor of surgery at Howard University Medical School in the 1940s, Drew trained nearly an entire cohort of Black surgeons.[1] The irony of Drew's discovery of a way to preserve whole blood was that it came at a time in American history when the blood of Black donors was routinely segregated from that of whites, an unfortunate politicizing of his intellectual and scientific property. Although Clarlene Drew Jarvis was still a child when her father died, she was clearly the heir to a divided legacy.

Born around 1940, Jarvis earned the baccalaureate degree in psychology at Ohio's Oberlin College in 1962. She quickly followed through, earning the master's degree, also in psychology, at Howard University in 1964.[2] She served as a psychology instructor at Montgomery College in Rockville, Maryland (1964– 1965), where she taught introductory psychology, mental hygiene, and social psychology courses to undergraduates. During this time, she also secured a position as a school psychologist with the District of Columbia Public Schools (1964– 1967), where she developed and administered tests to children in need of counseling and assessed students' environments to determine specific factors impacting on their development.[3] She then entered a graduate program at the University of Maryland at College Park, where she completed the Ph.D. in neuropsychology, in 1971. She also held a one-year psychology instructorship at Howard University (1970–1971).

Jarvis received a staff fellowship at the Laboratory of Neuropsychology of the National Institute of Mental Health from 1971 to 1976, where she conducted electrophysiological and ablation studies of the primate brain. She reviewed applications for pre-doctoral and postdoctoral awards in physiological psychology and comparative behavior as part of the National Institute of Mental Health's Biological Sciences Fellowships Initial Review Group from 1975 to 1976, then became a research psychologist at the Laboratory of Neuropsychology from 1976 to 1979, where she again conducted studies of the primate brain.[4] Both alone and with co-authors, Jarvis produced a number of research articles during the period between 1974 and 1982.

In the late 1970s, Jarvis began to move into politics and positions in the public sector. In 1979, she first took her seat on the Council of the District of Columbia, a position she still holds after having been elected to the Council five times as a representative of Ward 4. She chaired the Council's Committee on Housing and Economic Development from 1981 to 1990, and became chair of the Committee on Economic Development in 1990.[5] Jarvis was an elected delegate from the District of Columbia to the national democratic conventions in 1980, 1984, 1988, and 1992; she addressed the 1988 convention on the need for civil and human rights in the United States, including statehood for the residents of the District of Columbia. Jarvis became the national co-chair of the Clinton/Gore campaign in 1992. In 1993 she was appointed by President Bill Clinton to serve on the National Institute of Mental Health Advisory Council; she was also appointed by Donna Shalala, the Secretary of the Department of Health and Human Services, to serve on the Breast Cancer Task Force as well as the advisory board of the Women's Health Initiative of the National Institutes of Health.

Jarvis has received the Distinguished Service Award in 1994 from the Greater Washington Society of Association Executives; an honorary doctorate in humane letters in 1994 from Amherst College; and the Howard University Alumni Award for Distinguished Postgraduate Achievement in the fields of politics and community service in 1993.

Jarvis also has been active on the international level. In 1992, at the behest of the African American Institute, she went to Nigeria to conduct seminars, teach-ins, and workshops for Nigerian women on the issues of health, politics, business, and economic development. In 1988, at the behest of the Jewish Community Council, she went to Israel to become more acquainted with Israeli society, including the kibbutz system, housing problems/solutions, politics, and socioeconomic issues.

She has maintained interest in the medical and scientific fields by serving on the boards of directors of the National Museum of Health and Medicine (1990 to present) and the Ronald Reagan Center for Emergency Medicine of the George Washington University Hospital (1993 to present). She has been vice-president of the executive committee and chair of the transplantation subcommittee of the Washington, D.C., Chapter of American Red Cross, which sets policy for the organization's bone marrow, tissue, and blood bank programs; she has also served as the national spokesperson for the National Marrow Donor Program (1993 to present). Jarvis serves on the advisory panel of the Minority Institutions Science Improvement Program for the National Science Foundation, which reviews funding applications from minority institutions to improve science instruction. She also serves on the boards of directors of the Pennsylvania Avenue Development Corporation (1990 to present), the Economic Development Finance Corporation (1986 to present), the Private Industry Council (1986 to present), and the Girl Scouts of America.[6]

NOTES

1. Charles Richard Drew was born in 1904. He directed the Plasma for Britain program during World War II, and became director of the first American Red Cross blood bank program. He was a professor of surgery at Howard University Medical School from 1941 until his death in 1950. See Hamilton Bims, "Charles Drew's 'Other' Medical Revolution: Famed Pioneer in Blood Preservation Trained Vanguard of Black Surgeons," *Ebony*, February 1974 , pp. 88–93.

2. Charlene Drew Jarvis, written communication to author, including curriculum vitae, 29 January 1996.

3. Ibid.

4. Ibid.

5. Ibid.

6. Ibid.

MAE C. JEMISON
Physician, Astronaut

In 1987, when the National Aeronautics and Space Administration (NASA) announced its newest list of astronaut candidates, physician Mae C. Jemison suddenly found herself catapulted into the national spotlight.[1] Media attention intensified when Jemison rocketed into space on September 12, 1992, becoming the first Black woman in history to leave Earth's gravity well. As part of the crew of the space shuttle Endeavour, she participated in a one-week mission to study the effects of microgravity on humans and animals.[2] For a number of years after her historic flight, Jemison garnered more attention than any other Black woman scientist, and was touted as a role model for Black youngsters, particularly females. Sadly, her national prominence came at a time when various voices in the Black community had already invested much criticism in the space program,[3] which may have lessened her impact as a role model for Black girls. Today, perhaps due to Americans' declining interest in the space program in general, Jemison's face is no longer as instantly recognized as it ought to be.

Born in 1956, in Decatur, Alabama, Jemison was raised in Chicago, Illinois. Her parents, janitorial maintenance supervisor Charlie Jemison and schoolteacher Dorothy Jemison, reportedly moved their family to Chicago so that their three children could attend better schools. A social worker uncle added to her parents' focus on education, and fostered Jemison's girlhood interest in the sciences. An indefatigable reader, she soon began to haunt the local library.[4] As she recalled in a 1989 interview, "I went through a phase, when I was probably from 10 to 14 years old, where I read lots and lots of astronomy books, not science fiction, but actual astronomy books. . . . And I was always encouraged by my teachers and parents to be involved in science, but also in a number of other areas."[5] A few years later, a seemingly routine high school science-class trip to a local university sparked Jemison's interest in biomedical engineering. She wound up a well-rounded young woman, who balanced honor roll achievements at Chicago's Morgan Park High School with performing in the school's modern dance troupe and cheering on the pom-pom squad. By graduation from Morgan Park in 1973, she had earned a National Achievement Scholarship for college.[6]

Jemison entered California's Stanford University in the fall of 1973. As she had done in high school, she purposefully worked at being a well-rounded student. Majoring in both chemical engineering and African American studies, she was also active in dance and theater productions; she even, according to one source, represented Stanford at a 1976 festival in Jamaica, British West Indies.[7] In a 1991 interview, Jemison recalled her wide-ranging interests:

> Science is very important to me, but I . . . stress that you have to be well-rounded. One's love for science doesn't get rid of all the other areas. . . . someone interested in science is interested in understanding what's going on in the world. That means you have to find out about social science, art, and politics.[8]

Jemison graduated from Stanford in 1977, with two bachelor's degrees—a B.S. in chemical engineering, and a B.A. in African American studies.[9]

Degrees in hand, the twenty-one-year-old Jemison entered New York City's Cornell University Medical School. Again, she continued to demonstrate wider interests: during the summer of 1978 she did volunteer work with Cambodian refugees at a camp in Thailand; in 1979, with a grant from the International Travelers Institute, she participated in health studies in Kenya.[10] She completed the M.D. degree in 1981, interned at the Los Angeles County/University of Southern California Medical Center through July 1982, and worked briefly as a general practitioner with a medical group in Los Angeles.[11]

In 1983, Jemison began a 30-month posting as a staff physician with the Peace Corps in West Africa, caring for Peace Corps volunteers and State Department personnel in Sierra Leone, overseeing the medical health care program for volunteers in Liberia, and participating in research projects on schistosomiasis, rabies, and the hepatitis B vaccine for the National Institutes of Health and Centers for Disease Control. She also supervised medical personnel, provided oversight on laboratory conditions, developed curricula for public health and safety guidelines for volunteers, and helped write self-care health manuals for the indigenous people.[12]

Jemison was already something of a role model, albeit relatively unsung, when she returned to the United States in 1985 and went to work for a Los Angeles-based health maintenance organization. Within a matter of months she had applied to NASA's astronaut training program and enrolled in night classes in engineering at the University of California at Los Angeles.[13]

After a brief postponement in the selection of astronaut candidates in the aftermath of the 1986 Challenger space shuttle disaster, NASA notified Jemison in 1987 that she had been chosen. She completed the one-year mission specialist training program in August 1988.[14] In a 1989 interview, she expressed concern about the attitudes of some African American spokespersons toward NASA. She pointed out that the space program provided benefits for all Americans, particularly advances in communications, medicine, and more sophisticated knowledge about the state of the environment:

> Some might say that the environment is not a Black issue, but I worked in Los Angeles and I saw more Black and Hispanic children with uncontrolled asthma as a result of pollution. Just as many of us get sick from those types of things, and, in fact, we have more problems with them because many of us don't have the [easy] availability [of] health care.[15]

She also seemed sensitive to those critics who insisted that she would not fare well in the NASA environment, which they viewed it as a bastion of white male interests:

> One runs into things that discourage you all the time. . . . It depends on the way you react to it. One can react to it and say, "Oh, I got the feeling that somebody didn't want me there," or "They weren't very enthusiastic for me being around, so, therefore, I'm going to give it up." But for me, it just means that I have to keep trying and I have to work on it because this is something I want to do. This is something I know I'm capable and qualified to do.[16]

Although Jemison officially left NASA after her historic flight, she continues to make public appearances on behalf of the space program. Her outlook on various issues, coupled with her education and experiences, make Jemison a valuable role model, even though she has in the past expressed discomfort with the idea:

> The thing that I have done throughout my life is to do the best job that I can and be me. . . . And that's really all I can do. In terms of being a role model, what I'd like to be is someone who says, "No, don't try to necessarily be like me or live your life or grow up to be an astronaut or a physician unless that's what you want to do."[17]

NOTES

1. Four Black male astronauts had already preceded Jemison into space: Ronald E. McNair, who later died in the *Challenger* disaster; Guion S. Bluford Jr.; Frederick D. Gregory, a nephew of pioneering medical researcher Charles Richard Drew; and Charles F. Bolden Jr. (See "The Future-Makers: They Clone Cells and Smash Atoms in Search of Tomorrow's World Today," *Ebony*, August 1985, pp. 62–64, 66.) In fact, Blacks have a long history of being involved in the space program through the military, NASA, or their work with the aerospace corporations that supply NASA hardware and expertise. See "Long Island Negroes Built Guts of Missile," *New York Amsterdam News* (City Edition), 8 February 1958. James L. Hicks, "Negroes in Key Roles in U.S. Race for Space: Four Tan Yanks on Firing Team," *New York Amsterdam News* (City Edition), 8 February 1958, pp. 1, 15. "Negroes Who Help Conquer Space: Over 1,000 Negroes Are in the Satellite, Missile Field," *Ebony*, May 1958, pp. 19–22, 24, 26. "Negroes Aid in Space Research," *Chicago Defender*, May 10, 1958. "Space Computing Mathematician," *Ebony*, August 1960, p. 7. "Rocket Fuel Researcher," *Ebony*, September 1960, p. 6. "Her Paper Helped to Track Astronaut: Math Expert Who Aided Spaceman Is 'Thrilled,'" *New York Amsterdam News*, 24 February 1962, pp. 1, 20. Enoc P. Waters Jr., "They Helped Track Glenn in Orbit," *Afro-American*, 3 March 1962, pp. 1–2. Steven Morris, "How Blacks View Mankind's 'Giant Step,'" *Ebony*, September 1970, pp. 33–36, 38, 40, 42.

2. Christine Lunardini, "Mae C. Jemison," in *Black Women in America: An Historical Encyclopedia*, ed. Darlene Clark Hine, Elsa Barkley Brown, and Rosalyn Terborg-Penn, vol. 1 (New York: Carlson, 1993; Bloomington: Indiana University Press, 1994), pp. 633–635. See also Marilyn Marshall, "Child of the '60s Set to Become First Black Woman in Space," *Ebony*, August 1989, pp. 50, 52, 54–55. "Monument Recognizing Black Aviators Is Unveiled," *Jet*, September 1990, p. 34. "1988 Essence Awards," *Essence*, October 1988, pp. 59–60. "Space Is Her Destination," *Ebony*, October 1987, pp. 93–98. Maria C. Johnson, "Upward with Worldly Lessons," *Greensboro News and Record*, 28 January 1991, p. B5.

3. See, for example, "A Plan for Pioneer Eleven," *Ebony*, February 1974, p. 98. The Morris article, "How Blacks View Mankind's 'Giant Step'," first recounts negative views of space exploration among African Americans, at some length, and then does features on Grumman Aerospace Corporation's Carl Echols, a vehicle project engineer who worked on the lunar landing model, and aeronautical engineer O. S. (Ozzie) Williams, the company's expert in liquid fuel rockets; Thomas E. Jenkins, a deputy assistant director for the National Science Foundation and a management officer in NASA's space flight program who helped plan and organize NASA's initial space effort; Donald Gothard, an engineering supervisor for AC Electronics Division of General Motors who helped design the Apollo guidance equipment; "cosmo-chemist" George W. Reed, who studied lunar rocks at Argonne National Laboratories; and many more.

4. Jessie Carney Smith, "Mae C. Jemison," in *Notable Black American Women,* ed. Jessie Carney Smith, vol. 1 (Detroit: Gale Research, 1992), pp. 571–573.

5. "Space Is Her Destination," pp. 93–94.

6. Ibid.

7. Smith, "Mae C. Jemison," p. 571.

8. Johnson, "Upward with Worldly Lessons," p. B5. Also cited in Smith, "Mae C. Jemison," p. 571.

9. Smith, "Mae C. Jemison," p. 572.

10. Johnson, "Upward with Worldly Lessons," p. B5. See also Smith, "Mae C. Jemison," p. 572.

11. Johnson, "Upward with Worldly Lessons," p. B5.

12. Smith, "Mae C. Jemison," pp. 571–572.

13. "Space Is Her Destination," pp. 93–94.

14. Smith, "Mae C. Jemison," p. 572.

15. Marshall, "Child of the '60s," p. 54.

16. "Space Is Her Destination," pp. 93–94.

17. Ibid., p. 94.

KATHERINE COLEMAN GOBLE JOHNSON

The Ultimate Payoff—Putting America in Space

Katherine Coleman Goble Johnson

When mathematician and space scientist Katherine Coleman Goble Johnson began her career at the National Advisory Committee on Aeronautics (NACA), the forerunner of the National Aeronautics and Space Administration (NASA), Blacks were kept in a segregated office, and women were merely tools used to perform calculations. Yet because she was already in position when the space race became a Cold War priority, Johnson managed to carve out a measure of success in the predominantly white, male field of space science—a field so new that much of the early work involved seat-of-the-pants calculations and hunches.

Katherine Johnson was born in 1918 in White Sulphur Springs, West Virginia, and her father had a tremendous commitment to his children's education. Since White Sulphur Springs did not have a high school, when the eldest of the Coleman family's children approached high school age, her father moved the entire family—which included her two brothers and one sister—to Institute, where the children attended the laboratory high school at West Virginia State College.[1] According to Johnson,

> My father was really determined that we would all have good educations. So beginning the year my oldest brother was of high school age, we went to Institute every September and every June we came back to White Sulphur Springs.
>
> The summer before we went to Institute for the first time, my father went there to find a house to rent for us. He went to see the president of the College, and he told my father that it was impossible, that he'd never find a house to rent in the city. My father told him to wait and see, and as luck would have it he found a distant cousin of my mother's and she agreed to rent us a house.
>
> During the eight years that the four of us were in high school there, we only got to see my father about once a month during the school year. I know it was a real sacrifice for he and my mother—being separated like that for most of the year. He worked at the Greenbriar Hotel, a very elite establishment in White Sulphur Springs, while my mother stayed and kept house for us in Institute.
>
> I had a very, very interesting childhood, but, oh my, education was the primary focus in our family.[2]

After completing high school, Johnson entered West Virginia State College. One of Johnson's teachers at West Virginia State—Angie Turner King—had a tremendous influence on her, both in high school and in college.[3] King was "a wonderful teacher—bright, caring, and very rigorous," Johnson recalled.[4] Institute was an unusual town in those days: all Black with a significant number of very well-educated people.[5]

As do many of the other women profiled in this volume, Johnson has focused on the strong degree of mentoring students experienced at historically Black schools:

> Our teachers made such a difference—all my teachers and professors were very supportive and nurturing. One of my other favorites was Professor John F. Matthews, a scholar of Haitian literature, who had been the secretary to President Tubman, in Liberia. [Matthews] spoke at least seven languages fluently. He was head of the Romance Languages Department and a real hero of mine. I adored the ground he walked on. He was a wonderful mentor to me,

just as he was to many students at State who spent many hours reading the books in his library and discussing what they had read with him.[6]

In addition to King and Matthews, Johnson had other mentors, whom she considered equally important:

> James Carmichael Evans, was one of my math teachers in college—his wife had taught me math in the eighth grade—and because they didn't have children at that time, I became a kind of child to them. I was always at their house, and he was the sort of person who was always teaching, even at home. To please him I always had to do my very very best, and he always knew if it wasn't my very best effort.
>
> At that time, I was very interested in French and English studies [with] Professor Matthews, but Professor Evans said, "I know how good you are in French, but you *will* also major in mathematics." And then another of my math teachers said, "If you are not [signed up and] in my math class on Monday morning, I'm going to come and find you." So I really had no choice but to do a double major, which is what I did.[7]

Johnson earned the B.S. degree in mathematics and French in 1937.

Immediately after her graduation, Johnson began her career as a mathematics teacher. It was during the Depression, she recalled, so permanent teaching positions were hard to come by. Between 1937 and 1953, Johnson taught mathematics, and occasionally French, at a number of schools. She taught for a time in Marion and Morgantown, West Virginia, before moving to Virginia, where she worked part time at a high school in Newport News. Between 1938 and 1940, Johnson also pursued graduate studies in mathematics and physics at West Virginia State.[8]

In 1953, Johnson went to work as a mathematician at NACA, which had its research facility and headquarters at Langley Field, Virginia.[9] In 1958, in response to the Soviet success with Sputnik, Congress, at the behest of President Dwight D. Eisenhower and with the support of Senate Majority Leader Lyndon B. Johnson, passed the National Aeronautics and Space Act. In October, Johnson and the other employees of NACA became employees of NASA.[10]

Before the age of electronic computers, NACA employed hundreds of women to perform calculations for research engineers. Known as "human computers," these women, the majority of whom had mathematics degrees, performed the time-consuming job of doing calculations on wind tunnel experiments using slide rules and mechanical calculators. Historian of science Margaret Rossiter described their job as "women's work"[11]—males with similar educations were classified as "professionals" but the women were "subprofessionals."[12] Nevertheless, the women generally voiced few complaints about the discrimination, because they were doing the kind of work they loved and were being very well paid compared to what they might earn in other venues.[13]

When Johnson went to work at NACA, Black mathematicians were segregated in their own office and loaned out to wherever they were needed.[14] Assigned to the segregated office, Johnson was soon loaned out to the Flight Research Division and was "never sent back."[15] That department became the nucleus of the space project. "We were pioneers of the space era. We worked in secret for about

three years, often without knowing exactly what the total thrust of our work was.
. . . You had to read *Aviation Week* to find out what you'd done," she said. "The
Russians were already attempting to move into space at that point, so our efforts
were militarily strategic."[16]

When NACA became NASA in 1958, Johnson, along with the agency's other
employees, was assigned to it. As she recalled,

> Everything was so new—the whole idea of going into space was new and daring.
> There were no textbooks, so we had to write them. We wrote the first textbook by
> hand, starting from scratch. People would call us and ask, "what makes you
> think this or that is possible?" and we would try to tell them. We created the
> equations needed to track a vehicle in space. I was lucky that I was working with
> the division that worked out all the original trajectories, because I guess that's
> what I am remembered for.[17]

At that time, according to Johnson, women—regardless of their training or the
jobs they were doing—were not permitted to attend briefings. Johnson, eager to
learn all she could, felt locked out:

> These were such intelligent men, they knew so much, and I always loved
> intelligence, and so I'd ask what had gone on in the briefing—I'd listen and
> listen, and ask questions. Then, of course, I'd ask why I couldn't go myself, and
> eventually they just got tired of answering all my questions and just let me in [to
> the briefings].[18]

In addition to being kept out of briefings, women were generally not permitted to
put their names on their research:

> We needed to be assertive as women in those days—assertive and aggressive—
> and the degree to which we had to be that way depended on where you were. I
> had to be. In the early years at NASA women were not allowed to put their
> names on the reports—no woman in my division had had her name on a report.
> I was working with Ted Skopinski and he wanted to leave and go to Houston . . .
> but Pearson, our supervisor—who was not a fan of women—kept pushing him to
> finish the report we were working on. Finally, Ted told him, "Katherine should
> finish the report, she's done most of the work, anyway." So Ted left Pearson with
> no choice; I finished the report and my name went on it, and that was the first
> time that a woman in our division had her name on something.[19]

Johnson's ground-breaking report *Determination of Azimuth Angle at Burnout
for Placing a Satellite over a Selected Earth Position* would become the theoretical basis for launching, tracking, and returning vehicles in space.[20] Her computations were used in tracking the trajectories of the May 1961 Mercury flight
of astronaut Alan Shephard and the February 1962 Mercury flight of astronaut
John Glenn, as well as trajectories of a number of other manned and unmanned
flights.[21] She also developed simple emergency navigational methods that astronauts could use in the event that their spacecraft lost contact with ground control.
A few years later, in preparation for the Apollo moon landing project, she cal-
culated the trajectories for putting a spacecraft in lunar orbit, sending its lunar
lander to the moon's surface, returning it to the spacecraft, and returning the
mission to earth.[22]

Despite their mathematical formulas, Johnson recalled that much of what NASA scientists and engineers did at that time was based on hunches—she remembers that one of the astronauts said, "I'll take Kate's hunches anytime."[23] Because the various NASA teams were working in new, untried areas, Johnson said initiative and the ability to create what-if scenarios were important components in their work. By the mid-1960s, a close colleague, Al Hammer, had begun asking, "What do you do when things go wrong? What do you do when the computers go out?"[24] She and Hammer began to think about the possible problems, and to work up various back-up projections. Computers were still relatively new at that point, Johnson said, so the pair began to think about space flight scenarios involving a computer failure:

> [We were at Langley, in Virginia, but] the computers we were using were out in California, and there was the time differential. . . . We worked mostly at night, so we could communicate with the computers. Most of the time we worked sixteen hour days, or we'd work eight hours and then come back eight hours later. One morning I woke up in my car by the side of the road—I had fallen asleep behind the wheel, and I told myself I had to cut back. . . .
>
> Finally, the government accountants caught up with us and said they couldn't pay us the overtime [or] give us compensatory days off, and said we had to stop working so hard.[25]

In 1970 the Apollo 13 Mission experienced a fuel tank explosion en route to the Moon; many systems aboard the craft, including the on-board computers, had to be shut down. Johnson and Hammer were among the experts called in to monitor the flight, which was successfully returned to Earth.

During the late 1960s and early 1970s, Johnson analyzed data gathered by tracking stations around the world during various manned and unmanned missions. In the mid-1970s she began research on new navigation procedures to determine more practical ways to track objects in near-earth orbit and interplanetary space vehicles.[26] One of the special projects she was involved with was the Earth Resources Satellite, which was used to locate underground minerals and other essential resources.[27] "I had a wonderful, wonderful career at NASA," she said. "I don't imagine everyone can say that, but it's true in my case. I have always loved the idea of going into space—I still do."[28] Citing NASA's record of "glorious successes," Johnson says she is dismayed at the present lack of public and official support for the space program: "People want the payoff today—they want the paycheck now—and that's not the way things work, not worthwhile things that you want to do right because so much depends on them."[29]

Although Johnson started at NACA/NASA as a "human computer," a combination of luck and her own assertiveness, aggressiveness, and ability allowed her to expand the scope of her work into that of aerospace technologist and researcher.[30] She is a 1967 recipient of the Group Achievement Award presented to NASA's Lunar Orbiter Spacecraft and Operations Team, and the 1970 recipient of a similar award presented to the Apollo team. Katherine Coleman Goble Johnson co-authored 21 technical papers during her career. She retired in 1986, after 33 years with NASA and its forerunner agency.[31] She is the mother of three daughters, one of whom—Joylette Gobal Hylick—also became a mathematician.[32]

NOTES

1. Katherine Coleman Goble Johnson, telephone interviews with author, 6 March 1996, 27 May 1996, 7 June 1996. Johnson, written communications to author, including letter, résumé, personal statement ("What Goal in Scientific Achievement Did I Set for Myself?"), and assorted articles, 30 May 1996.

2. Johnson, telephone interview, 7 June 1996.

3. Johnson, telephone interview, 6 March 1996. See the entry for Angie Turner King in this volume.

4. Ibid.

5. Among Johnson's childhood acquaintances was Margaret James Strickland Collins, who also went on to a career in science (ibid.). See the entry for Collins in this book.

6. Johnson, telephone interview, 7 June 1996. Johnson and Matthews kept in touch even after she graduated from college and moved on. Many years afterward when Matthew retired, he moved to Hampton, Virginia, where Johnson lived, and he became a visiting scholar at Hampton Institute (now College). Margaret S. Collins was also befriended and mentored by Matthews. Johnson mentioned that her sister had been taught by Margaret Collins's father, Professor James.

7. Johnson, telephone interview, 7 June 1996. According to Johnson, James Carmichael Evans later went on to become a U.S. Assistant Secretary of Defense.

8. Johnson, résumé.

9. NACA was formed by an act of Congress in 1915. The agency was an outgrowth of the work of Samuel Pierpoint Langley, an astronomer with an interest in the development of heavier-than-air craft who had published *Experiments in Aerodynamics* (Washington, D.C.: Smithsonian Institution, 1891). Langley's work was supported partially by the Smithsonian, but it languished after an experimental failure in 1903. The start of World War I in Europe added new urgency to the idea's strategic nature. NACA originally had a committee of twelve members appointed by the president of the United States—two members from the Army; two from the Navy; one each from the Smithsonian, the Weather Bureau, and the Bureau of Standards; and five public members. All were to be "acquainted with the needs of aeronautical science, either civil or military or skilled in aeronautical engineering or its allied sciences." The agency's headquarters was located at Langley Field, Virginia. See A. Hunter Dupree, *Science in the Federal Government: A History of Policies and Activities* (Baltimore: Johns Hopkins University Press, 1986), pp. 283–288. See also Johnson, telephone interview, 27 May 1996.

10. Nicholas Booth, *The Encyclopedia of Space* (New York: Mallard Press, 1990), pp. 17–18.

11. Margaret Rossiter, *Women in Science: Struggles and Strategies to 1940* (Baltimore: Johns Hopkins University Press, 1982).

12. Kirk Saville, "What Was the First Computer Model?—A Woman," *Newport News Daily Press*, 14 December 14, 1990, pp. A–3, A–16. (My thanks to Katherine Johnson for providing me with a copy of this article.) See also, Johnson, interviews, 6 March 1996, 7 June 1996. The picture Saville paints of the way the women computers at NASA worked is strikingly similar to the work of the women astronomers at the Harvard Observatory, as described by Rossiter. According to Saville, "bent over long columns of figures, with magnifying glasses and mechanical calculators, the women whose job title was 'computer,' crunched numbers into formulas and equations" (Saville, "What Was the First Computer Model?" p. A–16).

13. Computer Rowena Becker, for example, was interviewed by Saville in 1990: "her decision to go to work as a human computer in January 1942, was easy. With a degree in math, she could look forward to earning $550 a year teaching in North Carolina. As a

computer, she started off at $1,620 a year." The salary was based on a six-day work week, and even women without mathematical degrees earned good money as computers— $1,440 a year. Men with similar qualifications were paid more (Saville, "What Was the First Computer Model?" p. A–16). Helen Wiley, another woman interviewed by Saville, joined NACA in 1941 and retired in 1973. She said, "They'd beef a little at it, [but] There really wasn't a lot of dissatisfaction. . . . I liked it right from the start . . . I must say, I was never bored. I was never unhappy" (ibid.). However, fifty years later, at least one of the women, Vivian Adair, did voice a complaint: "I was hired as a sub-professional. . . . A man with the same degree as I had was hired as a professional; but no, I was a sub-professional [and he was paid more]" (ibid.).

14. Johnson, interview, 6 March 1996. Laura Bateman, hired in 1943, was the first Black woman mathematician/computer at NACA. In an interview with Saville, she said, in regard to Blacks being hired, "I think everybody in Hampton and Newport News thought the world was coming to an end" ("What Was the First Computer Model?" p. A–16).

15. Johnson, interview, 6 March 1996.

16. Ibid.

17. Ibid. For a glimpse of other Black scientists who worked in the early days of the space program see Shawn D. Lewis, "She Lives with Wind Tunnels," *Ebony*, August 1977, p. 116 (an article on engineer/mathematician Mary Winston Jackson). See also Steven Morris, "How Blacks View Mankind's 'Giant Step,'" *Ebony*, September 1970, pp. 33–36, 38, 40, 42. "The Future-Makers: They Clone Cells and Smash Atoms in Search of To-morrow's World Today," *Ebony*, August 1985, pp. 62–64, 66. Enoc P. Waters Jr., "They Helped Track Glenn in Orbit," *Afro-American*, 3 March 1962, pp. 1–2. James L. Hicks, "Negroes in Key Roles in U.S. Race for Space: Four Tan Yanks on Firing Team," *New York Amsterdam News* (City Edition), February 8, 1958, pp. 1, 15. "Long Island Negroes Built Guts of Missile," *New York Amsterdam News* (City Edition), 8 February 1958. "Ne-groes Aid in Space Research," *Chicago Defender*, 10 May 1958. "Negroes Who Help Conquer Space: Over 1,000 Negroes Are in the Satellite, Missile Field," *Ebony* May 1958, pp. 19–22, 24, 26. "Rocket Fuel Researcher," *Ebony*, September 1960, p. 6. Ma-rianna W. Davis, "Black Women Achievers as Mathematicians, Engineers, and Space Technologists," in *Contributions of Black Women to America*, vol. 2 (Columbia, S.C.: Kenday Press, 1982), pp. 453–474.

18. Johnson, interview, 6 March 1996. The paper provided the means for ground con-trollers to tell, at any given time, the exact location of the spacecraft and thereby insure its safe retrieval.

19. T. H. Skopinski and Katherine G. Johnson, *Determination of Azimuth Angle of Burnout for Placing a Satellite over a Selected Earth Position*, Technical Note D–233 (Springfield, Va.: National Aeronautics and Space Administration, Scientific and Tech-nical Information Branch, 1959). According to Johnson, interview, 7 June 1996.

20. Skopinski and Johnson, *Determination of Azimuth Angle*. See also Johnson, rés-umé. "Her Paper Helped to Track Astronaut: Math Expert Who Aided Spaceman Is 'Thrilled,'" *New York Amsterdam News*, 24 February 1962, pp. 1, 20. *Black Contributors to Science and Energy Technology* (Washington, D.C.: U.S. Department of Energy, Of-fice of Public Affairs, 1979), p. 11. Davis, "Black Women Achievers," pp. 453–474.

21. Johnson, interview; see also "Her Paper Helped to Track Astronaut."

22. "NASA Scientists" (undated; from Johnson's personal collection). The publication venue of this article is unknown, but was most probably a local newspaper in Virginia.

23. Johnson, interview. Unfortunately, Johnson cannot remember which astronaut made this statement.

24. Ibid.

25. Ibid.

26. *Black Contributors to Science and Energy Technology,* p. 11.

27. Davis, "Black Women Achievers," pp. 453–474.

28. Johnson, interview, 6 March 1996.

29. Ibid.

30. In the 1960s and the 1970s, as NASA began to move to full electronic computerization, many of the human computers were phased out—some, like Johnson, had already become technologists, while others became technicians using the new computers. One of the early women computers, Kathleen Wicker, recalled in a 1990 interview that in the late 1950s when NASA officials first told her that she would begin working with the new Bell computer (an early model that performed calculations using a series of relay switches similar to an old telephone switchboard) she didn't realized that they were talking about a machine. "At that time," Wicker said, "I hardly knew what a computer was—I thought [it was] a *woman*" (Saville, "What Was the First Computer Model?" p. A–16).

31. Johnson, résumé. Johnson also received special achievement awards in 1970, 1980, and 1985. See also Davis, "Black Women Achievers," pp. 453–474. For a complete list of Johnson's published research papers, see the appendix.

32. Johnson, interview, 7 June 1996. Since 1977, information on Johnson's life and work has been housed at the Afro-American Historical and Cultural Museum in Philadelphia, Pennsylvania, as part of the permanent collection.

EUNICE L. JONES*

According to a 1956 *Ebony* magazine article, lab technician Eunice L. Jones was the "the only woman employed in the Buffalo, New York, water filtration plant."[1] As an assistant to the plant's associate chemist, Jones's job was to analyze and test "water samples to control the purification of Lake Erie drinking water," as well as testing "the fluoride salts content of water treated to reduce dental cavities in children."[2] According to the article, Jones majored in science at Buffalo University, and had worked as a computer for the Bell Aircraft Corporation. The article does not make clear if she completed her degree.[3]

NOTES

1. "Water Plant Tester," *Ebony,* March 1956, p. 5.

2. Ibid.

3. Ibid. She is also described as "the sister of Buffalo University trustee and city councilman Leland Jones."

SINAH E. KELLY*

In the 1940s, Radcliffe College graduate and analytical chemist Sinah E. Kelly was involved in developing methods for mass-producing penicillin at the U.S. Department of Agriculture.[1] Kelly had taken advanced course work in chemistry at New York University (the time frame is unknown), and began to work on penicillin mass-production during World War II. She had transferred to the USDA's

Fermentation Division Laboratory in Peoria, Illinois, from the Picatinny Arsenal. Kelly's supervisor, Dr. L. B. Howard, chief of the Bureau of Agricultural and Industrial Chemistry at the USDA, said that Kelly "performed important analyses."[2]

NOTES

1. "Mass Production of Penicillin Aided by Woman," *Afro-American*, 25 January 1947, p. 15.

2. Ibid. The remainder of the brief article devotes almost as much space to the history of penicillin and its link to agriculture as to Kelly's role in its production: "Penicillin was discovered by a British scientist nearly 20 years ago. But up until a few months before our entrance into World War II, British scientists had not been able to produce the drug in significant quantities. . . . The production of penicillin is linked up with agriculture in that it is produced by a mold that lives on a diet composed almost completely of agricultural products — corn steep liquor and milk sugar" (ibid.).

ANGIE TURNER KING

An Amazing Legacy of Scientists

Angie Turner King
Photo courtesy of the West Virginia State College Archives

Chemist and mathematician Angie Turner King has produced a remarkable legacy: two of the women in this volume—Margaret Strickland Collins and Katherine Coleman Goble Johnson—were her students, as was mathematician/ physicist Jasper Brown Jefferies (West Virginia State, 1933), who worked on the Manhattan Project in Chicago from 1943 to 1946 before becoming a professor of physics.[1] Yet during her entire career, King appears to have been fairly isolated: "I didn't have time to think if there were other Black women scientists with Ph.D.s," she said; "it never really entered my mind. I didn't think it was unique, although I guess it was. I just worked hard, and tried to do my best."[2] Other than her master's thesis and doctoral dissertation, King did not produce any published research. What she did produce, however, was an intellectual legacy. Out of the seventy-two former students who responded to a variety of questions for the West Virginia *State High School Reunion Booklet*, twenty-seven of them (seventeen women and ten men) chose King as their favorite teacher. Interestingly, according to the brief biographical updates each provided, at least twenty of the students who mentioned King went on to graduate school.[3]

Born in a segregated little hamlet in MacDowell County, West Virginia, in 1905, King was the grandchild of Virginia slaves who "were given land, a steer, and a log cabin when they were freed."[4] She had an extremely difficult early life. Her mother died when she was eight years old, and for a time she was sent to live with her maternal grandmother, a woman whom King has described as "so light-skinned she was almost white."[5] "My grandmother called me the 'black bitch,'" she recalled. "This was just her way; she never hit me, but she called me black bitch. . . . I slept upstairs in a cabin and in the winter time, when it would snow, I'd wake up with snow on my bed. I had it tough but it hasn't bothered my mind."[6] Eventually, she was able to live with her father again, and although he was illiterate, he insisted she go to school: "Even when I didn't want to attend he made me go. He encouraged me to do well in school."[7] She graduated from high school in 1919, at age fourteen. "I had good grades in high school and my teachers encouraged me to attend college," she said, but apparently both she and her teachers were unaware of scholarship opportunities.[8]

King attended the Bluefield Colored Institute for teacher training for a number of years, before transferring to West Virginia State College. She worked to pay the tuition at both schools: "I worked for my education—there were no scholarships in those days. Mostly I waited tables, washed dishes, anything like that I could get, to pay for school."[9] In 1927 she earned a B.S, cum laude, from West Virginia State in mathematics and chemistry. She began her teaching career at State High School, the laboratory high school at West Virginia State College, and attended Cornell University during the summers—paying for the tuition herself. She earned her master's degree in mathematics and chemistry in 1931.

King taught for eight years at the laboratory high school, and then was offered a position at the college. The first order of business, she recalled, was to "get a chemistry lab fixed up, so the students would know what a real laboratory was like."[10] She married in the early 1950s, and then attended the University of Pittsburgh, again paying the tuition herself, and was awarded her Ph.D. in mathematics and chemistry in 1955.

King retired from West Virginia State in the late 1980s, but she continues to live on the campus grounds, in the house she has occupied for years, surrounded by a collection of ceramic frogs given to her over the years by more students than she can count.

NOTES

1. See entries on Collins and Johnson in this volume. Jasper B. Jeffries later taught physics at North Carolina A&T University (1946–1949) and mathematics at Westchester Community College (1963–1971), where he was department chair from 1971 until his retirement. Vivian Ovelton Sammons, *Blacks in Science and Medicine* (New York: Hemisphere, 1990), p. 132.

2. Angie Turner King, interview with author in Institute, W.Va., 18 March 1996.

3. *West Virginia State High School Reunion Booklet, 1994–6*, Reunion Booklet Committee: Ethel H. Andrews, Helen S. Reynolds, Ann B. Turpeau (Institute, W.Va.: West Virginia State College Alumni Association, 1996). (Courtesy of Margaret S. Collins.) It should be noted that although only the earliest respondents had King as a high school teacher, nearly all of those who responded for the booklet also attended the college, where the majority had King as a professor of either math or chemistry.

4. King, interview.

5. Ibid.

6. Ibid.

7. Ibid.

8. Ibid.

9. Ibid.

10. Ibid.

REATHA CLARK KING
Chemist, College President

Reatha Clark King's life moved from a sharecropper's farm to the corporate boardroom, and her career encompassed research, college teaching, and both college and philanthropic organizational administration. Born in Pavo, Georgia, in 1936, and raised in nearby Moultrie, King grew up picking cotton and doing the heavy field work that was typical for a child born into a sharecropper family in the hardscrabble milieu of 1940s rural Georgia.[1] Her parents, who had almost no formal education, separated when she was young and she was raised by her mother. Education, particularly for those children who did well in school, was seen as a means to escape the poverty of southern farm life, according to King.

> When I was very young, my parents, neighbors, teachers, and church people constantly encouraged my two sisters and me to make good use of our brains to get ahead, to use our abilities which we had demonstrated in school, to study our books, and to make something out of ourselves. We heard this counsel over and over in the 1940s as we were growing up. . . . The strong encouragement from my family and community to use [my] academic ability to overcome disadvantages was like some kind of indoctrination. . . . This . . . undoubtedly strengthened my later confidence in my ability to succeed in the sciences.[2]

King recalled that she was well aware that the "destiny" of Black children "was to make something out of ourselves, so we could get better work than our uneducated parents had."[3] The desire of her family, as well as the local community, produced—not only for King but for other women of similar backgrounds—an environment in which they learned to please adults by succeeding in school. According to King,

"Making something out of yourself" was the language others used to motivate [us, when we were young]. So "making something out of ourselves" became our destiny as Blacks growing up in segregated south Georgia. Members of our family and community knew that getting an education was the pathway to a good job—a job with better pay than the substandard wages that others in our community were paid. . . . The idea that you as a young person could succeed with your goals became *so* indelibly imprinted in your head, [that] the encouragement you received helped even if it did not direct you to a specific career choice.[4]

Despite the atmosphere of encouragement to do well in school, King said there was "absolutely no encouragement"—either at home or in the segregated schools—for Black girls to consider a career in the sciences:

Becoming either a hairdresser, teacher or a nurse represented the reality of a better job for young Black women. Women in these jobs were our main role models both in person and in our books. So career counseling to women to be scientists just did not happen. . . . [But] even with the narrow career counseling that young Black women received in these years, I feel that there was still some positive benefit from this counseling for strengthening my self-image.[5]

Because teaching, home economics, and nursing were seen as the types of career choices that would enable a girl like King to escape the sharecropper life, she entered Clark College in 1954, intending to major in home economics.[6]

At Clark, King's knowledge about other career opportunities gradually broadened. Then, as was the case for so many other women profiled in this volume, a mentor played a pivotal role in directing her talents toward a career in science. "From Dr. Alfred S. Spriggs, who had become the Chair of the Chemistry Department at Clark College in 1955, I learned about the work of research scientists, [and] what would be required for me to become a research chemist, [and] something about the pay and the nature of the work."[7] King changed her major to chemistry, and recalled that it was the "perfect major" for her:

Both the subject matter and methodology were interesting and challenging; the laboratory and lecture sessions were exciting; and my fellow students in chemistry were both serious students and fun to work with . . . [and] the atmosphere at Clark was one of rigorous study in the sciences and mathematics.[8]

King completed her B.S. in 1958, received a Woodrow Wilson fellowship for graduate studies, and entered the University of Chicago. By her own account, the "intense atmosphere of research and scholarship" at Chicago—particularly the level of interaction with faculty and other students, not only in chemistry, but in physics, mathematics, computer science, and geology, as well—greatly enhanced her intellectual development and provided a solid preparation for her later research.[9] In collaboration with others she had already published two articles on thermochemical reactions before she completed her doctorate in 1963.[10]

Immediately after receiving her Ph.D., King obtained a research position at the National Bureau of Standards (NBS) in Washington, D.C. During her tenure at NBS, King continued the research she had begun as a graduate student on the thermochemical properties of fused salts, fluorine, and other inter-halogen compounds. She also engaged in work on the Laves phases and thermochemical studies on certain metals, and she conducted measurements on the levels of heat produced in the formation of oxygen difluoride.[11] Her career as a researcher was somewhat brief—only five years in addition to her graduate work—but she described it as "intense."[12] The senior staff at the NBS impressed upon the younger scientists the need for precision of measurements and accuracy. "As a measurement laboratory," King said, "the NBS was the only one of its kind in the United States and the work done at the Bureau had to be of the highest quality possible."[13]

A steady although not a prolific researcher, King has indicated that her publications can be better evaluated by the depth of the studies, rather than the length of the list. Her article on the heat of formation of oxygen difluoride, for example, took three years to complete, and involved the development of new techniques for flame calorimetry, which were later used for studies on fluorine and chlorine trifluoride with hydrogen.[14] By age thirty-one, however, King had all but ended her research career. She turned to college teaching and administration.

From 1968 to 1977, King taught chemistry at York College of the City University of New York, where she was also an associate dean. From 1977 to 1988, she was president of Metropolitan State University in Minneapolis–St. Paul, Minnesota.[15] King took over as president of Metropolitan State University only six years after its inception, and she described it as an "extraordinary institution—one that was young and had a mission to function in creative ways to provide educational services."[16] She left Metropolitan State in 1988 to become executive director of the General Mills Foundation and a vice-president of the General Mills Corporation in Minneapolis, a position which she currently holds.[17]

King places great emphasis on the degree to which her scientific training prepared her for her later roles as a university president and as a foundation executive: "Young people, especially, often fail to realize that a scientific education is flexible and that it can prepare one to function in varied kinds of careers."[18]

NOTES

1. Reatha Clark King, "Becoming a Scientist: An Important Career Decision," *Sage* 6 (fall 1989): 47–50; see page 47.

2. Ibid., pp. 47–48.

3. Ibid.

4. Ibid.

5. Ibid., p. 48.

6. Ibid., p. 47. King's older sister, in fact, majored in nursing at Dillard University in New Orleans.

7. Ibid., p. 48.

8. Ibid., p. 49.

9. Ibid.

10. Ibid. See the appendix.

11. Ibid., p. 48. Some of the metals she worked on in the thermochemical studies were gallium, copper, brass, gold, silver, sodium, copper, and zinc alloys. See the appendix for a representative list of King's publications.

12. Ibid., p. 49.

13. Ibid.

14. Ibid., pp. 49–50.

15. Ibid., p. 50.

16. Ibid., p. 50. King stated that when she took charge of the university its enrollment "increased from 1,600, in 1977, to 6,000 when [she] left in 1988; the graduates increased from 1,200, in 1977, to 5,500, in 1988" (ibid.).

17. Ibid.

18. Ibid.

FLEMMIE PANSEY KITTRELL

An Extraordinary Life

When you are an extraordinary scholar, people don't care what color you are.[1]

—*Flemmie P. Kittrell*

As one of the most nationally and internationally prominent scientists among the early women doctorates, nutritionist Flemmie Pansey Kittrell provides a study in contrasts. Modest and self-effacing yet determined and dynamic, she developed a plethora of home economics and nutrition projects around the world, strongly supported higher education for women, and helped Howard University gain an international reputation. She described her field, the science of home economics, as a combination of nutrition science, chemistry, biology, and the psychology of child and family development.

Kittrell began her professional career as a high school teacher; then, in 1928, she became director of nutrition at Bennett College in Greensboro, North Carolina. Twelve years later, she moved to Hampton Institute in Virginia where, from 1940 to 1944, she was professor of nutrition, director of the department of home economics, and dean of students. For the next twenty-nine years (1944–1973), she held a full professorship at Howard University, was chair of the home economics department, and pioneered in the development of international cooperation in the home sciences, most notably in Africa and India. Toward the end of her tenure at Howard, the Department of Home Economics was restructured and renamed the Department of Human Ecology. Concurrent with her tenure at Howard, Kittrell expanded the scope of her work to the international level, and she became widely known as an authority on nutritional and sanitary standards in Third World countries. After her retirement in 1973, Howard University named her Professor Emerita of the Department of Human Ecology, a position she held until her death in 1980.[2]

Flemmie Kittrell was born to James Lee Kittrell and Alice (Mills) Kittrell in 1904, in Henderson, North Carolina; she was the eighth of nine children (four girls and five boys). Her father was the grandson of a slave who had lived on a plantation owned by a white family named Kittrell, and, following accepted custom, he had taken their surname as his own.[3] Her mother's grandmother was also a slave.[4] By the time of her birth, Kittrell's parents owned a small farm, which they worked with the help of her older siblings. In those days, being a child in a southern Black farm family meant hard work—child labor was needed to keep a farm going. Kittrell, however, was different. She was always small for her age, and exhibited a gift for "book learning" from the moment she began attending school: "I remember very well my first year in school. I had a very good teacher. She was so good and so pretty that when I was promoted to the second grade, I cried. It was really that teacher who gave me my great love for school."[5] Kittrell always acknowledged that this early teacher, followed by a succession of other teachers and mentors, had a tremendous impact on her life. As she recalled, "I was not too alert when it came to working on the farm, but when it came to going to school, I was always quite willing to get there on time."[6] Although it was usual for farm children to drop out of school at around age eight or nine to work, Kittrell's parents continued to allow her to attend because of her early success there.

Henderson was a small rural farm community and as such did not have a high school. Once again Kittrell's parents sacrificed her services as a farm hand, instead sending her, at about age twelve, to the Hampton Academy in Virginia, which was part of Hampton Institute, a historically Black college.[7] Her family could not afford to pay for her high school education, however; Kittrell had to work to pay for it herself:

> There were many young people who couldn't afford to go to school. I had to work in order to make ends meet. [Hampton Academy] called it a "work-year," where you worked a year and earned money, and that money was put aside for your tuition. You were also able to get additional jobs later on if you needed them, but the work-year was a solid background investment for money, and students who went through the work-year, profited a great deal. They don't have that any more, but it was necessary then.[8]

According to historian Merze Tate, the philosophy driving this type of work-study program was to train the hand, the head, and the heart—as opposed to just training the mind. This philosophy was later refined by one of Hampton's earlier graduates, Booker T. Washington, who had earned his tuition money cleaning rooms.[9]

At Hampton Academy, Kittrell had thought about studying political science, economics, or history. Her path into the study of home economics is a good illustration of the importance of teachers and role models. Just before Kittrell graduated from high school, one of her teachers, Blanche Rollonson, called her in to discuss her options in college. As Kittrell recalled,

> [Mrs. Rollonson] wanted to know what I had decided to do for the future. When I told her my choices, she said, "But have you thought of home economics?" I said, "I don't think I'd like that." And she said, "Why not?"

And I said, "Well, I just don't think I'd like it." I didn't have a good reason,
except I thought the home was just so ordinary, women know all about it
anyway.
   But she said, "Well, you read this book and come back and let me know what
you think about it." It was a book on the life of Ellen H. Richards.[10]

The book Rollonson gave her was a biography of Ellen H. Swallow Richards,
whom historian Margaret Rossiter has said "almost single-handedly created the
field of home economics."[11] Richards had graduated from Vassar in 1870, and
then from the Massachusetts Institute of Technology with a degree in chemistry
in 1873. Between 1880 and 1920, according to Rossiter, "She propagandized for
[home economics], ran demonstration projects, raised money, performed many
chemical analyses, wrote several handbooks, trained and inspired her co-workers,
and organized its main activities and professional associations."[12] Rossiter also
says that "as a student in the male world of MIT [Richards] had learned how to
make a place for herself and other women by capitalizing on woman's traditional role and using it as an entree into the sciences." This aspect of Richards's career
struck a familiar chord in Kittrell:

> Ellen Richards was one of the outstanding women who went to MIT, and
> strangely enough, she had the same experiences and hardships that some
> of us women seem to have today, in having equality with men. She made a
> terrific contribution and now she is cited at MIT as one of [its] outstanding
> graduates. . . .
>   So that's how I got into home economics at Hampton.[13]

Clearly, Rollonson saw Richards as a role model, as would Kittrell. She majored
in home economics at Hampton Institute, graduated with a bachelor of science
degree in 1928, and immediately secured a position as a high school teacher at
Bennett College's laboratory high school.[14] In the following year, 1929, she was
awarded a Julius Rosenwald Fund scholarship to enter the master's program at
Cornell University. A General Education Board fellowship (1933–1935) enabled
her to complete her doctoral studies.[15]
   At Cornell, Kittrell met her most influential mentor, Ethel B. Waring, who
would become her lifelong friend. More than forty years after her initial experience with Waring, Kittrell recalled,

> If I had to look back and write down what has impressed me the most as a
> student, it would probably be Ethel B. Waring. . . . She was my beloved
> professor, the most humane person I've ever known. . . . When I was one of her
> students in graduate school, [she gave us a formula] dealing with how to observe
> [the] behavior of humans and how to not [rely on] what you didn't observe. . . .
> She taught us how to observe and how to read.
>   [She helped] me to observe objectively and not be prejudicial, because
> prejudice means having your thoughts get right too soon.[16]

Waring's formula for conducting research in home economics was, according to
Kittrell, both crucial and simple because it took into account the broadness of a
field that had to interrelate its various disciplines:

[Waring] said there are four basic elements that people must have if they are going to work well in an interrelated way. This interrelated way would have to begin first of all with just a plain love for people. If one loves a person it means one can love humanity; unless one can love humanity, one cannot love even one person. The second principle is to have respect. The third point is to help people as they need it in the areas where they really want help.[17]

Kittrell did not fully articulate Waring's fourth point, but based on her various recounts of it, it would have been to recognize both the wisdom and resources of "poor people around the world." Kittrell continued to expand on the ways in which Waring's formula had influenced her own work:

I have worked on projects [around the world] and have been amazed to find that we did not need to help people much if we loved them and they believed in us. If we encouraged them with respect, and as they developed more self-respect, they would come up with all kinds of creative ideas. In this way, a team made up of a person of the poverty group and one who was more affluent and educated worked for common goals and moved ahead together.[18]

According to Ethel Waring's son, Dana Bushnell Waring, Kittrell and his mother maintained their friendship and their "mutual concern for the overall development and welfare of families in this country and abroad" until Waring's death in 1976.[19]

Throughout her long career, Kittrell was actively involved in developing and implementing various projects in home economics in the United States, the Indian subcontinent, and Africa. Her research interests included nutritional standards for poor children, early child development, family life and parent support, and sanitary food conditions in Third World countries.

At Hampton Institute

From 1940 to 1944, Kittrell was professor of nutrition, director of the department of home economics, and dean of students at Hampton Institute. In practice, Kittrell primarily worked as the dean of women. As World War II escalated, however, the Hampton campus—because of its proximity to Hampton Roads Channel, which the Navy considered strategic—changed almost overnight into the center of a military training program. "That was a very unusual experience," Kittrell recalled, "and I had a very high regard for the armed forces because the men who were in charge of the program were outstanding."[20] Educational and training opportunities for women also increased due to the war effort, and this served to heighten Kittrell's own awareness of her career potential in science.

Sometime in early 1943, at the behest of Howard University's president, Mordecai Johnson, Dean Charles H. Thompson began trying to recruit Kittrell to head Howard's home economics department. This apparently came as a direct result of Thompson's wife's having known Kittrell at graduate school at Cornell.[21] Kittrell, however, was modest about her abilities. She told Thompson that Howard was a very sophisticated university and she didn't think she was up to its standards.[22] Despite Thompson's assurances to the contrary, Kittrell said she "didn't take [his offer] very seriously."[23] As Thompson continued to write to her, her confidence in herself grew, and she began to consider leaving Hampton.[24] In late 1943, she went to Howard to meet with President Mordecai Johnson, ostensibly to decide on the direction the home economics program would take were she to assume the chairmanship. She also wanted to assure herself firsthand of Johnson's commitment to what she perceived as her field's goals:

> President M. Johnson was enthusiastic, of course, and he wanted me. And so President Johnson said, "Well, now, Miss Kittrell, we would like to have you come because we think home economics is a most important field."
>
> And I said, "Mr. President, I have seen your building and your facilities, and it doesn't show that you think this way."
>
> And he laughed and said, "Well, daughter, what would you suggest I do?"
>
> And I said, "Mr. President, I think you need a new building; you need a new plant. . . . Home economics . . . is a very important field dealing with the science of living and it draws on the natural sciences and the humanities and the arts, and therefore I think . . . you really need better facilities."
>
> "Well," he said, "daughter, if you come, we will get that building."[25]

Kittrell, who was surely not working with the best facilities at Hampton, was not reticent about mentioning the poor state of Howard's home economics facilities. She also expressed very clearly her definition of home economics as combining the natural sciences, the humanities, and the arts. Johnson did promise to have a new home economics facility built if she would accept his offer.

For the time being, however, Kittrell decided to remain at Hampton Institute. She had only been there three years at that point, and was earning $4,000 per year.[26] When she had arrived as dean of women in 1940, morale was low and enrollment was down.[27] Although enrollment and morale had risen by 1944, Kittrell still documented a number of problems at Hampton. And, perhaps more importantly, she felt it "wasn't good to leave the Navy and Army program," which she clearly viewed as vital to the nation's war effort and which offered her a strong measure of intellectual stimulation.[28]

When Kittrell suggested the possibility of her moving to Howard a year later, Johnson and Thompson readily agreed. Thompson was clearly a supporter of Kittrell's: "Now, Miss [sic] Kittrell, when you come, remind the president that he said that if you'd come, you would get that [new] building."[29]

At Howard University

In September 1944, Kittrell assumed the position as head of the Department of Home Economics at Howard University. During her first year there her salary was the same as it had been at Hampton—$4,000 per year—but Howard offered other benefits. At Hampton, she had been the dean of women and the lone home economist; at Howard, however, she moved into an established department that had already instituted a master's degree program, and she would have the opportunity to build a staff.[30] During the 1945–1946 academic year, in addition to Kittrell herself, the Howard home economics department included an associate professor, a professional lecturer, and three instructors; its annual budget was $14,775.[31] At that point, the promised new building for home economics remained just a promise. According to Kittrell, "The first year passed, of course when the war was still going on; the second year passed, and we were getting a new start; and the third year we had a new [federal] secretary of education who wanted to cut the budget."[32] Nevertheless, her position at Howard gave her an opportunity to build a curriculum that transcended the traditional home economics model and was fashioned instead along a research-oriented model. Kittrell's program offered new ideas, and gave increasing attention to the important role of women in making significant contributions in a changing world. Kittrell herself became the role model for this new form of home economics.

Kittrell's position at Howard also afforded her the opportunity to develop her research interests at the international level. She had first gained international attention in 1947 to 1948 when she conducted a six-month nutrition survey under the auspices of the U.S. Department of State that studied the general living conditions, family life, and food consumption of natives of Liberia, Africa. Her findings, published in the booklet *Preliminary Food and Nutrition Survey of Liberia* (1947), showed that ninety percent of the Liberian people experienced "hidden hunger"—in which people may feel their stomachs are full, even though they are seriously malnourished—as a result of subsisting largely on diets of rice and cassava dishes.[33] Kittrell also found that many African countries had shortages of meat and fish, resulting in a diet that was dangerously low in protein. "We worked out diets that mothers could feed their children to make them healthy," Kittrell said. "In my lectures, I suggested ways in which people could get more protein from the food supplies available around them," particularly by broadening and varying their daily intake of local grains, vegetables, and fruits, and "I touched on a lot of other subjects that are universal in appeal—food in general, clothing, shelter, and children."[34] Her report contained a number of recommendations —both immediate and long-range—to remedy Liberia's malnutrition problems; these included refinement of agricultural production, development of the fishing industry, and enlargement and increased professional training of personnel within Liberia's Agricultural Bureau. The publication of Kittrell's report on hidden hunger resulted in a series of international programs aimed at eliminating the condition.

Kittrell's growing international reputation had an impact on president Mordecai Johnson and the trustees of Howard University. Soon after her return from Liberia, the home economics department's budget was substantially increased. For the academic year 1948–1949, the department's budget nearly tripled, to $39,198.28. Kittrell's salary rose to $5,500 and the size of her staff also increased dramatically.[35]

In 1949, Kittrell represented the United States as its official delegate to the International Congress of Home Economists in Stockholm, Sweden. That year she also proposed that Howard University permit the home economics department to set up a nursery school, which would be used as a research site for her growing program in early child development. This was approved, and the department's budget for the 1949–1950 academic year was increased by $4,500.

Despite these advances, however, funding at Howard was becoming an increasing problem. Kittrell was busy trying to fine-tune her new program and develop a satisfactory design for the building that would someday house it, but she "didn't dare say a word [about the new building] to the president at that time."[36] During the post-war period, the nation experienced a recession, and the financial situation at Howard worsened—department heads were called in and told that their budgets had to be cut.[37] Kittrell fretted, "Where is the [new] building going to be, with cutting our budget?"[38] She also worried about having to cut staff; she suggested to Dean William Stuart Nelson that rather than letting anyone else go, she would take a leave of absence. He initially rejected that idea, but Kittrell persisted, telling the dean that a leave would give her a "new perspective."[39] It was then that her relationship with Ethel Waring and Cornell University came into play once again:

> [The] very day I made that decision, I had a long-distance call from Cornell University asking me would I be interested in a post in India and [they] said that Mrs. Hansa Mehta was there [at Cornell]. Her daughter was in the College of Home Economics at Cornell and wanted to talk with me, and so Mrs. Mehta got on the telephone and said she'd be down to Washington. She came the next day—this was in the midst of the discussion with Dr. Nelson—and [she] asked if I would come to India to help them get their home economics program started. At the time they were going to pay me from their funds, because the Maharajah [of Baroda] had left a lot of money to be turned over to their state for a university.[40]

Kittrell, however, had another idea for funding the trip:

> I got to thinking that Fulbright ought to have something, but I couldn't accept the idea that Fulbright would come through, so I just said to Mrs. Mehta, "Yes, I would like to do that."
> At the same time I knew that we had to reduce our staff. So what did we do? She signed me up and I signed the papers and so forth. I had not been there long enough to look for more counsel. I just went right ahead [and signed], I usually do that.
> So I inquired downtown if the Fulbright people would accept my application for home economists to go to India and they said, "Miss Kittrell, today we have just gotten word that that program would be approved for India, and you will be the first home economist to ever go to Asia under the Fulbright."[41]

Mordecai Johnson and Dean Nelson approved the plan; Kittrell would receive $2,700 of her yearly salary, in addition to the Fulbright fellowship. She arrived in India in August 1950 and stayed for a full year.[42]

In the Third World

India, then newly independent, was in the process of redesigning its colleges and universities.[43] Given the enormous size of its population, new emphasis had to be given to the various facets of the "home sciences," as home economics was referred to there. Kittrell's job was to help set up a college of home science—the first of its kind in India—at the University of Baroda. She was to help select and train faculty, design a curriculum, teach courses in foods and nutrition, develop a research program in foods and nutrition, and assist in creating a home science extension service that would provide basic nutritional and sanitary education to the people of rural India.[44] She worked directly with India's vice-chancellor of education, Mrs. Hansa Mehta (who was also vice-chancellor of the University of Baroda), and another woman, Raj Kamary Amrit Keun, India's minister of health. Kittrell later credited these two women as having had "a profound influence on education [in India] and particularly in this area of home science."[45] Before she returned to the United States in the fall of 1951, Kittrell had also organized the first All-India Home Science and Nutrition Conference.[46]

As a 1951 letter from Dorothy M. Pearson, a professor of nutrition at the Women's Christian College, in Madras, India, to the Committee on Missionary Education, a Canadian religious/missionary organization, reveals, trained nutritionists and home economists were desperately needed in India:

> We were told that if we do not get heavy rains by the 15th of this month all domestic supplies will be cut off in this City. . . . Four years of drought in this area have had terrible consequences. Central India also has a serious scarcity this year.
>
> Had a committee meeting of the Madras State Women's Food Council at Government House. We cooperate with them to the extent of going with them to villages one day a week and visiting ration-card distributing centres twice a week to help the illiterate women with their cards and ration problems.
>
> It has been very difficult getting research and extension work built up. [But] as a result of our work there is a great demand for our ["Nutrition Notes"] charts. We get many requests for recipes for new foods and also requests to determine the approximate values of diets.
>
> Research work is also developing. Miss Theophilus, of our staff, has started research to fill in the gaps of our knowledge of the nitrogen balance of South Indian women. This is in furtherance of work done by Dr. [Emily] Mason. Other workers have done a great deal on men but not on women.
>
> In my area, Mrs. Rowlands has just completed the writing of her M.Sc. thesis on the effect of cooking on the ascorbic acid content of some South Indian vegetables. As a research student I have one girl, M. Lalitha, who is getting her first [research] experience in an attempt to develop an infant food with materials available in India that will be as good as pabulum, we hope.[47]

Kittrell's correspondence with nutrition scientists around the world was voluminous—it seems as though she was in contact with almost everyone who was working on health and nutrition problems in Third World countries.[48]

Occasionally, Kittrell's correspondence reveals that she ran into funding problems. For example, Mrs. Mehta had invited Kittrell to return to India in 1953, and she clearly wanted to be there to teach the advanced courses in foods and nutrition to the senior students, who had started their training with her in 1950. As Kittrell wrote,

> Because time was of great importance to Baroda, and because I had to meet a deadline for asking for a leave of absence from Howard, I had to come to a decision. At my request, Mr. Charles N. Keating, Director of the Security Division-Mutual Security Agency, arranged for a conference with me. I explained my total situation to him and my need to be in Baroda then. I was

led to believe that my coming now would not affect my receiving Point Four Aid if my clearance came through. I left for India on August 4 by Pan American, paying my own fare. I am hoping that I will be reimbursed for this as I was in 1950 when I came over under the Fulbright program.[49]

Her request for reimbursement was rejected, as this excerpt from a reply letter explains: "The Government is so large that it has to have definite rules and one of them is against retroactive payments to compensate a person for what they may have done voluntarily for the Government and not under Government orders. . . . there is absolutely no way in which the Government can make an advance."[50]

In 1952, Kittrell went to Zaire, under the auspices of the U.S. State Department, to help organize the Congo Polytechnic Institute's School of Home Economics. Then for two years, 1953 to 1955, she returned to India, under the auspices of the Agency for International Development (AID).[51] During that period, she did a great deal of work to develop India's home science extension program. "We invited many women throughout the country, through the government of India, to come to Baroda to study extension methods in home economics. And in turn we went to Japan and to Hawaii, because Japan had some outstanding [extension] programs there."[52]

A 1955 letter from John P. Ferris, chief of the South Asia Division of the Department of State's International Cooperation Administration, to Howard University President Mordecai Johnson provides a good indication of Kittrell's influence, her stature in federal and international circles, and her importance to America's image abroad:

> Dr. Kittrell, her initiative and dedication overcomes all obstacles and resulted in such a significant contribution to home science in India; to improvement in the academic standards of India institutions of higher learning; to the role of women in Indian life; and to the cause of the United States in India.
>
> It is our considered opinion that Dr. Kittrell's earlier assignment in India as Fulbright professor in 1950–1, and her more recent work under ICA sponsorship, were largely responsible for the Government of India's increased interest in improving and expanding home science work in India and for recognition of the importance of that field to India's Development Program. . . .
>
> One of the results of this increased interest on the part of the Government of India has been requests that the United States provide additional assistance in home economics. Now not only is Baroda University profiting by help from American teachers, but six other institutions as well, under a contract with the University of Tennessee, will have assistance in upgrading their teaching facilities.[53]

Ferris ended the letter by saying, "Our Mission Director has called Dr. Kittrell one of our most outstanding ambassadors of good will in India."[54] Kittrell's international reputation was so important to Howard president Mordecai Johnson and to the university that he continued to grant her leaves of absence.

In 1957 and 1958, the State Department sent Kittrell to East and West Africa to lecture and conduct nutrition research; in 1959 she was sent to West and Central Africa, again to lecture and conduct nutrition research. In 1959, she was named the official U.S. delegate to the United Nations Geneva Conference for

Non-governmental Organizations, and in 1960 she participated in a nutrition survey of India and Thailand for the Food and Agriculture Organization of the United Nations. The State Department then sent her on a cultural exchange tour to Guinea in 1961.

In addition to her government-sponsored trips, Kittrell also made a number of foreign trips sponsored by various religious organizations and private aid foundations, most notably her visits to Leopoldville (Belgian Congo), Rhodesia (now Zimbabwe), and Mozambique in 1961 for the Methodist Church. She continued to serve as a consultant to the Congo's Home Economics Program from 1961 to 1972.[55]

Expanding the Howard Program

THE HOME ECONOMICS BUILDING

While she worked abroad, Kittrell continued to do research in both nutrition and child development, and she published her findings. She also continued to develop her department at Howard University, which now included an expanded master's program.[56] In particular, Kittrell lobbied for the promised new home economics building, which, due to the funding situation at Howard, was continually put off. Forced to make do, she struggled to carve out more room and better facilities for home economics. By 1947, she was actively engaged in a letter-writing campaign to colleagues asking for their help in supporting her idea that home economics needed its own separate building, and should not have to share a facility with the biology department.[57] A tireless campaigner, she wrote to whoever she thought might help advance the cause of her program, and frequently to men in a position of power at Howard University. Her letters reveal a savvy woman, who knew what types of information might produce the hoped for results—she continually stressed her department's rising enrollment of both home economics majors and elective students.[58]

Ironically, during her 1953 to 1955 return to India, a new building for the College of Home Science at the University of Baroda in India was already nearing completion,[59] but it was clear that a new building at Howard would not be built in the immediate future. Nevertheless, she pressed for additional facilities and was not loath to mention her own and the department's (largely through her efforts) growing national and international reputation: "Since our new Home Economics Building will not be immediately forthcoming, it is earnestly hoped that these suggestions will be carried out to help us make continued progress and to be worthy of the confidence that we have received nationally and internationally."[60]

Kittrell continued her international work, often taking extended leaves of absence from Howard. The university's president, Mordecai Johnson, was due to retire sometime around 1958. According to Kittrell,

> One day [President Johnson] sent me a note and asked me to come over and talk with him, because we still didn't have the building [and he was due to leave in two years]. [When I went to his office] he said, "You have been very patient with me and you know you haven't even called me to inquire on that building, but

you have reminded me every year when you sent your budget in, and I haven't forgotten it. Now, I want to get your opinion on something.

"You know the [U.S.] Congress will only give one building a year, and now we have two buildings on the drawing board, the physical education building for men and the home economics building. The physical education building for men, [will cost] three million dollars. The home economics building, [will cost] two and a half million dollars. If you were president of Howard University, which would you choose?"

I said, "The home economics building."

"Well, daughter," he said, "that is what I have chosen."[61]

The University's board of trustees and the federal government approved construction two days before Johnson retired.[62]

During Kittrell's 1961 trip abroad, construction finally started on the home economics building at Howard University. In February 1963, twenty years after Johnson's original promise, Kittrell's home economics building was dedicated. Howard's new president, James M. Nabrit, officiated, and anthropologist Margaret Mead, then associate curator of ethnology at the American Museum of Natural History, gave the keynote address, in which she highlighted the changes that the home economics field had undergone:

For many years there was a general feeling that home economics belonged in Land Grant colleges and the girls who studied home economics could marry the boys who studied agriculture. . . .

Especially as it was represented in the liberal arts colleges, home economics was treated as something that was very necessary to have to improve the poor, the underprivileged, [and] the tasteless, but at the same time we seemed to have a whole group of people in the country who said they didn't need it. . . . Possibly it was the rootedness of home economics in our Land Grant colleges and in the traditions of equality and opportunity that made training any person in a servant position, even though they were to be treated with dignity and paid properly, repugnant. Possibly it was just the onsweep of the times and the gradual elimination of domestic workers in the United States. . . .

The change has been marked most by the shift in home economics from an interest in food and textiles to an interest in children. . . . Home Economics has shifted from its knowledge of skills and techniques, knowledge of foods and textiles, knowledge of specific professional skills, all of which are still greatly needed in far greater numbers than before . . . to include children and human relations more importantly than it ever did before. . . .

[This new building] means a great deal to all the other colleges in the country that are thinking about just what they are going to do with home economics and whether home economics is quite the thing they want now or the kind of thing that goes on at Vassar where there's a straight split between child development and child psychology. Child development has children in it and galoshes, zippers, all sorts of complications, while child psychology has just books in it.

Now we have a world today in which every woman, no matter how privileged, no matter how wealthy, virtually has to apply the skills the home economists have been teaching. . . . she can no longer replace herself with the labor of the uneducated and the poor. Every woman—the farm woman, the wealthy city woman, the professional woman, who has a job of her own and does need help at home—every kind of woman in this country, is now up against the same sort of problem and needs the same kind of help.[63]

For Kittrell, the dedication of the new building was perhaps a bittersweet time, because by 1961 the science of home economics was coming under attack from a number of quarters. In her introductory remarks, Kittrell acknowledged this:

[The dedication] mark[s] a new milestone in the profession and comes at a time when there was much confusion in the minds of some educators as to the place and function of Home Economics today in education, and especially at the university level.[64]

As Mead's dedication address had indicated, there was a move afoot in the colleges and universities to make home economics more "scholarly," to distance it from its Land Grant college background. Such a move was anathema to Kittrell.

HUMAN ECOLOGY

As shown in the speech Kittrell delivered at the 1972 Convocation of the All-India Home Science Association, one of the most important features of the science of home economics to her was that it should always be viewed, and practiced, as a *composite* field.[65] She described it as drawing its content from the "basic fields of knowledge," which included, but was not limited to, chemistry, bacteriology, biology, physiology, psychology, physics, mathematics, fine art, religion, philosophy, economics (including consumer choices), political science, and sociology. "It focuses on the all around growth of the individual, the family, and the community. [It] is unique in that it is the only composite field of integrated subject matter performing this specific function."[66]

Kittrell objected to the changes she saw occurring in her field—changes that had resulted in the new term "human ecology." She protested the narrowing of the field's focus in an era of increased specialization and compartmentalization. Kittrell longed for the well-rounded approach of the generalist:

Now they call it human ecology, but then it was home economics. Home economics really had a broader base than human ecology, because human ecology can be broken into so many different areas, and people are apt to not see it as a whole. . . .

I really prefer home economics to human ecology, [especially] after I have seen how many different interpretations people can give it.

But in home economics, those of us who have been educated in that area, particularly in research, know that if you have a home economics program that is valid, it has to draw its basic content from the major discipline. The people who propose to work in the field must have a strong background in all the fields of education, not as specialists, but firmly rooted in the core material.

Piecemeal specialization, [with] no one to integrate it into a whole is useless. . . . Now there are many aspects of home economics where it's all right to become a specialist [at some point]. But a specialist is not much good unless the specialist knows about all other parts that make a whole.[67]

In addition, and equally important, Kittrell felt the lack of *home*—both the word and the concept—in "human ecology." She firmly believed that the home ought to be central to their field: "the home is the child's first school and parents are [a child's] first teachers." In an era of rapid change, she said, "the family and the home must remain the cornerstones from which learning flows, to enable society to handle change in a progressive way."[68]

Helping people—uplifting their lives and living conditions—was the central foundation of Kittrell's philosophy of home economics. She thought too much specialization would prove disastrous, and that practitioners would become so "bookish" they would not know how to relate to the very people the field was originally designed to help. She believed in a basic give-and-take series of interactions, in which each side learned from the other, as she explained in a 1968 speech:

> We experts don't really need to do much for people. Many of the experts put out the light that is there. There is a lot of light among poor people. If we would observe and help as needed, we could stimulate their abilities. This gives the person a sense of satisfaction, it also gives a person of the poverty group a desire to continue to make choices, and make choices because knowing that back of these choices is support on the part of the person with whom he is working—the professional.[69]

Following Ethel Waring's guidelines, Kittrell thought practitioners ought to maintain a high level of respect for their clients, regardless of their circumstances. Without this respect, practitioners would fail to recognize people's own ability to find solutions. And Kittrell felt that increased specialization would hamper this relationship; a loss of the important focus on home and family would result in ordinary people feeling ill at ease dealing with the professionals.

As historian Margaret Rossiter has shown, the increased entry of men into the home economics field during the 1960s provided the major impetus for the movement away from the field's more traditional focuses.[70] As Margaret Mead had mentioned in her 1963 speech, increased efforts at professionalization had resulted in, for example, the split between child development studies and child psychology. Kittrell worried that the field was losing its understanding of what home economics meant. She found the movement to "upgrade" the field from its old roots in the Land Grant colleges was ill-conceived:

> There are so many people who are eager to do a specialized thing that they forget the total meaning of "home economics." I would suggest that the folks who would want to be interested in this further would read Ellen Richard's life story in regards to human ecology and home economics. They're the same, those meanings are really the same, but [the term] human ecology, being so much more sophisticated, [makes] people think it's so far removed from the home, and they prefer it to be something else. But the home is really the center of all our learning, it will one day come into its own, after we learn how *not* to be sophisticated, but to be really sincere and scientific in our working with people. In the future, home economics will come into its full understanding among people, and we will have to have this as one of the basics if we're going to have a good life.[71]

Despite her expressed preference to the term "home economics," and her antipathy to the term "human ecology," Kittrell was a pragmatic woman. When it suited her larger aims, she appeared willing to compromise for the sake of the program at Howard University. In 1970, she wished to elevate Howard's Department of Home Economics to the level of a school within the University's graduate school. To this end, she herself proposed changing her department's name:

> Because the name Home Economics has not been fully understood by the public in terms of its true function and nature, an alternate name can help to reach this goal.
>
> A School of Human Ecology would offer greater creativity and flexibility in curriculum design at the undergraduate level . . . [and] give opportunity for greater involvement in basic human problems and needs.[72]

It seems clear that Kittrell's capitulation was merely window dressing. Her primary interest was to ensure the continued growth of her program. Because of what she termed the "rigid requirements of the College of Liberal Arts," it was increasingly "difficult for students to find space for electives and supporting courses for majors."[73] Clearly, she was aiming to increase enrollment in home economics courses: "A *school* would offer greater opportunities for admission and retention of students, and at the same time it could provide for individual projects that would be unmanageable within a department that has to fit its program into a larger liberal arts program."[74] Then, without batting the proverbial eyelash, she would forcefully reiterate her definition of home economics, as her philosophy of home economics remained largely unchanged over the years.

For Kittrell, home economics was the science of living—a means to instruct all people toward the attainment of a better life. Society needed strengthening, and the family was the locus where this could best be accomplished:

> We want to keep in mind constantly our major objectives. We want to work for the best development of Negro women, and *all women*, in areas of their social, economic, homemaking, spiritual, health, and civic life. We must not only know the scientific and spiritual facts as they relate to each of these, but we must be able to know and have good techniques in bringing about the desired results in learning.[75]

DAY CARE PIONEER

Kittrell was a pioneer in research on day care facilities. As early as 1949 she had begun to develop the nursery school program and research facility at Howard University. The nursery school program functioned both as a training ground for students and as a laboratory for research in human development. Kittrell's carefully controlled studies of child development were instrumental in providing the research material needed to support the development of the federal Head Start Program; her influence as a member of the executive committee for the White House Conference on Children and Youth in 1960 was also critical.

In the 1960s, Kittrell had again pushed for the expansion of the department's research in child care, early child development, and parent training. By 1962, the nursery school had been expanded into a day care program. During this period, her correspondence reveals that she tried, unsuccessfully, to interest a number of philanthropic organizations in providing funds for the program.[76] Then in 1963–

1964, the Children's Bureau of the U.S. Department of Health, Education, and Welfare provided a substantial grant; in 1965, the day care program, under Kittrell's direction, received an $83,000 grant from the Children's Bureau to continue research already begun on the effects of day care on culturally deprived children and parents.[77]

After Retirement

Kittrell continued to develop her home economics department during the final years of her tenure as chair. In the 1970–1971 academic year, there were 46 students enrolled in the home economics master's program; fifteen of that number were from foreign countries. Nineteen students were awarded the master's degree in June 1970.[78] In the fall of 1971, the Howard home economics department began to offer a doctoral degree in nutrition, and in the following year it also instituted a doctoral program in child development and parent education.[79] In 1972 Kittrell traveled to Zaire on a research grant sponsored by the Agency for International Development to gather material on family ecology in that country for a book.

Kittrell retired from Howard University in 1973, and she became a visiting postdoctoral research fellow at the College of Human Ecology at Cornell University, where her interests centered around the need for research on the development of an individual from infancy through old age. She also continued her work as a senior research fellow at the Moton Center for Independent Studies in Philadelphia from 1976 to 1977.[80] In 1977, she made her final trip to India, as an invited participant in the American Lecture Series sponsored by the U.S. State Department, and also as a participant in the twenty-fifth anniversary celebration of the College of Home Science at Baroda University.

Her science brought Flemmie Pansey Kittrell a life that few individuals have the opportunity to even contemplate, let alone practice. She traveled to many countries—Bangladesh, Ethiopia, Ghana, India, Japan, Kenya, Liberia, Morocco, Mozambique, the Netherlands, Nigeria, the Republic of the Congo and the Republic of Zaire (Leopoldville and Brazzaville), Rhodesia, South Africa, Sweden, Switzerland, and Uganda.[81] In each country, she lectured on the needs of women and children—their nutritional needs, their educational needs, and their human needs. Often, she happened to be in a country at a pivotal time in its history—India during the first years of its independence from Britain, the Congo during a dangerous period of political and civil unrest, Kenya during the Mau Mau uprisings, Bangladesh during its struggle for independence from Pakistan, and South Africa during the worst days of apartheid and human rights violations. Often she was the representative of the U.S. State Department or other government agencies, but at other times she was sent by church missionary groups or the United Nations.[82] Despite political and civil turmoil, she was welcomed everywhere she went because she brought something unique—scientific knowledge and a love and respect for people. In each place she was both a teacher and a student, and she brought the lessons she learned home to the United States.

Kittrell knew world leaders, and knew them well enough to call upon them for help whenever she felt she needed it.[83] And she knew ordinary people — people from many lands, many of them girls and young women. She was a forceful advocate for women's education, both at home and abroad:

> No country can go higher than the performance of its women. If women are kept down, the men will keep themselves down. When one educates a man, one educates an individual; when one educates a girl, one educates a whole family. Without advanced women, a country cannot survive. It is good for parents to encourage their daughters to go to school, and when they graduate give them work which will make them glad they had finished the school and make others want to go.[84]

Often Kittrell helped sponsor African women when they were ready to come to the United States to further their educations in science. Howard University benefited as well, because many Africans came to Howard to study:

> As I look back over my African experience, in regard to Howard, Howard got so many of these African men and women students. . . . the African people felt so much at ease at Howard, and then they went back home to carry on the work they wanted to do — we were able to see the fruits of our labor in that the graduates went back.[85]

"I'd like to think of myself as an international citizen," Kittrell said in a 1977 interview. "I have enjoyed knowing that human beings all over the world respond in predominantly the same ways to the problems they meet."[86] In each country she was able to make a real contribution and make an important difference to the lives of its people. During the scientific revolution of the seventeenth century, natural philosophers frequently expressed the idea that science was a way of praising God — praising Him for His infinite workmanship through the diversity of the Creation. For Kittrell, however, science was not only a way of praising God, but a way of doing God's work by feeding mothers and children and making them healthy. This religious aspect of her science was reflected in every project she undertook:

> I have been interested in the church all my life, not from the emotional side, but the doing side. I think as a youngster I learned so much from the church that helped me as I grew up that I just kept on with it, and I think it has paid off. . . . I really can't describe how thrilling it was to see [once malnourished] children develop into glowing, happy youngsters.[87]

Flemmie Kittrell died in 1980.[88] The American Association of Family and Consumer Sciences has established a fellowship for American and foreign minority women in her name.[89]

NOTES

1. Flemmie P. Kittrell, transcript of an interview by historian Merze Tate, at the home of Mrs. Orris Robinson, 31 Argyle Terrace, Washington, D.C., 29–30 August 1977, p. 42. This interview can be found in the archives of the Women's Oral History Project, in the Gerda Lerner Collection (Schlesinger Library, Radcliffe College, 10 Garden Street, Cambridge, Mass.) and at Fisk University (Nashville Tenn.)

2. Flemmie P. Kittrell, curriculum vitae, taken from the "Proposal for a Program Leading to the Doctor of Philosophy Degree in Nutrition in the Department of Home Economics, Howard University," prepared by Kittrell, department chair, 6 May 1971 (Flemmie P. Kittrell Collection, uncataloged material, Moorland-Spingarn Research Center, Howard University, Washington, D.C.). See also Kittrell, Tate interview; James M. Jay, *Negroes in Science: Natural Science Doctorates, 1876–1969* (Detroit: Belamp, 1971), pp. 41, 60; and "Nutritional Chemist," in *Contributions of Black Women to America*, ed. Marianna W. Davis, vol. 2 (Columbia, S.C.: Kenday Press, 1982), pp. 443–444.

3. Kittrell, Tate interview, p. 1

4. Ibid.

5. Flemmie P. Kittrell, "Interrelated Services for the Planned Community," talk presented at an international conference on home economics, at the University of Vermont, 9 August 1968 (Kittrell Collection, uncataloged material, Moorland-Spingarn Research Center).

6. Kittrell, Tate interview, p. 1.

7. Hampton Institute was founded in 1868. According to Kittrell, "it wasn't a missionary type of school, but it was founded for the freed slaves; ten years later, the Indians [Native Americans] came to Hampton Institute, and the first buildings on the campus for Indians were called the wigwams and Winona Hall" (Tate interview, p. 2). Merze Tate adds that "the founder of Hampton Institute was Samuel Chapman Armstrong, the son of missionaries in Hawaii. They were, in a sense, the first technical agents to go out from the United States, to carry handicrafts and agriculture and work with the hand as well as the mind with other people" (ibid.).

8. Kittrell, Tate interview, p. 3.

9. Ibid., p. 2. According to Tate, "when Mahatma Gandhi wanted to establish basic education in India, he studied and used the concept of training the hand, head and heart that [Booker T.] Washington had learned at Hampton and later had implemented at Tuskegee" (ibid.).

10. Ibid., p. 4.

11. See Margaret Rossiter, *Women in Science: Struggles and Strategies to 1940* (Baltimore: Johns Hopkins University Press, 1982), pp. 68–70, 77–79, 82, 91, 97–98, 121, 195, 238–239. "The founder of 'home economics,' and one whose leadership and character touched her contemporaries deeply, was Ellen Swallow Richards, Vassar 1870 and MIT 1873. Between 1880 and 1920 she almost single-handedly created the field—she propagandized for it, ran demonstration projects, raised money, performed many chemical analyses, wrote several handbooks, trained and inspired her co-workers, and organized its main activities and professional associations. . . . As a student at MIT she had learned how to make a place for herself (and other women) by capitalizing on woman's traditional role, or, as she put in 1871, 'Perhaps the fact that I am not a radical or a believer in the all powerful ballot for women to right her wrongs and that I do not scorn womanly duties, but claim it as a privilege to clean up and sort of supervise the room and sew things, etc. is winning me stronger allies than anything else'" (ibid.). Rossiter cites Carolyn L. Hunt, *The Life of Ellen H. Richards* (Boston: Whitcomb and Barrows, 1912), p. 91. This is most probably the biography Kittrell read.

12. Rossiter, *Women in Science*.

13. Kittrell, Tate interview, p. 4.

14. Laboratory schools were then a common feature of colleges which offered degrees in education or teacher training.

15. Kittrell, curriculum vitae.

16. Kittrell, Tate interview, pp. 16–17.

17. Kittrell, "Interrelated Services for the Planned Community."

18. Ibid. This paragraph immediately follows the previously cited quote (note 17) in the source document.

19. Dana Bushnell Waring, memorandum to Dr. Flemmie P. Kittrell, 4 December 1976; also Waring, telephone conversation with author, 17 April 1995. A portion of the Kittrell-Waring correspondence is housed in the Kittrell Collection at the Moorland-Spingarn Research Center; the remainder is housed in the Ethel B. Waring Collection at Cornell University, Ithaca, New York.

20. Kittrell, Tate interview, p. 8.

21. Ibid.

22. Ibid.

23. Ibid.

24. Ibid.

25. Ibid., pp. 8–9. Johnson repeatedly addressed Dr. Kittrell as "daughter," and he apparently always addressed her in this manner; this is perhaps revealing with regard to women's place in his hierarchy.

26. Ralph P. Bridgman, President, Hampton Institute, Hampton, Va., letter to Flemmie Kittrell, 3 May 1944 (Kittrell Collection, Moorland-Spingarn Research Center). "I am continuing your appointment at Hampton Institute as Dean of Women for the Year 1944–45 at a salary of $4,000 for 12 months beginning July 1."

27. Flemmie P. Kittrell, "Annual Report to the Trustees at Hampton Institute," 1 September 1944 (Kittrell Collection, Moorland-Spingarn Research Center, Box 4, uncataloged).

Kittrell said that "the enrollment of women students in 1940 was smaller than it has been in several years. The total number of freshmen women admitted was 93. This is a very significant fact because in 1940 other Negro colleges reached an almost all-time high in their women's enrollment. The instability in Hampton's program and general unrest were part of the problem. Parents were sending their daughters to colleges that seemed to offer better education and guidance" (ibid.).

28. Among the problems she mentioned in her annual report were lack of what she considered proper courses, poor food service, and poor pupil counseling services (Kittrell, Tate interview, p. 9).

29. Ibid., p. 9.

30. The first M.S. in home economics was awarded in 1943. Information from Flemmie P. Kittrell, "Program for the Implementation of a School of Human Ecology," Howard University, Washington, D.C., 8 June 1970, p. 1 (Kittrell Collection).

31. "Department of Home Economics Budget, 1945–6" (Kittrell Collection). The budget included salaries for Kittrell, professor and head of department, $4,000; one associate professor, $3,000; one professional lecturer, $2,000; two instructors, at $2,000 each; and one instructor at $1,800 (ibid.). All faculty members were women, and Kittrell was the only one with a doctorate.

32. Kittrell, Tate interview, p. 9.

33. See the appendix.

34. Biographical sketch, *Africa Feature*, IPS/Africa, June 1963.

35. "Department of Home Economics Budget for 1948–1949" (Kittrell Collection). The breakdown was Kittrell, $5,500; associate professor Margaret Brainard, $4,500; one instructor at $3,800; one instructor at $3,250; three instructors at $3,000 each; the department secretary at $2,168.28; and a housekeeper (10 months) at $1,408. The budget

also provided for an additional instructor in 1950 at $3,250. Total salaries were $32,876.28. The total budget with supplies, equipment, and expenses was $39,198.28 (ibid.).

36. Kittrell, Tate interview, p. 9.

37. Ibid.

38. Ibid.

39. Ibid.

40. Ibid. Note: Mrs. Hansa Mehta was then vice-chancellor of education for India and also a member of the United Nations Human Rights Commission. See Flemmie P. Kittrell, "University of Baroda Establishes Home Economics in Higher Education," *Journal of Home Economics* 44 (February 1952): 97–100.

41. Kittrell, Tate interview, p. 9.

42. Ibid., p. 10. Mrs. Lydia J. Rogers took over as acting head of the Department of Home Economics at Howard. See Hettie Pearson, "Scrapbook 2, Howard University Home Economics Department: The Open House and Career Conference, April 13–14, 1951 (Senior Day)" (Kittrell Collection).

43. India became an independent country in 1948.

44. The B.S. core-curriculum included introduction to the social sciences, biology, world literature, fine and applied arts, personal hygiene and physical education, current topics, general chemistry, child development (including observation and participation in the nursery school), and general psychology. Initially, four major fields of study were offered: child development and family life, foods and nutrition, household management, and home economics teaching. Later, the extension service program was added, and later still, an advanced degree program (Kittrell, "University of Baroda").

45. Kittrell, Tate interview, p. 5.

46. The two-day conference was held August 4–5, 1951. According to Kittrell, its purpose was fact-finding and establishing criteria for selecting course content, evaluating homemaking programs, and recruiting of teachers. The fifty-four delegates came from all parts of India and Ceylon, from eleven colleges and universities and eight secondary schools. Kittrell knew many of the women personally. Among the delegates were several American-trained women: Surjeet Chopra, Dr. Rajammal P. Devadas, Mrs. Sucy Koshy, and Mrs. Kumudini (Pandit) Karandikar, all of whom studied at Ohio State University; Angani Mehta (the daughter of Mrs. Hansa Mehta), Hazel Manoranjan Sadoc, and Theodora E. Bryce of Cornell University; and Florence Theopilus, Department of Physiology, Vassar College (Kittrell, "University of Baroda"). Dorothy M. Pearson wrote a letter with more details: "One of the epic-making events this year was the first Home Science Conference to be organized in India at Baroda in August by Dr. Kittrell of Howard University. She did a magnificent job." Dorothy M. Pearson (Professor of Nutrition, Women's Christian College, Madras, India) to the Committee on Missionary Education (514 Wesley Building, Toronto 2B, Ontario, Canada), 13 November 1951. This letter was forwarded to Kittrell by the Committee (Kittrell Collection).

47. Pearson, letter, 13 November 1951.

48. See the letters on file in Kittrell Collection.

49. Flemmie P. Kittrell, letter to George Dolgin (Deputy Chief, Indian Branch, Asian Development Service, Technical Cooperation Administration, United States Department of State, Washington, D.C.), 12 August 1953. A copy of this letter was sent to Henry E. Niles, Deputy Director of TCA for India, at the American Embassy in India (Kittrell Collection).

50. Henry Niles (U.S. Deputy Director of Technical Cooperation Agency for India, U.S. Foreign Service) letter of reply to Flemmie P. Kittrell (then at Maharaja Seyajireo University, Baroda), 13 August 1953 (Kittrell Collection).

51. This was a part of the International Cooperation Program (also known as the Point Four Program) of the U.S. Department of State.

52. Kittrell, Tate interview, p. 5. The 1957 Japan trip, in which she led a group of twenty-two Indian women, was sponsored by the U.S. Department of State.

53. John P. Ferris (Chief, South Asia Division, International Cooperation Administration, Washington, D.C.) letter of appreciation to Dr. Mordecai W. Johnson (President, Howard University), 1 November 1955 (Kittrell Collection).

54. Ibid.

55. Flemmie P. Kittrell, curriculum vitae. Political unrest erupted during Kittrell's first visit to the Congo (now the Republic of Zaire) in 1961. The situation there got so bad that the United Nations had to send in troops to restore order.

56. See the appendix.

57. Flemmie P. Kittrell, letters to Dr. P. Mabel Nelson (School of Home Economics, Iowa State College, Ames, Iowa) and Miss Jessie Harris (Dean, School of Home Economics, University of Tennessee, Knoxville), 11 August 1947 (Kittrell Collection). Kittrell wrote, "It has been suggested recently by some of the members on our committee that we have our Department combined in a building with biology. This seems a most unsatisfactory arrangement, and I have been asked to work up justifications for a separate building. This I have done, but it would have greater significance if I could get letters from colleges and universities who have moved progressively forward in home economics thinking throughout the years. . . .

"Would you kindly write me a short letter saying why home economics at Howard University should be in a building specifically designed, arranged, and equipped to promote home economics in the following areas: foods and nutrition; child development; nursery school and parent education; teacher education; clothing and textiles; institutional management and cafeteria; home management house. Research will be carried on in all of these areas in cooperation with other schools and colleges of the University."

58. See, for example, Flemmie P. Kittrell, letter to Mr. Julian A. Cook (Coordinator of the Building Program, Howard University), 22 January 1952: "We look forward to an enrollment of approximately 325 students. This will include the graduate program which is increasing at a more rapid pace than I had anticipated. The courses in home crafts are being increasingly elected by students throughout the university, and the anticipated enrollment in that area will be approximately 40 students per quarter beginning next year. We limited enrollment this quarter because of lack of space. In Home Economics Education (student teaching) we need, in keeping with the present progressive trends, a workshop type of classroom where students can work with projects in Foods and Nutrition, clothing and textiles, child development, home furnishings and household equipment. This arrangement should be a very special type of room—semi-classroom and semi-laboratory. In addition, we need: for nutrition research, a small animal laboratory to be located at the top of the building; a general purpose room for living room activities, and for meetings of groups for demonstrations and the like. This is a must, a room for approximately 350 people, large enough for the coming together of all home economics majors, and for demonstrations and lectures. . . . It is necessary to have considerable play space adjourning the Nursery School, and I hope the building will not be crowded into too small space and that it will be located so that we will not be in danger of constant fear of cars running down the children."

See also Kittrell, letter to Dean J. St. Clair Price (College of Liberal Arts, Howard University), 25 March 1953: "I am stating herewith some information concerning serious problems and needs in the Department of Home Economics. You will note that we have made substantial gains in enrollment and also that our Nursery School has grown from a Child Care Center of six in 1944, when I came to Howard, to an educational project now involving thirty-five children. Our physical space has not increased to take care of minimum needs. Our Child Development Program not only serves all of our students in Home Economics as a basic area in general education, but it also serves the medical students in the Department of Pediatrics, nurses at Freedman's Hospital, students in Psychology and education, and trainees who come from various foreign countries, and the United States Territories. We need immediate relief from physical pressure.

"We are asking that, beginning in September, the whole first floor be given over to the Home Economics department—a place for psychological testing of the children, an isolation room, office space for the staff for records. We cannot do an education job of significance unless we have these aspects provided. We need a class room on the first floor —it is impossible to conduct all of our instruction in the one lecture room provided on the third floor."

59. Kittrell, Tate interview, p. 5.

60. Kittrell, letter to Dean J. St. Clair Price.

61. Kittrell, Tate interview, p. 10.

62. Kittrell, Tate interview, p. 10.

63. Margaret Mead, "Keynote Address." The complete text of Dr. Mead's speech is in the Kittrell Collection.

64. Flemmie P. Kittrell, "Introduction," in *Excerpts from the Dedication Ceremonies of the Home Economics Building*, Howard University, 1–2 February 1963 (Kittrell Collection).

65. Flemmie P. Kittrell, "The Home and the Total Family Must Be Our Operational Base for Launching Medical, Social, Nutritional and Economics Education," Convocation of the All-India Home Science Association, Travandrum, South India, 1972 (Kittrell Collection).

66. Ibid.

67. Kittrell, Tate interview. See also Kittrell, "The Home and the Total Family."

68. Ibid.

69. Kittrell, "Interrelated Services for the Planned Community."

70. Margaret W. Rossiter, *Women Scientists in America: Before Affirmative Action, 1940–1972* (Baltimore: Johns Hopkins University Press, 1995), pp. 165–185.

71. Kittrell, Tate interview, pp. 28–29.

72. Flemmie P. Kittrell, "Program for the Implementation of a School of Human Ecology," pp. 1, 3.

73. Ibid.

74. Ibid.

75. Flemmie P. Kittrell, talk or report to the Trustees at Hampton Institute, 1 September 1944 (Kittrell Collection). I have added the emphasis in this passage.

76. See for example F. D. Patterson (Director, Phelps-Stokes Fund) letter to Dr. William Stuart Nelson (Dean, Howard University), 5 June 1956; and Patterson letter to Flemmie P. Kittrell, 15 June 1956, regarding an appointment he had made for her with Dr. Lindley F. Kimball, vice-president of the Rockefeller Foundation, and also an appointment with Mr. Chester Bowles of the Ford Foundation. Also see Kittrell letter to Dr. Maxwell Hahn (Executive Vice President, Field Foundation), 12 November 1957. All rejected her request for funding (Kittrell Collection).

77. "Budget for a Grant of $83,041.00 from the Children's Bureau, Department of Health Education and Welfare (Project NO. D185 (01) for Renewal of Research Entitled 'A Group Day Care Program for Culturally Deprived Children and Parents' (GS-RPG 1083A) under the Direction of Dr. Flemmie P. Kittrell, Professor and Head of the

Department of Home Economics, Beginning Jun 1, 1965 and Terminating Automatically May 31, 1966" (Kittrell Collection, document dated 8 June 1965). The budget lists the following salaries: principal investigator (Kittrell), $6,665; head teacher, full-time 12 months, $8,500; research assistant, full-time 12 months, $7,700; assistant teacher, full-time 12 months, $6,500; and parent educator, full-time 12 months, $8,000. Note that, with one exception, Kittrell assigned herself the lowest salary.

78. Kittrell, "Program for the Implementation of a School of Human Ecology," p. 3. The countries included India, Formosa, Korea, Nigeria, Liberia, the Philippines, and the Caribbean areas.

79. Ibid.

80. Kittrell, Tate interview, p. ii.

81. Kittrell was in Nigeria, Liberia, Ghana, Kenya, and Uganda in 1957; she visited Kenya and Uganda again in 1958 and 1959; she was in Guinea and the Congo in 1961; and she returned to the Congo in 1962. With regard to Uganda, Kittrell said, "I was particularly interested in their program, especially at the university. They had a very outstanding university [in Uganda], and many of the professors there were women, in the upper ranks. . . . They were outstanding scholars. At that time [when I was there], there were a great many Indian women there, who were in leadership positions. But you know, after Amin [Idi Amin DaDa] came [1971] in he ordered the Asians to leave the country [1972], and that was the beginning of the trouble that we now face" (Kittrell, Tate interview, p. 11).

82. Kittrell's trip to Bangladesh was a part of the Airlift of Understanding mission: "When they got their independence . . . There were seventy of us who volunteered to go, and we paid our own way, of course, to see and to offer any sympathetic understanding we could as individuals, not representing our government, but as individuals" (Kittrell, "Tate Interview," p. 27). Her other remarks make it clear that this trip was directed by the Methodist Church; subsequently, Kittrell participated in a number of annual international development conferences sponsored by the church to develop programs to assist Bangladesh (ibid.). The Women's Division of the Methodist Church's Board of Missions sponsored her Congo, Mozambique, Ethiopia and Rhodesia trips in 1961, and her visit to the Congo in 1962. The purpose of these church-sponsored trips was to "uplift the educational status of women in those countries." See the Biographical Sketch, *Africa Feature*, IPS/Africa, June 1963.

83. Among the world leaders Kittrell knew personally, were Prime Minister Indira Gandhi of India, President Lyndon B. Johnson of the United States, and Tom Mboya, the Kenyan political leader and General Secretary of the Kenya Federation of Labor. Mboya was supposed to succeed Kenyan President Jomo Kenyatta but he was assassinated in 1964.

84. Flemmie P. Kittrell, "Consultation on Home Economics," a talk presented at Elizabethville, Congo, 2 August 1961 (Kittrell Collection).

85. Kittrell, Tate interview, p. 11. Kittrell also sponsored the higher education of a number of women from other countries, as well as women from the United States (Kittrell Collection, letters).

86. Kittrell, Tate interview.

87. Kittrell, Tate interview.

88. Obituary, *Washington Post*, 5 October 1980; also see the *Afro-American*, 11 October 1980.

89. Flemmie P. Kittrell Fellowship for Minorities, American Association of Family and Consumer Sciences, 1555 King Street, Alexandria, Va. 22314 (Reference Service Press, 1995, via the Internet).

MARGARET MORGAN LAWRENCE
Psychiatrist, Researcher on Black Children

Psychiatrist Margaret Morgan Lawrence made considerable contributions to the mental health field through teaching, clinical work, and research. Her research interests have included adaptation to trauma, and ego strength. She was born in New York City in 1914 to the Reverend Sandy Alonzo Morgan, an Episcopal minister, and Mary Elizabeth (Smith) Morgan, a schoolteacher.[1] Although the Morgans lived near Richmond, Virginia, their first child, a son, had died in a segregated hospital. Rather than risk a similar event, late in Mary Morgan's next pregnancy, they traveled to New York and stayed with family until shortly after their daughter's birth at Sloan's Hospital for Women. The family then returned to Virginia, but soon moved to Mississippi, eventually settling in Vicksburg. According to a biography written by Lawrence's daughter, Sarah Lawrence Lightfoot, Reverend Morgan's ministry was nurtured by Black parishioners and white philanthropists. Nevertheless, Vicksburg was rigidly segregated, and Lawrence was well aware of her boundaries. Every other summer, she and her mother would visit New York City, where they stayed with relatives for extended periods.[2]

By the time Lawrence was fourteen, she had already decided to become a doctor, and she persuaded her parents to let her continue her high school education in New York. She lived with her maternal aunts and attended Wadleigh High, one of two classical high schools for girls in the city. All the teachers at Wadleigh were white, but Lawrence was an excellent student and they were very supportive. With the help of Wadleigh's dean, Anna Pearl MacVay, Lawrence was accepted to three schools: Hunter College, Smith College, and Cornell University.

Lawrence chose to attend Cornell and was awarded a scholarship by the National Council of the Episcopal Church. When she began attending classes in 1932, she was the college's only current Black undergraduate. She supported herself by working first as a housekeeper and later as a laboratory assistant. She graduated in 1936. Although she had excelled at Cornell, she was denied admittance to Cornell Medical College "because another black, a man, admitted before had 'not worked out.'"[3]

With the encouragement of Madeline Ramee, an administrator with the National Council of the Episcopal Church, which had awarded her undergraduate scholarship, Lawrence applied to and was accepted at Columbia Medical School. She entered in the fall of 1936, becoming the third Black to be admitted. She was one of ten women in a class of 104 who graduated in 1940. According to Jessie Carney Smith, Lawrence was rejected for pediatric residency at Babies Hospital in New York because there were no accommodations for a black female intern; and she was rejected at Grasslands Hospital in Westchester, New York, because she was married. She finally secured a two-year pediatric internship at Harlem Hospital in 1940.

In 1942, Lawrence received a Rosenwald Foundation fellowship to study at Columbia University's School of Public Health, and in 1943 she earned the M.S. in public health. During her studies for the master's degree, Lawrence worked with Benjamin Spock, who focused her interests on children and their emotional concerns; she considers him to have been a major influence on her career.[4]

Lawrence began her career as an instructor in pediatrics and public health at Meharry Medical College in Nashville, Tennessee, in 1943. When she left Meharry in 1947 to begin a National Research Council research fellowship at New York's Babies Hospital, she had risen to the rank of associate professor. In 1948, under a United States Public Health Service fellowship, she began psychoanalytic training at the Columbia University Psychoanalytic Center, which was affiliated with the New York Psychiatric Institute; she was the first Black ever admitted to the Institute. Her decision to enter psychiatry was supported by Charles Pickett, of the American Friends Service Committee; Pickett introduced her to Viola Bernard, who encouraged Lawrence to enter the Columbia Psychoanalytic Center, and served as her mentor.

In 1951, Lawrence completed a fellowship in child psychiatry and pediatric consultation with the Council of Child Development Center and was awarded a Certificate in Psychoanalytic Training from the Columbia Psychoanalytic Center.[5]

She spent a number of years at Harlem Hospital, directing the Therapeutic Developmental Nursery; she was an associate clinical professor of psychiatry at Columbia College of Physicians and Surgeons from 1963 to 1988; she was the director of the Child Development Center from 1969 to 1974; she was one of the founders of the Rockland County Center for Mental Health; and she served on the New York State Planning Council for Mental Health in the 1970s and early 1980s. She retired from Columbia University as an associate clinical professor of psychiatry in 1984.

In 1938, during her second year in medical school, Lawrence married Charles Radford Lawrence II, who was then studying sociology at Morehouse College in Atlanta. For most of his career, Charles Lawrence was a professor of sociology at Brooklyn College. They were both very active in the Episcopal Church, and he served for a time as president of the House of Deputies of the General Convention of the Episcopal Church in the United States. At various times during their careers, their research interests coincided. In 1973, they took a sabbatical leave to study the strengths and the pathologies of school children in East and West Africa, a continuation of her research on Black children who had been identified as "strong" by their teachers in schools in Georgia and Mississippi. The couple had three children: Charles Lawrence, a professor of law at Stanford University; Sarah Lawrence Lightfoot, a sociologist at Harvard University's School of Education; and Paula Lawrence Wehmiller, principal of the Wilmington Friends Lower School in Delaware.[6]

NOTES

1. Sarah Lawrence Lightfoot, *Balm in Gilead: Journey of a Healer* (Reading, Mass.: Addison-Wesley, 1989). See also Joanna Chapin, "Interview with Dr. Margaret Morgan Lawrence," *Association for Psychoanalytic Medicine Bulletin* 28 (March 1989): 116–122. *Who's Who among Black Americans* (Northbrook, Ill.: Who's Who among Black Amer-

icans Publishing, 1985), p. 504. *Who's Who of American Women* (Chicago: Marquis Who's Who, 1974), p. 552. "The American Negro in College, 1939–1940," *The Crisis,* August 1940, p. 233.

2. Lightfoot, *Balm in Gilead.* See also Chapin, "Interview with Dr. Margaret Morgan Lawrence," pp. 116–122; *Who's Who among Black Americans,* p. 504. *Who's Who of American Women,* p. 552; "American Negro in College," p. 233.

3. Susan Brown Wallace, "Margaret Morgan Lawrence," in *Notable Black American Women,* ed. Jessie Carney Smith, vol. 1 (Detroit: Gale Research, 1992), pp. 658–660.

4. Ibid., p. 659.

5. Ibid., p. 660.

6. Ibid.

KATHERYN EMANUEL LAWSON

Researcher at Sandia National Laboratory

Chemist Katheryn Emanuel Lawson was another beneficiary of Cold War science research. Born in Shreveport, Louisiana, in 1926, she attended segregated public schools. She graduated from high school in 1941, as salutatorian of her class, and then attended Dillard University in New Orleans where she earned the baccalaureate degree, cum laude, in the natural sciences in 1945. She completed the master's degree in organic chemistry at Tuskegee Institute in 1947.

During the next four years (1947–1951), Lawson held a series of temporary appointments at historically Black colleges as an assistant professor of chemistry: Bishop, Savannah State, Talladega, and Grambling. Between 1951 and 1954, she moved from the rank of assistant to associate professor of chemistry at Central State College.[1] Lawson married chemist-bacteriologist Kenneth Lawson at some point during this period, although the exact date is not available.[2]

In 1954, Lawson received an offer to study for the doctorate at the University of New Mexico in Albuquerque. She received an assistantship, and Kenneth Lawson got a job as a laboratory technician at the city's sewage plant while he resumed his own graduate studies a the University of New Mexico.[3] He completed his degree a year later, and she completed her doctorate in physical chemistry with an emphasis on radiochemistry in 1957. Her dissertation was "Behavior of Indium at Tracer Concentrations."[4]

During the next year (1957–1958) Lawson was a staff member in biochemistry at the Veterans Administration Hospital in Albuquerque. Then, in 1958, she obtained a position at the Crystal Physics Research Division of Sandia Corporation, an ordnance engineering laboratory under contract to the Atomic Energy Commission.[5] At Sandia Laboratory, Lawson's research involved studying the optical properties of transition metals—copper, nickel, iron, silver, gold—in their combinations with other elements and their associated electrical and magnetic properties, using infrared spectrophotometers and x-ray secondary emission spectrometers.[6] By the mid-1960s, according to a Sandia Corporation spokesperson,

Lawson was working on "systems that were expected to shed light on the validity of the Crystal Field Theory, which deals with ions and their interchangeable relationships."[7]

Lawson is the mother of two sons, the first of whom was born while she was in graduate school.

NOTES

1. Vivian Ovelton Sammons, *Blacks in Science and Medicine* (New York: Hemisphere, 1990), p. 149. See also "Physical Chemists," in *Contributions of Black Women to America*, ed. Marianna W. Davis, vol. 2 (Columbia, S.C.: Kenday Press, 1982), pp. 444–445. "Scientific Couple Finds Success in Albuquerque: Chemists Kenneth and Katheryn Lawson Say Desert Town Is Ideal Community for Them," *Ebony*, June 1965, pp. 67–70, 72, 73.

2. The *Ebony* article seems to put the date at around 1954: "At the time, she was a chemistry instructor at Central State College in Wilberforce, Ohio, and her husband (they had been married only one month) was a graduate student there, having obtained a degree in biology the previous year" ("Scientific Couple," p. 67).

3. Ibid., p. 68.

4. Sammons, *Blacks in Science and Medicine*, pp. 444–445.

5. "Scientific Couple," p. 67.

6. "Physical Chemists," p. 445.

7. "Scientific Couple," p. 68.

LILLIAN BURWELL LEWIS

Student of Ernest Everett Just

Zoologist and endocrinologist Lillian Burwell Lewis studied with Ernest Everett Just at Howard University. Born in Meridian, Mississippi, in 1904, Lewis earned her baccalaureate degree in biology at Howard in 1925, two years after Roger Arliner Young and Marguerite Thomas Williams.

Lewis began her career as an associate professor of zoology at South Carolina State A&M College in 1926, moving up in rank to associate professor by the time she left in 1929.[1] It is likely that Just, her former mentor, had asked South Carolina State President R. S. Wilkinson to hire Lewis.[2] Lewis studied for her master's degree at the University of Chicago; although exact information is not available, it is most probable that she attended during the summers, because during the academic years 1929–1931 she was teaching at Morgan State College. She received her M.S. in 1931. According to Kenneth Manning, at that time Just tried to secure research funding for a number of his former students, among them Lewis (then Burwell).[3] The funding crisis was exacerbated by the Depression, and apparently Just was unable to divert any of his Julius Rosenwald fund grant money to her.[4]

Lewis moved on to an associate professorship at Tillotson College, where she remained until 1947. During her tenure at Tillotson, she completed the Ph.D. in zoology at the University of Chicago in 1946, becoming the first Black woman to earn a natural science degree at that school. She was elected to the Sigma Xi honorary society and Sigma Delta Epsilon, a national honorary scientific society for women.[5] Her dissertation was "A Study of the Effects of Hormones upon the Reproductive System of the White Pekin Duck."[6]

After earning her doctorate, Lewis secured a full professorship at Winston Salem State University, a position she held from 1947 until her retirement in 1971.[7]

NOTES

1. Vivian Ovelton Sammons, *Blacks in Science and Medicine* (New York: Hemisphere, 1990), p. 153. See also *American Men and Women of Science: The Physical and Biological Sciences*, 12th ed. (New York: Bowker, 1971–), p. 3697.

2. Kenneth W. Manning, *Black Apollo of Science: The Life of Ernest Everett Just* (New York: Oxford University Press, 1983), p. 234. Manning states that Just "persuaded R. S. Wilkinson, president of South Carolina State College to do his utmost to hire three of his students: [Leona] Gray, [Caroline] Silence, and Dorothy Young." Lewis was hired, so it is probable that Just interceded on her behalf.

3. Ibid., pp. 233–234.

4. Ibid.

5. "The American Negro in College, 1946," *The Crisis*, August 1946, p. 243.

6. See also *American Men and Women of Science*, p. 3697. "The American Negro in College, 1946," p. 243.

7. *American Men and Women of Science*, p. 3697.

RUTH SMITH LLOYD*

Born in Washington, D.C., in 1917, anatomist Ruth Smith Lloyd earned the baccalaureate at Mount Holyoke College, then a predominantly white institution, in 1937. She completed her master's degree at Hood University in 1938, and her Ph.D. at Western Reserve University in 1941. Her dissertation research on the macaque, an Old World monkey, was titled "Adolescence of Macaques (Macacus Rhesus)."[1] Lloyd was an assistant instructor in physiology at Howard University's College of Medicine in 1941, and was an instructor of zoology at Hampton Institute in Virginia during the academic year 1941–1942. Between 1952 and her retirement, she moved from technician to instructor, to assistant professor, to associate professor of physiology and anatomy at Howard University.[2]

NOTES

1. *American Men and Women of Science: The Physical and Biological Sciences*, 14th ed. (New York: Bowker, 1971–), p. 3030. Harry W. Greene, *Holders of Doctorates among American Negroes: An Educational and Social Study of Negroes Who Have Earned Doctoral Degrees in Course, 1876–1943* (Boston: Meador, 1946), p. 193. James M. Jay, *Negroes in Science: Natural Science Doctorates, 1876–1969* (Detroit: Belamp, 1971), p. 60.

2. *American Men and Women of Science*, p. 3030. Greene, *Holders of Doctorates*, p. 193. Jay, *Negroes in Science*, p. 60.

BEEBE STEVEN LYNK
Science Forerunner

Little substantive information is now available on Beebe Steven Lynk, because the University of West Tennessee — the venue most likely to house records on her — no longer exists. Born in Mason, Tennessee, in 1872, Lynk was the daughter of Henderson and Judiam (Boyd) Steven. She earned the bachelor's degree at Lane College in Jackson, Tennessee, in 1892, and in the following year married physician Miles Vandahurst Lynk.[1]

By all accounts, Miles V. Lynk was devoted to education. In addition to an M.D. degree, he earned an M.S. (Walden University, 1900), a second M.S. (Agricultural and Mechanical College, 1901), and an L.L.B. (University of West Tennessee, 1902). He also founded, edited, and published the *Medical and Surgical Observer* (1892), the first medical journal in the United States issued by a Black man, and he authored at least three books.[2] In 1900, Miles Lynk founded the University of West Tennessee, a school for Blacks.

In 1901, Beebe Lynk embarked upon the study of pharmaceutical chemistry at the University of West Tennessee, and earned the Ph.C., a two-year degree, in 1903.[3] In the same year she graduated, she became a professor of pharmacy and chemistry at West Tennessee.[4] Like Josephine Yates,[5] another early Black woman forerunner, Beebe Lynk wrote and took an active part in the Black women's club movement. Her book *Advice to Colored Women* was published in 1896. A member of the National Federation of Woman's Clubs, she also served as its treasurer.

Although Lynk was a forerunner in science, it is not clear if she had any effect as a teacher or mentor of other Black women interested in the sciences. She married a successful, well-educated man who shared his strong interests in education; she became a science teacher at the college level; and she was active in the Black women's club movement and had a strong interest in writing. There may have been articles written about her in the Black popular press of the time, and this might have given her a certain visibility as a role model, particularly with regard to her activities in the women's club movement and her writing career.

NOTES

1. Miles Lynk earned the M.D. at Meharry Medical College in 1891, according to Frank Lincoln Mather, ed., *Who's Who of the Colored Race: A General Biographical Dictionary of Men and Women of African Descent* (Chicago, Ill., 1915; reprint, Detroit: Gale Research, 1976). Miles Lynk's books include *Afro-American School Speaker and Gems of Literature* (Jackson, Tenn.: M. V. Lynk, 1896); *The Black Troopers, or The Daring Heroism of the Negro Soldiers in the Spanish-American War* (Jackson, Tenn.: M. V. Lynk, 1899; reprint, New York: AMS Press, [1971]); and *The Negro Pictorial Review of the Great World War: A Visual Narrative of the Negro's Glorious Part in the World's Greatest War* (Memphis, Tenn.: Twentieth Century Art Company, [1919]).

2. *Who's Who of the Colored Race*, p. 181; see also *Who's Who in Colored America: A Biographical Dictionary of Notable Living Persons of Negro Descent in America* (New York: Who's Who in Colored America, 1927–), p. 348.

3. *Who's Who of the Colored Race*, 1915 ed., p. 181; see also *Who's Who in Colored America*, p. 348. In the early 1900s, when Beebe Lynk earned the Ph.C., the degree re-

quired two years of study. Although she earned this degree after her bachelor's, it was considered a pre-bachelor's degree that was required if one wished to become a pharmaceutical chemist (pharmacist) or perhaps if one wished to teach chemistry and pharmacy, as Lynk did. For example, Meharry Medical College awarded a Ph.C. to Xenophon Lamar Neal in 1927; Neal was listed as an assistant in the chemistry department at Spelman College (*Spelman College Catalogue, 1944–6*, April 1945, p. 24; Special Collections Archives, Woodruff Library, Atlanta University Center, Atlanta, Ga.).

4. Information is not available on how long she held that position.

5. See the entry on Josephine A. Silone Yates in this volume.

CAROLYN R. MAHONEY
Balancing a Mathematics Career with Life

Carolyn R. Mahoney

If a person enjoys serving others, and wants a quality of life that he or she has a good measure of control over, then I would strongly recommend higher education.
—Carolyn R. Mahoney

Mathematician Carolyn R. Mahoney is vice-president of academic affairs at California State University at San Marcos. An effective teacher and a much-sought-after public speaker, she has presented her research—combinatorics, graph theory, and matroids, as well as mathematics pedagogy—in the United States and in the People's Republic of China.

Born in 1946, the sixth in a line of thirteen children, Carolyn Mahoney grew up in what she has described as a "poor but happy family," in Memphis, Tennessee.[1] Her father, Stephen Boone, had only completed the fifth grade in school, but her mother, Myrtle Boone, had completed one year of college.[2] Her grandmother, a seamstress who took in sewing, lived with the family and stayed at home with the children while their mother worked. The family, Mahoney recalled, was "proud that it was never on welfare."[3] Although she says she didn't read a great deal as a young child, she did demonstrate an early interest in mathematics, puzzles, and problem solving. "I always loved mathematics—I can remember dreaming about solutions to math problems in eighth grade."[4] Her interest was encouraged by the nuns in the Catholic schools she attended. She recalled that

> although these schools were segregated, they were considered the best for black kids—[as a result] many students were middle class blacks, who may or may not have been Catholic. . . .
> The nuns and the Catholic school environment of support, encouragement and expectation is the single most important factor [in my success]. My family believed very strongly in eduction. My mother, grandmother, and the nuns were important role models in the sense that women can work.[5]

On the cusp of her teen years, her family's fortunes changed, Mahoney said, due largely to her father's drinking and gambling. Her parents separated and eventually divorced. The family, sans father, was forced to move to a lower-middle-class neighborhood where it was "probably the least wealthy family," and where they were viewed by long-time residents as "blockbusting."[6] Despite their changing finances, however, the Boones

> were viewed in the neighborhood as being "smart" in school; so, there was strong expectation that I would do well. My older brothers were gifted athletes as well as good students; my older sisters were prom queens and cheerleaders as well as good students. I was not considered as pretty as they, but I was considered much smarter in school than they were.[7]

A top student, Mahoney recalls having had a great deal of support at Father Bertrand High School, from which she graduated in 1964. She credits the Catholic school environment with encouraging "all students to perform at their highest potential," recalling that she received particularly strong encouragement from the nuns—especially from Sister Kilian, B.V.M.—who had no doubt she would do well in college and so arranged scholarships for her.

For her first three years, Mahoney attended the Catholic, all-female Mount St. Scholastica College in Atchison, Kansas, where she was mentored by Sister Malachy Kennedy, O.S.B. Mahoney then returned to Memphis and in 1970 completed her B.S. degree in mathematics at Sienna College, also a Catholic, all-female school.[8] She believes that attending an all-female college—along with all-Black Catholic elementary and high schools—had a major impact on her ability to pursue a career in mathematics.[9]

Mahoney entered Ohio State University for graduate study, and received a series of National Science Foundation fellowships, followed by teaching assistantships awarded by Ohio State. Although she does mention professors such as Dijen Ray-Chadhiuri and Tom Dowling, Mahoney says almost every professor she came in contact with at Ohio State provided some measure of mentoring or acted as a role model. Several people were especially encouraging, and were, she says, "always there for me." These included Joan and Tim Leitzel, Frank Demana, Bogden Baishanski, Joe Ferrar, and L. Allayne Parsons.[10] Perhaps due to the nature of mathematics, Mahoney says she was "essentially a foreign student [in graduate school] as most of my friends were Canadian or Chinese. . . . I was definitely in a severe minority situation. [But] . . . enough people came forward to help that the negative effects of racism were minimized for me."[11]

Mahoney completed her master's degree and doctorate in mathematics, with a concentration in combinatories and graph theory, at Ohio State (1972 and 1983, respectively), and then taught for five years (1984–1989) at Denison University in Granville, Ohio, and then for two years (1987–1989) at Ohio State. During this period (1986–1989) she also served on the test development committee of the College Board exams. In 1989, she became the first mathematician selected for the faculty at the new California State University at San Marcos.[12]

Mahoney has said that her technical research is important, but not prolific. Her contributions have focused on various open problems in graph theory and combinatories, on making progress in answering two old problems (questions) related to independent set numbers of a class of matroids and unit distance graphs. Neither problem has been fully solved. She has also said that finding collaborators has been much more problematic because she is a Black female. In the past several years, however, "finding time" has been one of her biggest challenges:

> When I was a faculty member, I taught courses and did research. I tried to schedule my days so that I had big blocks of time to do research. In the last several years, my time has been taken up more and more with administrative tasks.[13]

Mahoney also has a strong, active interest in educational reform at all levels, but says that until she was in a position to "strongly influence hiring," most of her colleagues were white males. She now enjoys a diverse set of professional colleagues, and is well aware of their potential as mentors and role models:

> I like teaching and coaching students to learn. I like facilitating faculty to accomplish these goals. . . . Over the years I have been encouraged and mentored by many men and women, some of whom were unaware of their influence and some of whom I was not totally aware of at the particular moment in my life history.[14]

Mahoney points to the many efforts underway by professional organizations, such as the Mathematical Association of America, the Association for Women in Mathematics, and others, as well as numerous efforts by faculty and departments, as having facilitated a "cultural change that is taking place so that the overall environment for women is improving." She believes the "most critical key to continued success is increasing the number of female administrators and tenured faculty."[15]

Cognizant of the fact that networking and continued support have been crucial in her career, Mahoney tries to provide strong mentoring to her own students, both male and female. "I can't imagine anyone succeeding without it," she said. Mahoney is married, with three daughters "to inspire and support," and says it is difficult to imagine anyone succeeding without financial support in the form of scholarships or fellowships, hard work, lots of luck—and in the case of married women, without a supportive husband and family. At this point in her career, Mahoney says she enjoys "dividing my time between administrative tasks and research and educational reform activities. . . . I understand and embrace my position and ability to influence positive changes in education and in people's lives."[16]

NOTES

1. Carolyn R. Mahoney, personal communications with author, including questionnaire of 5 January 1996, e-mail message of 5 November 1995. Mahoney, personal interview with author, GWSE Conference, Washington, D.C., 20–23 October 1995.

2. Mahoney, questionnaire. See also Patricia Clark Kenschaft, "Carolyn R. Mahoney," in *Black Women in America: An Historical Encyclopedia*, ed. Darlene Clark Hine, Elsa Barkley Brown, and Rosalyn Terborg-Penn, vol. 2 (Brooklyn, N.Y.: Carlson, 1993; Bloomington: Indiana University Press, 1994), pp. 743.

3. Mahoney, questionnaire.

4. Ibid.

5. Ibid.

6. Ibid.

7. Ibid.

8. Ibid.

9. Ibid.

10. Ibid.

11. Ibid.

12. Ibid.; see also Kenschaft, "Carolyn R. Mahoney," p. 743.

13. Mahoney, questionnaire.

14. Ibid.

15. Ibid.

16. Ibid.

SHIRLEY M. MALCOLM
Directing Changes in Science

Shirley M. Malcolm

[Minority and women scientists] may not look like past generations [of scientists] but they are those generations' heirs.

—*Shirley M. Malcolm*

Ecologist Shirley Mahaley Malcolm heads the American Association for the Advancement of Science's (AAAS) Directorate for Education and Human Resources Programs, is a member of the board of trustees of the Carnegie Foundation of New York, and until May 1998, served as a member of the National Science Board, the policy-making body of the National Science Foundation. She is a member of the President's Committee of Advisors on Science and Technology.[1] She oversees development and implementation of a wide range of programs designed to increase the participation of women, minorities, and the handicapped in the sciences and related fields. Strategically situated in Washington, D.C., she has testified before congressional and senate committees, has presented her views at public hearings, and has written and lectured—particularly within the scientific and intellectual community—on the complex issues related to increasing the participation of those who have historically felt they were outside the mainstream in the sciences.

Malcolm brings to her task the viewpoint of one who remembers quite well being on the outside, looking in. During her undergraduate studies at the University of Washington from 1963 to 1967, she was the lone Black zoology major in a very large department. Later, from 1967 to 1968, she was one of only three minority graduate students in the biological sciences at the University of California at Los Angeles. UCLA then had few female science faculty, and no minority faculty; there were only a handful of minority students in the labs and study sections Malcolm supervised as a teaching assistant.[2] "I thought these small numbers of minorities and women were peculiar to the institutions I attended. I found out later that the small numbers were not at all atypical."[3] When she surveyed the situation again in 1989, she found, to her dismay,

> In many colleges and universities now, more than twenty years later, it is as though time has stood still. We see more women faculty and students, but . . . the increased numbers of women are largely a by-product of their increased participation in higher education rather than their increased attraction to science and mathematics.[4]

Malcolm has worked tirelessly to change the status quo, yet acknowledges "there are entrenched ideas that are hard to combat."[5]

Malcolm's training as a scientist, as well as her experience as a science facilitator, has provided her with the authority to scoff at critics who claim that minority and female students don't have what it takes to make it in science. They then state that the very absence of women and minorities in science is evidence of their specious claim. She states emphatically,

> There is no evidence to support biological determinism . . . nor any genetic basis for some group's participation or non-participation/performance or achievement in science and mathematics. Complex social and cultural factors, the opportunity to learn, and the availability of resources have likely interacted to produce the current distribution of groups in the science fields. The existence of programs which can alter [the so-called] "inevitable" support the idea that there are many interventions we must exhaust before we are allowed to claim biology [as a reason].[6]

Instead of allowing critics to use nature as an (albeit questionable) excuse, Malcolm wants to let *nurture*—in the form of education, encouragement, opportunity, and support—foster the pursuit of science literacy and science careers in people from groups with historically weak traditions in science and its related fields. She has a good chance of achieving this goal—chemist Gloria Long Anderson, of Atlanta's Morris Brown University, describes Malcolm as "a fighter, a go-getter."[7]

In order to pursue her goal more fully, Malcolm forsook her own once-promising career in research to become a science facilitator and administrator. As she explained in a recent interview,

> I have a role in translation and bridge-building that's perhaps unique—the kind of things that are integral for one community to know about another. It is

essential that this be done today, and it involves the efforts of many people. . . . Yet . . . there is also the prevailing movement in some circles to divide and conquer. Those of us who span the different groups must find our natural allies, particularly in a policy climate that is terribly ugly. We must reach out to those who think there's value to diversity—find innovative things to emphasize the positive effects of our influence. I try to speak up for my friends who are Hispanic, Asian and Native American so that what has happened to African-American men cannot continue to happen to women and other minorities. . . . Somebody has to say these things.[8]

Malcolm believes America is sitting on a storehouse of knowledge: "We are right in the midst of a community and yet many don't see the problems in it—the Emperor is buck naked but they don't see it."[9]

Shirley Mahaley Malcolm was born in Birmingham, Alabama, in 1946, into what she herself has described as an "intact lower-middle-class family that was better off than many."[10] Her mother, Lillie Mae Funderburg Mahaley, had completed college training as a schoolteacher, although she stayed at home during much of the children's younger years; her father, Ben Lee Mahaley, who had completed two years of college under the GI Bill, worked as a meat-packer.[11] Although the family was fortunate enough to have a television during her childhood, Malcolm was already reading when she entered kindergarten at age four. As she recalled,

My family made this possible. We had books, and I read anything that was there. My reading was quite eclectic. I read a lot of mythology—when my grandfather died he left a bunch of books on Greek and Roman myths—and I read them all. My mother bought a set of encyclopedia, which I read faithfully, and I especially liked mysteries and fiction, and biographies of people like Marie Curie.[12]

Malcolm went directly from kindergarten to second grade, and she stressed the fact that her parents encouraged her. Later, her teachers would also provide strong encouragement.

Malcolm grew up in an all-Black community, where, she said, "one would have to travel a long distance to find white people."[13] She attended segregated schools, showed early signs of being gifted in math and science, and credits her sixth-grade science teacher, Mr. Smoot, with providing both interest and encouragement, not only to the gifted children but to all his students. According to Malcolm, Smoot attended National Science Foundation workshops and came back to the classroom energized—"that got us excited about the possibilities, in spite of our lack of staff and [equipment]."[14] "Other things seemed so subjective," Malcolm recalled, "but with science and math it felt like you could really know what the real deal was."[15] Miss Goddard, her seventh-grade teacher, also provided strong mentoring:

In those days, you could only teach if you were not married. Miss Goddard was married to the profession, and worked very hard to bring learning to all of us. There were more than 35 kids [in my class] and she worked endlessly, trying to push us forward.

I was lazy in some areas, but she tried to give me a sense of what excellence

was. Miss Goddard said, "Just because you are smart and can get by doesn't
mean you can sit back and take it easy. You have to start doing things for *you.*
Find someplace within you where you do it, whether anyone appreciates it or
recognizes it or not." Hers were words to live by.[16]

At some point during seventh grade, Malcolm remembers Goddard saying:
"Why am I holding you back? Just keep on [in your math book] until you run out
of things to work on."[17] Like everyone else in her community—parents, teachers,
classmates, neighbors—Goddard expected her to do well. "I was marked to succeed," Malcolm believes. "I would have had to perform an unnatural act to not
succeed."[18]

Malcolm's thoughts turned toward choosing a college, and she began to picture herself pursuing a career in medicine. Meanwhile, the impact of larger
events had begun to create changes. As Malcolm recalled,

> My home community *was* the civil rights struggle—remember, my generation
> had to study for the literacy test to vote. As a ten-year-old I helped. . . . We
> were on the cusp of the revolution and maybe to a certain extent that shaped
> who I am, along with Montgomery and Sputnik and the science education
> revolution.[19]

She graduated from Birmingham's George Washington Carver High School in
1963, with a perfect 4.0 grade point average.

Because she was only sixteen, Malcolm's parents were understandably reluctant to have her go off entirely on her own. Her mother had already returned to
teaching to help pay college costs for Malcolm's older sister, Sandra, who was
then, serendipitously, living with an aunt and uncle in Seattle, Washington. The
University of Washington had an excellent medical school with a highly rated
pre-med program; with close family nearby, Malcolm's parents felt better about
her living away from home. Additionally, as Malcolm recalled, Birmingham was
not among the safest places to be at that time: "Those were tough times in Birmingham. Mom's church, the 16th Street Baptist Church was bombed. . . . it was
a good time to be away."[20]

She entered the University of Washington, and while taking required pre-med
courses—organic and inorganic chemistry, physics, mathematics, biology—she
suddenly became aware of how unique she was:

> In spite of the fact that students from majors all across a campus of more than
> 25,000 converged in these classes, all too often I was the only Black student in
> my science classes, and one of only a few women. I never had a Black science
> professor at any of the three major research universities I attended.[21]

Even during the 1960s, according to Malcolm, medical schools maintained
quotas on the number of women admitted, so females had to perform better across
the board academically than their male classmates. Nevertheless, Malcolm was
a determined pre-med student until she encountered another mentor:

> Dr. Alan Kohn taught zoology and was my advisor. He said, "Why not academic
> science?" When he said that, it sort of gave me permission. [It was a] kind of gift

he gave me. [He was] the type of person who was always accepting of people, whatever different gifts they had.[22]

Malcolm completed her baccalaureate degree in zoology in 1967. She then entered the University of California at Los Angeles (UCLA).[23] Although her parents had paid for most of her undergraduate education, she did receive some financial relief by securing a resident assistantship for part of her time at the University of Washington and later earned an undergraduate teaching assistantship. At UCLA, she secured a teaching assistantship; at one point she had also applied for a minority fellowship, but because her grades were so competitive she earned a Regents Fellowship. She completed her master's degree in zoology in 1968.[24]

Malcolm taught high school for a few years, then returned to Birmingham for a while before deciding to pursue doctoral studies at Pennsylvania State University. Although the idea of studying for the doctorate had been in the back of her mind for some time, she made her actual decision too suddenly and too late in the year to qualify for funding. Essentially, she paid for Penn State by herself, as she recalled:

> I was a walk-on [at Penn State]. I eventually walked into the lab of H. B. Graves, [the man who would become] my mentor, to see if he would take me on. I had come for one of the major labs but he said. "You've had more animal behavior courses than we offer here."
>
> A good old boy from Mississippi who was not much older than I was, Graves was a wonderful mentor, just terrific. He taught animal behavior, and was brilliant, stimulating, and exciting. If I have anything now it's really because of him. He focused me on the research, and formed both an intellectual family and a family-type family. "Whatever you do I'll be proud of you," he said.[25]

While at Penn State, Malcolm also developed the kind of friendship with a fellow student, Vena Case, that can be so crucial to graduate students, allowing them to bounce ideas off one another and "talk shop."[26] She completed her Ph.D. in ecology in 1974.

Malcolm taught biology for a brief period at the University of North Carolina, before she accepted a position at the National Science Foundation's Minority Institutions Science Improvement Program (NSF/MISIP), which had been set up to help accelerate science development in schools with high minority and/or disadvantaged student enrollments.[27]

Malcolm's career with the American Association for the Advancement of Science has moved from research assistant to staff associate to project director of the AAAS's Office of Opportunities in Science, where she organized conferences and prepared publications that examined issues of minorities and women in science. During this period in Malcolm's career, it became abundantly clear to her that despite the science ideal of awarding position based solely on talent, differential opportunities routinely had been extended in science, independent of talent, throughout much of history. As Malcolm wrote in a 1990 article,

> Examples abound of where the tools needed to do science, the recognition within science, the opportunity for the study of science have all been handed out

based on considerations of race and gender. . . . But just as this country changed over time so did the community of scientists.

As the U.S. responded to the challenge raised with the launch of Sputnik, opportunities opened up which Black Americans seized. Black as well as white teachers attended summer institutes to improve their skills and content knowledge. They took advantage of opportunities to study science offered by the National Defense Educational Act and the National Science Foundation programs.

Often Black teachers attended all white institutions in the South in their summer science programs long before these institutions were officially integrated, since the NSF which funded these institutes insisted that they be integrated. Perhaps this speaks as strongly as anything about the power of federal policy and the need for federal support of equal opportunity goals.[28]

Yet, despite the national push toward science for all, underrepresentation among certain groups has made it all but impossible to play an effective game of catch-up, Malcolm believes.

The very fact of underrepresentation makes it difficult to change the numbers, for we are asking Black children (or girls, or Hispanics) to become something which they have not seen, which they may not believe it possible or important to be — a world peopled by exceptions.

Underrepresentation [also] makes it difficult to change the numbers because one can only become a scientist by the process of "apprenticeship" with a scientist. Someone must see within a student the potential to contribute that will inspire a commitment to mentor, to guide and to make the student into a scientist.[29]

Malcolm has long espoused the power of a strong communication network among minority women scientists, and notes that with but a few exceptions, minority women still remain outside prominent mainstream positions within science.[30] To help foster change, she sees her mission as "stirring the pot":

Our country has a lot to congratulate itself for, but those kinds of benefits have not extended to *all* people. What we owe to the American public is to solve problems. To keep us engaged with real people — the taxpayers who are the investors — we owe an accounting for what they have invested in us. I try to help stimulate and remind our [science] community of its responsibility for the next generation.[31]

If science is to fulfill its promise to future generations, if America is to maintain its technological and scientific hegemony, Malcolm says we must look increasingly to contributions from participants other than the traditional white male. She sees diversity within the community of scientists as a benefit to science:

I didn't, nor do other minorities — both majority women and minority men and women — in science leave our life experiences at the door of the lab. There is a set of experiences that define me, and this goes into my work.

Once, when I was doing my research I didn't feel it was going well, I didn't think it was going to work. I had to stop and try to look at the world the way my organism would. I suppose this is risky, in a way, but when you are at that edge you have nothing to lose.

In trying to look at the world the way my organism would, I tried to build on the way I had approached other problems in the past—when I had tried to see through other people's eyes. Black women are especially experienced at doing this—we not only look at the world through our own eyes, but we try to get into the heads of others and look at the world the way they would. . . .

 I think it makes a difference who does science and if we aren't aware of the difference diversity brings then we will be stuck in the same old rut. New types of people bring different perspectives, new knowledge to value.[32]

Believing that students from diverse backgrounds can do science, helping them to envision themselves as part of a scientific community by providing mentors and role models, and developing introductory courses that recruit students into the enterprise rather than weeding them out are among some of Malcolm's suggestions for bringing new blood into the scientific enterprise. In fact, Malcolm wants to make science more accessible to everyone:

 If this is our last shot at providing a future president or future congress-person with a view of science, what image do we want them to have? What do we want them to remember? I think we would agree that we want them to understand the quest, the dynamism and the energy of science, its place in our lives and its capacity to empower.[33]

In addition to numerous articles, Malcolm's publications include two books, *The Double Bind: The Price of Being a Minority Woman in Science* and *An Inventory of Programs in Science for Minority Students.*[34]

Malcolm is the mother of two daughters, and is married to a research physicist who works at Johns Hopkins University's Applied Physics Laboratory. Recently she heard one of her daughters asking her father why they had such an "upside down family"—"Mom watches Monday Night Football," their daughter said, "and you sew your own pants." Laughing, Malcolm recounted her husband's answer: "What makes you think this is upside down?" he asked.[35] Shirley Malcolm's philosophy of who ought to be inside the community of scientists would surely echo her husband's reply to their daughter.

NOTES

 1. Shirley M. Malcolm, interview with author, 10 January 1996, Omni Berkshire Hotel and J. Sung Dynasty Restaurant, New York City. Malcolm, personal communications by mail, including curriculum vitae, partially completed questionnaire, and "Research Universities in the United States," 6 January 1996. Malcolm, e-mail communications with author, 2 January 1996, 4 January 1996, 16 January 1996, 19 January 1996, 17 April 1996, 27 April 1996, 22 July 1998. I am quoting the 22 July 1998 e-mail. Malcolm's position on the National Science Board resulted from an appointment by President Bill Clinton, and was confirmed by the U.S. Senate. President Clinton also appointed her to serve on the executive committee of the National Science Board, and as a member of the president's committee of advisors on science and technology (Malcolm, curriculum vitae).

 2. Shirley M. Malcolm, "The Unfinished Agenda" (unpublished manuscript, 1989). I received this manuscript from Malcolm on January 4, 1996. The other minority students were all female: two Blacks, including Malcolm, and one Hispanic.

 3. Malcolm, "The Unfinished Agenda."

 4. Ibid.

5. Malcolm, personal interview.

6. Ibid. (emphasis added).

7. Gloria Long Anderson, personal interview with author, Morris Brown College, Atlanta, Ga., 19 November 1995.

8. Malcolm, personal interview.

9. Malcolm, personal interview.

10. Malcolm, questionnaire and personal interview. See also Yolanda Scott George, "The Status of Black Women in Science," *The Black Collegian*, May/June 1979, p. 70. "Ecologist," in *Contributions of Black Women to America*, ed. Marianna W. Davis, vol. 2 (Columbia, S.C.: Kenday Press, 1982), pp. 430–431.

11. Malcolm, questionnaire and personal interview. Lillie Mae Mahaley died in 1989; Ben Lee Mahaley is still living (Malcolm, e-mail communication, 22 July 1998).

12. Ibid. Malcolm's kindergarten schooling was not free—her family paid for it.

13. Ibid.

14. Ibid.

15. Ibid.

16. Ibid.

17. Ibid.

18. Ibid.

19. Ibid. Malcolm was referring to the Montgomery boycott and march.

20. Ibid.

21. Malcolm, "The Unfinished Agenda."

22. Malcolm, questionnaire.

23. Malcolm, e-mail communication, 22 July 1998.

24. Ibid.

25. Malcolm, questionnaire. H. B. Graves is now chief scientist at a Washington, D.C. law firm (Malcolm, e-mail communication, 22 July 1998).

26. Vena Case, who is white, is now head of the biology department at Davidson College (Malcolm, questionnaire).

27. "Ecologist," pp. 430–431; George, "The Status of Black Women," p. 70.

28. Malcolm, "Reclaiming Our Past."

29. Ibid.

30. "Ecologist," p. 431.

31. Ibid.

32. Malcolm, personal interview.

33. Malcolm, "The Unfinished Agenda."

34. *The Double Bind: The Price of Being a Minority Woman in Science* (Washington, D.C.: American Association for the Advancement of Science, 1976). *An Inventory of Programs in Science for Minority Students, 1960–1975* (Washington, D.C.: American Association for the Advancement of Science, 1976).

35. Malcolm, personal interview.

JESSIE JARUE MARK*

An Unsung Path-Breaker

Born in Apple Springs, Texas, in 1906, botanist and plant physiologist Jessie Jarue Mark became the first Black women scientist on the faculty at Iowa State University, a majority white institution.[1]

Mark earned the baccalaureate degree from Prairie View State College, a historically Black institution in Texas, in 1929; she earned her master's and doctoral degrees from Iowa State University in 1931 and 1935, respectively. Upon receipt of the master's degree, Mark became a professor at Kentucky State Industrial College in 1931, while continuing her studies for the doctorate. Her dissertation was "Relation of Root Reserves to Cold Resistance in Alfalfa."[2]

NOTES

1. Six years of research have yielded only the barest information on this path-breaking researcher. Mark worked for some time at the Iowa Agricultural Experiment Station while she was on the faculty at Iowa State University. To date, I have been unable to obtain the titles of any of her publications, but did find a reference to her work in A. R. Langrille, "Seasonal Variation in Carbohydrate Root Reserves and Crude Protein and Tannin in Crownvetch Forage *Coronilla varia* L.," *Agronomy Journal* 60 (1968): 415.

2. *American Men of Science: A Biographical Directory,* 7th ed. (New York: Science Press, 1953), p. 1167. Harry Washington Greene. *Holders of Doctorates among American Negroes: An Educational and Social Study of Negroes Who Have Earned Doctoral Degrees in Course, 1876–1943* (Boston: Meador Publishing, 1946), p. 129. James M. Jay, *Negroes in Science: Natural Science Doctorates, 1876–1969* (Detroit: Belamp, 1971), p. 60. Vivian Ovelton Sammons, *Blacks in Science and Medicine* (New York: Hemisphere, 1990), p. 161.

VIVIENNE LUCILLE MALONE MAYES
Mathematical Legacy

Born in Waco, Texas, around 1930, Vivienne Lucille Malone Mayes attended the city's segregated public schools, which she has described as "strictly separate and strictly unequal."[1] Despite this, Mayes has said, that "in every Black school I've attended there has always been at least one Black woman teacher or professor with whom I could identify."[2]

After her graduation from high school in 1948, Mayes enrolled as a mathematics major at Fisk University in Nashville, Tennessee. In 1950, her junior year, the university hired two mathematics professors—Lee Lorch as department chair, and Evelyn Boyd Granville—both of whom had a tremendous impact on Mayes. Granville was at Fisk only until 1952, and Lorch was there only until 1955. In that short time, however, they both mentored Mayes, as well as two other Black women students who went on to earn doctorates in mathematics, Etta Zuber Falconer and Gloria Conyers Hewitt.[3] Lorch also taught another woman, Joan Murrell Owens, who minored in mathematics and later earned a doctorate in geology/marine biology.[4]

According to Mayes, Lorch had a particular gift for recognizing and nurturing talented students. Almost immediately after his arrival at Fisk, he hired her as his grader, even though she was still an undergraduate. When she had completed the baccalaureate degree in 1952, he convinced her to remain with him at Fisk for graduate work. She received an assistantship and earned the M.A. in mathematics in 1954. Lorch encouraged her to continue on for the doctorate, but money was a problem.[5]

After earning her master's degree, Mayes took a position as an associate professor of mathematics at Paul Quinn College, where she remained until 1961. Then, with an American Association of University Women fellowship (1964–1965), she enrolled at the University of Texas at Austin.

Although women had outnumbered men in mathematics classes at Fisk University and at Paul Quinn College, the situation was quite different at the University of Texas. In her first class, Mayes was the only Black and the only woman: "For nine weeks thirty or forty white men ignored me completely. . . . It seemed to me that conversations before class quickly terminated if it appeared that I was listening. . . . My mathematical isolation was complete."[6]

Mayes's sense of isolation at the University of Texas was compounded even further when one professor adamantly refused to allow Blacks to attend his classes. She was not permitted to hold a teaching assistantship because of her race, and she could not meet off campus with her advisor and other math majors because the coffee shop they favored did not, by Texas law, serve Blacks.[7] Mayes didn't realize until after the law had been changed that women were rarely included in such conversations, no matter what their color.[8] She completed her Ph.D. in 1966; her dissertation, "A Structure Problem in Asymptotic Analysis," was done under the direction of Don E. Edmondson.[9]

In the fall of 1966, Mayes became an assistant professor of mathematics at Baylor University, a majority white institution in Waco, Texas. In 1967 and 1968, she received Research College Teachers summer study grants, which helped her prepare material for publication, and she quickly moved up in rank to full professor.[10] In 1971, she was named Baylor University's outstanding faculty member of the year by the student congress.[11]

NOTES

1. Vivienne Malone Mayes, "Black and Female," *Association for Women in Mathematics Newsletter* 5 (1975): 4–6. See also Vivian Ovelton Sammons, *Blacks in Science and Medicine* (New York: Hemisphere, 1990), p. 165. "Mathematicians," in *Contributions of Black Women to America*, ed. Marianna W. Davis, vol. 2 (Columbia, S.C.: Kenday Press, 1982), pp. 457–458. Patricia Clark Kenschaft, "Black Women in Mathematics in the United States," *American Mathematical Monthly* 88 (October 1981): 599–600. Kenschaft, "Black Women in Mathematics in the United States," *Journal of African Civilizations* 4 (April 1982): 63–83. Kenschaft, "Blacks and Women in Mathematics," paper presented at the Science and Technology Seminar, City University of New York, 3 March 1983, in *Educational Policy Seminar Papers* (New York: City University of New York, Office of Special Programs, 1986), pp. 22–23. Virginia K. Newell, et al., eds., *Black Mathematicians and Their Works* (Ardmore, Penn.: Dorrance, 1980), pp. 192–157, 290. *National Faculty Directory* (Detroit: Gale Research, 1988), p. 2448. Note: none of the above sources list a date of birth for Mayes.

2. Mayes, "Black and Female."

3. See my entry on Evelyn Boyd Granville in this volume.

4. Lee Lorch is discussed in the Granville entry. Also see the entry on Joan M. Owens in this volume.

5. Vivienne M. Mayes, "Lee Lorch at Fisk: A Tribute." *American Mathematical Monthly* 83 (November 1976): 708–711.

6. Mayes, "Black and Female," p. 5.

7. Ibid. See also Kenschaft, "Black Women in Mathematics in the United States," 1982, pp. 70–71. Blacks in Austin, mainly students from the university, picketed Hilberg's Cafe. Mayes said, however, that after the law had been changed and she was allowed inside she realized that sexism was also an obstacle—women, regardless of color, were rarely included in "shop talk" conversations (ibid.).

8. According to Mayes, there was only one other female student in her graduate courses (ibid.).

9. Sammons, *Blacks in Science and Medicine*, p. 165. See also "Black Women in Mathematics in the United States," 1982, pp. 70–71.

10. Sammons, *Blacks in Science and Medicine*, p. 165. See the appendix for a partial list of Mayes's publications.

11. Kenschaft, "Black Women in Mathematics in the United States," 1982, p. 71.

SHIRLEY MATHIS McBAY

Focusing on Minority Issues in Science

Mathematician and chemist Shirley Mathis McBay was born in Bainbridge, Georgia, in 1935. She earned the baccalaureate degree in chemistry, summa cum laude, at Paine College in 1954.[1] Between 1955 and 1963, McBay was a chemistry instructor at Spelman College. During this period, she attended Atlanta University, where she earned two master's degrees: her M.S. in chemistry in 1957, and her M.S. in mathematics in 1958.[2] In 1964, she was awarded a United Negro College Fund fellowship, sponsored by the IBM Corporation, to study for the doctorate at the University of Georgia.[3] She received the Ph.D. in mathematics in 1966; her dissertation, done under the supervision of Thomas R. Brahana, was "The Homology Theory of Metabelian Life Algebras."[4]

Immediately after earning the doctorate, McBay returned to Spelman College as an associate professor of mathematics.[5] By 1970, she was also chair of the mathematics department.[6] According to McBay's colleague, Etta Zuber Falconer, in the late 1960s "the science faculty [at Spelman College] began questioning the low production of science graduates and the 'perceived' low status of science at the College."[7] Falconer further stated that the "situation would have continued if the science faculty under the leadership of Dr. Shirley McBay had not presented a strong case."[8] By Falconer's own account, she, McBay, and Gladys Glass were the "pioneers who built the Mathematics Program."[9] As a result of McBay's leadership, in 1972 Spelman College created the Division of Natural Sciences, with McBay as its chairman, to develop and implement her plans for increasing the science emphasis at the college.[10]

To increase the number of students enrolled as science majors, McBay instituted a pre-freshman summer program for students interested in becoming science majors. The college also renovated the science building and had an annex built to house a biochemistry department. As student interest increased, a department of chemistry was created in 1976. McBay served as chairman of the Division of Natural Sciences until 1975, and also as associate academic dean at Spelman from 1973 to 1975.[11] She left Spelman in 1975 and took a position at the National Science Foundation, where she later became program director of the Minority Institutions Science Improvement Program.[12] In 1980 she became dean of student affairs at the Massachusetts Institute of Technology.[13]

According to friend and former colleague, chemist Gloria Long Anderson, McBay's advanced education and career—particularly during the years she was in Atlanta—were nurtured by Henry Cecil McBay.[14] Anderson also said that Shirley McBay always had a very strong interest in minority issues in science, and she reiterated Falconer's statement that McBay was "the one who really started the whole Spelman turnaround."[15] McBay is now president of the Quality Education for Minorities Network, in Washington, D.C.[16] Although little information is available on McBay's research interests, if any, she has been described as an effective teacher who recruited many Black women into science careers during her tenure at Spelman College.[17]

NOTES

1. Paine College is a historically Black college in Augusta, Georgia.

2. Vivian Ovelton Sammons, *Blacks in Science and Medicine* (New York: Hemisphere, 1990), p. 167. *Who's Who among Black Americans* (Northbrook, Ill.: Who's Who among Black Americans Publishing, 1985), p. 558. Virginia K. Newell et al., eds., *Black Mathematicians and Their Works* (Ardmore, Penn., 1980), p. 290. Etta Z. Falconer, "A Story of Success: The Sciences at Spelman College," *Sage* 6 (fall 1989): 36–38. Patricia C. Kenschaft, "Black Women in Mathematics in the United States," *Journal of African Civilizations* 4 (April 1982): 73. "Mathematicians," in *Contributions of Black Women to America*, ed. Marianna W. Davis, vol. 2 (Columbia, S.C.: Kenday Press, 1982), pp. 456–457.

3. Advertisement for the United Negro College Fund sponsored by the IBM Corporation. *Ebony*, November 1975, p. 10. The text of the ad reads, in part: "Shirley McBay's toughest equation: $E + S+ \$ = Ph.D.$ [where E stands for enthusiasm; S stands for skill; and $ stands for funds]. . . . But this was one equation Shirley McBay couldn't solve herself. Because it takes more than enthusiasm and skill to earn an advanced degree. It takes money, too. The United Negro College Fund recognized Shirley McBay's potential. But it did more than that. UNCF awarded her an IBM Faculty Fellowship grant. With the help of that grant, Shirley McBay went on to a doctoral degree" (ibid.). According to Sammons, McBay was also an assistant professor of mathematics at the University of Georgia during the period of her doctoral study (1965–1966), but this may be a misprint. It seems more likely that McBay was a teaching or research assistant during this period, as it was rare for someone without the doctorate to be awarded the rank of assistant professor at a major university.

4. Kenschaft, "Black Women in Mathematics," p. 73. See also Sammons, *Blacks in Science and Medicine*, p. 167, and "Mathematicians," pp. 456–457.

5. Kenschaft, "Black Women in Mathematics," p. 73. See also Sammons, *Blacks in Science and Medicine*, p. 167; "Mathematicians," pp. 456–457; and *Who's Who among Black Americans*, p. 558.

6. "Faculty and Staff Notes," *Spelman Messenger*, February 1971, p. 24.

7. Falconer, "A Story of Success," p. 36. According to Falconer, "there was an abun-

dance of evidence to support the premise the Spelman students were not seriously encouraged to pursue the sciences: the science building was dark and uninviting; there was a lack of emphasis on science and health careers and little recognition of scientists and their contributions to society; the role of women in science and engineering was not discussed in classes, or presented in College publications; major classes in the sciences beyond the freshman level had tiny enrollments; major classes were offered only in biology and mathematics—the only chemistry course was a service course for majors in home economics and physical education—the only physics course had been deleted from the curriculum before 1960; [and] science was perceived by the general student population as difficult and uninteresting" (ibid.).

8. Ibid.

9. Ibid.

10. Sammons, *Blacks in Science and Medicine*, p. 167.

11. Kenschaft, "Black Women in Mathematics," p. 73; Sammons, *Blacks in Science and Medicine*, p. 167.

12. "Mathematicians," pp. 456–457.

13. Kenschaft, "Black Women in Mathematics," p. 73.

14. Gloria Long Anderson, interview by author at her office at Morris Brown College, Atlanta, Ga., November 1995. Chemist Henry Cecil Ransom McBay (B.S., Wiley College 1934; M.S., Atlanta University 1936; Ph.D., University of Chicago 1945) was a professor of chemistry at Morehouse College in Atlanta from 1945 until 1981. It is not clear exactly how and where Henry McBay mentored Shirley McBay, but they were married and later divorced. I have been unable to obtain the dates regarding their marriage. For more information on Henry McBay, see the entry for Gloria Long Anderson in this volume.

15. Anderson, interview.

16. Ibid. This was confirmed via e-mail by Shirley Malcolm of the American Association for the Advancement of Science. McBay has not responded to repeated requests for information.

17. "Mathematicians," pp. 456–457. "Mention the name Shirley McBay as you talk with any young woman who attended Spelman College between 1955 and 1975, and there is immediately a response of body language and verbal expressions of academic respect. 'A great teacher, tough but fair,' they say with pride" (p. 456).

F. PEARL McBROOM

Research Physician

A specialist in heart disease research, physician F. Pearl McBroom was born in Louisville, Mississippi, in 1926. She earned a bachelor of arts degree at the University of Chicago in 1946, and a bachelor of science degree at Columbia University in 1949. McBroom earned her M.D. degree at the Columbia University College of Physicians and Surgeons in 1953.[1] She interned at Bellevue Medical Center in New York (1953–1954), and did her residency at Columbia University's Goldwater Hospital (1954–1955). After completing her residency, McBroom moved to California.

McBroom performed research in cardiology at the University of California at

Los Angeles Medical Center, where she was the first Black physician to receive an appointment. From 1957 to 1958, she was a Fellow in Cardiology at the University of Southern California.[2]

From 1958 to 1962 McBroom held a series of National Institutes of Health research grants, and in 1960 she developed a new method for observing changes in coronary blood vessel tissues that were affected by hardening of the arteries.[3] From 1962 to 1963, with her husband Dr. Marcus S. W. McBroom, she conducted a two-year study on the effects that varying levels of male and female hormones have on the incidence and severity of coronary heart disease (this research was supported by a $17,000 U.S. Public Health Service grant).[4] Between 1962 and 1985, McBroom preformed independent research in cardiovascular and preventive medicine.[5]

NOTES

1. Vivian Ovelton Sammons, *Blacks in Science and Medicine* (New York: Hemisphere, 1990), p. 167. Note that Sammons lists McBroom as a male; a 17 May 1996 telephone call to Dr. McBroom confirmed that she is, indeed, female. See also *Who's Who among Black Americans*, 4th ed. (Northbrook, Ill.: Who's Who among Black Americans Publishing, 1985), p. 558. *Who's Who of American Women* (Chicago: Marquis Who's Who, 1975), p. 624. "Outstanding Women Doctors: They Make Their Mark in Medicine," *Ebony*, May 1964, pp. 68–69, 72–74, 76. People, *Jet*, 11 February 1960, p. 18.

2. "Outstanding Women Doctors."

3. Sammons, *Blacks in Science and Medicine*, p. 167.

4. "Outstanding Women Doctors."

5. *Who's Who among Black Americans.*

DOROTHY McCLENDON*

Born in Minden, Louisiana, and raised in Detroit, Michigan, microbiologist Dorothy McClendon graduated from Detroit's Cass Technical High School, and earned a B.S. degree in biology at Tennessee A&I State University. She then took graduate courses at Wayne State University, Purdue University, and the University of Detroit.

Before she began a long career coordinating microbiology research for the U.S. Army Tank Automotive Command (TACOM) in the early 1960s, she taught public school in Phoenix, Arizona, and Eldorado, Arkansas. With TACOM, she was a specialist in the types of organisms that can contaminate solids and liquids and cause decay or a loss of integrity. McClendon's research ranged from developing methods to prevent fuel contamination by various microorganisms to developing chemical fungicides to protect a variety of stored materials without causing harm to military personnel.[1]

NOTES

1. "Microbiologist," in *Contributions of Black Women to America*, ed. Marianna W.

Davis, vol. 2 (Columbia, S.C.: Kenday Press, 1982), p. 431. Also, Hattie Carwell, *Blacks in Science: Astrophysicist to Zoologist* (Hicksville, N.Y.: Exposition Press, 1977), p. 44.

RUTH ELLA MOORE*
First Black Woman to Hold an Earned Natural Science Ph.D.

Although bacteriologist Ruth Ella Moore has the distinction of being the first Black woman in the United States to hold an earned doctorate in a natural science, little has been written about her. Born in 1903, she obtained the bachelor's degree from Ohio State University in 1926 and the master's degree from that school in 1927. In 1933, Ohio State awarded her the doctorate in bacteriology; her dissertation was "Studies on Dissociation of Mycobacterium Tuberculosis: A New Method of Concentration on the Tubercule Bacilli as Applied to Sputum and Urine Examination."[1]

From 1940 to 1951, Moore held the rank of assistant professor of bacteriology at Howard University Medical College in Washington, D.C.[2] In 1952, although she was still ranked only as an assistant professor, she was made head of the department of bacteriology, a position she held until 1957.[3] Although she had achieved the rank of associate professor, she stepped down as department head in 1957.[4] The records indicate a gap in her career during the years from 1959 to 1971, at which time she returned to Howard to teach bacteriology on a part-time basis.[5] During her tenure at Howard, Moore conducted research on blood grouping and Enterobacteriaceae.

NOTES

1. Harry W. Greene, *Holders of Doctorates among American Negroes: An Educational and Social Study of Negroes Who Have Earned Doctoral Degrees in Course, 1876–1943* (Boston: Meador, 1946), pp. 193–194. "The American Negro in College, 1932–1933," *The Crisis*, August 1933, p. 181 (picture only).

2. *Howard University Bulletin, Personnel Register: 1940–41*, 20, no. 4 (15 October 1940), p. 13. *Howard University Bulletin, Personnel Register*: 23, no. 2 (15 July 1943), p. 14. *Howard University Bulletin, Personnel Register: 1944–45*, 24, no. 3 (1 October 1944), p. 12. *Howard University Bulletin, Personnel Register: 1946–47*, 26, no. 3 (15 October 1946), p. 15. *Howard University Bulletin, Personnel Register: 1948–49*, 28, no. 2 (15 October 1948), p. 21. *Howard University Bulletin, Personnel Register: 1949–50*, 29, no. 2 (15 October 1949), p. 20.

3. Ibid. See *Personnel Register* for years 1952 through 1957.

4. Ibid. See *Personnel Register* for years 1958 through 1960.

5. Sammons indicates that Moore returned to part-time teaching at Howard in 1971. See Vivian Ovelton Sammons, *Blacks in Science and Medicine* (New York: Hemisphere, 1990), p. 176. This is most probably an error, because Moore would have been sixty-eight or sixty-nine years old at the time. Moore is not listed in the *Personnel Register* for 1971–1972, but this could have been a simple omission. See also *American Men and Women of Science: The Physical and Biological Sciences*, 12th ed. (New York: Bowker, 1971), p. 4370.

LENORA MORAGNE

Nutrition Expert

Nutrition scientist Lenora Moragne was born in Evanston, Illinois, in 1931.[1] She earned her baccalaureate degree in nutrition at Iowa State University sometime around 1953.[2] Between 1955 and 1957, she was the chief dietitian at Community Hospital in Evanston, Illinois. In 1959, she was awarded a research fellowship at Cornell University, and she earned her M.S. degree in nutrition 1963.[3]

Between 1965 and the end of the spring 1967 semester, Moragne was an assistant professor of foods and nutrition at North Carolina College. In the fall of 1967, she received a second research assistantship at Cornell, where she earned the Ph.D. in 1969; her dissertation was "Influence of Household Differentiation on Food Habits among Low-Income Urban Negro Families."[4] While working toward the doctorate at Cornell, Moragne was also a nutrition publicist for the General Foods Corporation in White Plains, New York (1968–1971).[5]

Moragne's most productive research years were between 1959 and 1963, during her time as a research assistant at Cornell University. Her primary focus was on the effects of temperature on foods;[6] her name appears first on all the publications, indicating that she was the principal investigator.[7] From 1971 to 1972, Moragne was a nutrition lecturer at Columbia University, and at the same time, she was a professor of foods and nutrition at Hunter College of the City University of New York, and an adjunct professor of foods and nutrition at Herbert H. Lehman College in the Bronx, New York. In late 1972, she joined the Food and Nutrition Service of the United States Department of Agriculture in Washington, D.C., where she became head of the Division of Nutrition Education and Training, a position she held until 1977. In 1974, she published two books—one was a junior high school nutrition textbook and the other was on infant care.[8]

Moragne was appointed to the staff of the Agriculture, Nutrition, and Forestry Commission of the United States Senate in 1977. In 1979, she became coordinator of nutrition policy for the Commission's Health and Human Services Division. She was also a member of the Board of Directors of the American Dietetic Association (1981–1984).[9]

NOTES

1. Vivian Ovelton Sammons, *Blacks in Science and Medicine* (New York: Hemisphere, 1990), p. 176. Information on Moragne's early life and schooling was not available. See also *Who's Who among Black Americans* (Northbrook, Ill.: Who's Who among Black Americans Publishing, 1985), p. 608. George L. Jackson, *Black Women Makers of History: A Portrait* (Sacramento, Calif.: Fong and Fong, 1977), pp. 198–199. Lenora Moragne, "Black Women Marketing Executives," *Black Business Digest*, February 1973.

2. None of the above-cited sources lists the exact date of her B.S. degree.

3. I determined the approximate date from the dates of the Cornell research fellowship (Jackson, *Black Women Makers*, pp. 198–199).

4. Sammons, *Blacks in Science and Medicine*, p. 176.

5. Jackson, *Black Women Makers*, pp. 198–199; *Who's Who among Black Americans*, p. 608.

6. See the appendix for a representative list of publications.

7. Jackson states that Moragne was an "Assistant Professor of Institution Management at Cornell University," although he doesn't give the exact dates; and *Who's Who among Black Americans* lists her as an assistant professor of foods and nutrition at Cornell University between 1961 and 1963, the years she was studying for the master's degree. I believe it is more likely that she was a research assistant at Cornell during this period, primarily because it does not seem likely that a major university like Cornell would have appointed a person with only the B.S. to a faculty position. With regard to her research, one of her collaborators at Cornell, Karla Longree, cited her work in a few publications, even as late as 1964 when Moragne had moved on (see the appendix).

8. See the appendix.

9. *Who's Who among Black Americans*, p. 608.

DIANE POWELL MURRAY

Crunching Numbers to Command Space Satellites

Diane Powell Murray

One of seven women profiled in the National Science Foundation film, *Science: Woman's Work*, mathematician Diane Powell Murray moved steadily up the research and corporate ladder at TRW, a major American developer of intelligence systems for the military and other venues. She served as manager of major intelligence support programs in the Integrated Engineering Division of the TRW Systems Integration Group, overseeing work contracts valued at over $75 million. She worked on the CAMS project, an intelligence-gathering system that could process information from over a thousand users at more than fifty sites worldwide. Her responsibilities with the company's Intelligence Support Programs have involved the developing of strategies and acquisition plans for new business pursuits, providing systems engineering, systems analysis, and applications development support to government customers, serving as the primary interface with government program management, and representing TRW in project contract negotiations.[1]

A 1975 graduate of Spelman College (B.A., mathematics), Murray earned the master of science degree in operations research at Cornell University in 1976. She began her career at the Eastman Kodak Company (January 1976 to June 1977). She then joined TRW in August of 1977, where she has worked over the last twenty years in areas of systems engineering, software development, and project management.[2]

Murray credits her mathematics training at Spelman as providing the crucial intellectual background for her success:

> One of the reasons math majors go into law, medicine, journalism and so on is that math trains you to be very analytical and it trains you to be very precise in your thinking. As a result, you have a better chance to excel in a variety of fields.[3]

Murray has participated in or managed projects ranging from the command and control of space satellites, to earthquake detection, to developing an architectural description of radioactive waste management. For three years (1977–1980), she was a member of the company's Missile Application Program, where she analyzed software requirements and studied algorithm designs of a trigonometric and geographic problem for determining "brother-killing" conflicts in the missile application. Between 1981 and 1986, she managed programmatic and technical specification and design reviews of a subcontracted synthetic seismogram generator, and did systems analysis and software development on strategic and space defense projects. By 1986 she had become the manager of a department of sixty professional and technical personnel involved in software engineering, executive control system software, software productivity, and reuse software development for a series of Department of Defense contracts. In 1990, she was providing technical support to the Earth Observing System Satellite and Information System; from 1991 to 1992 she supplied technical support to a development of an architectural description of the United States Department of Energy's Office of Civilian Radioactive Waste Management's Integrated Office Automation and Records Management System; and from 1991 to 1994 she was involved in software engineering to develop, integrate, and test a variety of U.S. government intelligence programs.[4]

In a National Science Foundation film made around 1980, Murray's husband, psychologist Garrett Murray, spoke about how domestic partnerships can be crucial to both personal happiness and professional success: "Many of the men I know . . . inevitably feel they will be neglected. . . . But this is not a problem—we *share* the household responsibilities."[5] Murray is the mother of one child, and has been actively involved in community and church service throughout her career. She has served as a district coordinator for the Boy Scouts of America, and has been a TRW-appointed member of the Carson California Child Guidance Advisory Committee, which developed strategies to solve a variety of problems confronting local school children and their families. She has also taught math in the Los Angeles Upward Bound Program, and participated in career day activities at local public schools, both in the Maryland/Washington, D.C., area and in California. She received the Candace Award in New Technologies from the National Coalition of 100 Black Women, and an award for excellence in science and technology from the Spelman College Alumnae Association. TRW has nominated her for the Black Engineer of the Year Award for Professional Achievement.[6]

NOTES

1. Diane P. Murray, personal communications: job description, résumé; summary of activities, 27 April 1996. The *Science: Woman's Work* video (VHS, 28 min.), produced by the National Science Foundation, is available through Media Design Associates, Inc., P.O. Box 3189, Boulder, Colo. 80307 (303) 443-2800. Oddly, no publication date appears either in the video itself nor on the jacket. Based on style of dress, male hair styles —which were more striking at that time—and various textual comments, I very loosely estimate the video was made around 1980. At one point, for example, a woman scientist talks about men being "male chauvinist pigs," which would appear to be a 1960s comment, yet at another point in the video the narrator makes a comment about the state of science and math at Spelman College in the "early 1970s."

2. Murray, personal communications.
3. Murray, from *Science: Woman's Work*.
4. Murray, résumé.
5. Garrett Murray, in *Science: Woman's Work*.
6. Murray, summary of activities.

MARION ANTOINETTE RICHARDS MYLES*

Botanist/plant physiologist Marion Antoinette Richards Myles was born in Philadelphia, Pennsylvania, in 1917. Information is scant on her early life, schooling, and career. She earned the baccalaureate degree at Pennsylvania University in 1937, and the master's degree at Atlanta University in 1939. From 1941 to 1943, she was head of the biology department at Philander Smith College, a historically Black institution. During her time at Philander Smith, she apparently attended Iowa State University during at least one summer session, and then in 1943, was awarded a research fellowship in plant physiology by Iowa State to work on her doctorate. Botanist/plant physiologist Jessie Jarue Mark was the only Black woman on the faculty at Iowa State at the time, so it is likely that Myles took some courses with her. Myles held the fellowship until 1945, when she was awarded the Ph.D.

Although her degree was in botany/plant physiology, Myles's career was somewhat eclectic. From 1945 to 1948, she was an associate professor of biology at Tennessee A&I State University, and from 1948 to 1951 she was an associate professor of agronomy there. She moved on to Fort Valley State College in 1951, where she was professor of botany and zoology and acting head of the science division until 1959. During this period, she was awarded a Carnegie Foundation research grant (1952–1954), and was also head of Fort Valley's Division of Science and Mathematics (1953–1954). From 1959 to 1961, she was a research associate in enzymology at Vanderbilt University; and in 1965, she became an assistant professor of pharmacology at the Medical Center of Mississippi. During her career Myles also did research at the California Institute of Technology, and participated in nuclear studies specializing in plant physiology at the Oak Ridge Institute in Tennessee.[1]

NOTE

1. James M. Jay, *Negroes in Science: Natural Science Doctorates, 1876–1969* (Detroit: Belamp, 1971), p. 179. *American Men of Science*, 11th ed. (New York: Bowker, 1968), pp. 3816–3817. *Leaders in American Science*, 2nd ed. (Nashville, Tenn.: Who's Who in American Education, 1955), p. 495. "Plant Physiologist," in *Contributions of Black Women to America*, ed. Marianna W. Davis, vol. 2 (Columbia, S.C.: Kenday Press, 1982), pp. 431–432.

ANN T. NELMS*

In 1954, Howard University graduate Ann T. Nelms was a "25-year-old nuclear physicist at the Bureau of Standards."[1] Nelms had been "cited for writing a book of technical mathematics physics formulas," which was "being used extensively by the Atomic Energy Commission to speed up its experiments."[2] It would be interesting to know the types of experiences Nelms had as an early female physics or mathematics major at Howard University and later during her professional career at the Bureau of Standards.

NOTES

1. "Atomic Mathematician," *Ebony*, May 1954, p. 5.

2. Ibid. The book was Ann T. Nelms, *Graphs of the Compton Energy-Angle Relationship and the Klein-Nishina Formula from 10 KEV to 500 MEV* (Washington, D.C.: U.S. Government Printing Office, 1953). Her other book is *Energy Loss and Range of Electrons and Positrons* (Washington, D.C.: U.S. Government Printing Office, 1956).

BARBARA J. ODEN
Zoologist at West Virginia State

Barbara J. Oden
Photo courtesy of the West Virginia State College Office of News Services

Zoologist Barbara J. Oden is currently vice-president for academic affairs at West Virginia State College. She earned the baccalaureate degree, with a major in biology and a minor in French, at Tougaloo College in 1953. She then taught high school biology in Utica, Mississippi from 1953 to 1955, when she entered the master's program at the University of Wisconsin at Madison; she received the M.S. degree in invertebrate zoology in 1957 with her master's thesis, "Cytoplasmic Streaming in the Axopodia of *Actinosphaerium eichornii*."[1]

Immediately after she had earned her master's degree, Oden became a professor of biology at West Virginia State University, a historically Black institution in Institute, West Virginia. In 1959, she received a National Science Foundation summer grant to study at the Oregon State University in Corvallis, Oregon; she received another summer grant in 1961 to study at the Theodore Franze Stone Laboratory at Ohio State University, in Columbus. From 1963 to 1964, she took sabbatical leave for further graduate study at Ohio State. In 1977, twenty-four years after earning the baccalaureate degree, she was awarded the Ph.D. in invertebrate zoology by the University of South Carolina at Columbia; the research for her dissertation, "Temporal and Spatial Variations among Fresh-Water Littoral Meiofauna in a Reservoir Receiving Thermal Effluents," was performed at Par Pond, in Aiken, South Carolina.

Oden continued as a professor of zoology at West Virginia State until 1987, when she was named assistant vice president for academic affairs at the college. She was named vice president for academic affairs in 1990.[2]

NOTES

1. Barbara J. Oden, curriculum vitae, received from Oden on a visit to West Virginia State College, Institute, W.Va., 18 March 1996.

2. Ibid.

JOAN MURRELL OWENS
Becoming a Scientist at Mid-life

Joan Murrell Owens, pictured with her late husband, Frank

Never give up your dream, in spite of the obstacles. There have been several points in my life where a door to my planned career closed and another, totally different door, opened. Though one door may seem to close, the other door can lead to the fulfillment of your dream.[1]

—*Joan Murrell Owens*

Geologist/paleontologist Joan Murrell Owens's path to a career in science was a most unusual one. To begin with, Owens was originally deterred from marine biology, the science of her choice. She earned her first baccalaureate degree in art, with minors in mathematics and psychology, and her first master's degree in guidance counseling, with a specialty in reading therapy. She did not embark upon the study of science until she was thirty-seven years old, later earning her second bachelor's and master's degrees as well as her doctorate in geology, with a minor in zoology. Owens's reentry into college at mid-life provides an early example of another group of women—those of all races who, in the aftermath of the women's liberation movement of the late 1960s, began to exercise their options for continued education. Although her degrees are not in marine biology per se, she is essentially both a marine biologist and a geologist/paleontologist.[2]

Owens was born in Miami, Florida, in 1933, the youngest of the three daughters of William Murrell, a dentist, and Leola (Peterson) Murrell, who had been a schoolteacher before her 1924 marriage. The family placed tremendous emphasis on education. William Murrell was descended from free Blacks in South Carolina, most of whom, including his siblings, aunts, and uncles, were college educated. He had been orphaned at age ten, and was raised by an aunt. Leola Murrell, who had been orphaned at age thirteen, was raised by an aunt and later by family friends. Despite being orphaned, both graduated from Florida A&M College (now Florida A&M University), which was where they met.

Leola Murrell taught her daughters to read before they began attending school, so like a majority of the women in this study, Owens was an early reader. "I particularly remember reading two children's books—one about three African-American sisters named Rowena, Tina, and Tot, and another about a Mexican boy and his family," she said, adding that her interest in reading was also reinforced by her oldest sister, whom she idolized: "I 'graduated' at about age nine or ten to what my sister was reading. One of my first books then was Booth Tarkington's *Image of Josephine.*"[3] William Murrell was an avid fisherman, and routinely took his wife and daughters on weekend fishing trips. Owens became fascinated with marine life, and at her parents' urging she complemented what she observed on the fishing trips with books. Jacques Cousteau's *Silent World* and Eugenie Clark's *Lady with a Spear* fueled her growing passion for marine biology.[4] "Both books fed my interest in marine organisms," Owens said, "and I was particularly struck by the Clark book because it was about a woman."[5]

Owens attended Miami's segregated public schools, where the only science textbooks available to Black students were outdated hand-me-downs from the white schools; the laboratory equipment was cobbled together in a piecemeal fashion.[6] As was the case with so many of the women in this study, access to a home library and family dinner-table discussions of schoolwork were vitally important. According to Owens,

> When I was in junior high school, I told my father at dinner one night that there were *eight* planets in the solar system. He was so appalled [at this misinformation] that he started buying the current editions of my text books to add to the books and periodicals we already had at home.[7]

By junior high school, Owens had already demonstrated a gift for the sciences, and her extended family and the close-knit Miami community also played an important role in nurturing her academic success:

> In the segregated South of the 1940s and 1950s when I grew up, the home, church, community and school joined forces to make you realize that your only "escape" from the oppression of segregation was to excel in school and "go on." ... My generation of family and friends was indoctrinated early [with the idea] that education was the only path to "freedom."
>
> All the people in our community knew each other, and the church, school and community jointly saw to the raising of the children. . . . There was a tremendous atmosphere of support for students who showed promise. People of your community knew your family and felt free to make sure you "toed the line." Financial support in my case came from my father, being a dentist and prominent citizen. But non-financial support came from those other sources—from the community itself. . . . They *expected* you to excel, and you responded to their loving pressure.[8]

Despite the lack of proper books and laboratory equipment in the segregated public schools, Owens described her teachers as both dedicated and competent. "Between them and my parents, I was well prepared when I entered college," she said, adding that the year she took the entrance examinations for Fisk University, she registered the highest score, and described biology and geology as her special sciences.[9]

There was never any question that Owens, her siblings, and her cousins would go on to college; it was merely a question of where they would go:

> In my family—and I mean that in the extended sense—it was *understood* that you would go to college. My oldest sister was even told *which* school she would attend! She was only 16 at the time, which my mother used as her reason for sending her to [all-female] Spelman. My next sister, however, who was also 16 when she graduated, but a few months older [than my oldest sister had been at graduation], said she would join the WACS rather than go to Spelman. This sank my mother's ship since she thought the WACS were prostitutes for the Army. Thus, I had no problems when I said I wished to go to Fisk as my sisters had.[10]

All of Owens's cousins, on both sides of the family, and all her close friends attended college. The number of close family members who attended college, particularly the large number of them who majored in the sciences and mathematics, is worthy of a brief digression.

As mentioned, Owens's father, William Murrell, was a dentist and a graduate of Meharry Medical College. His sister, Delia Murrell, was married to Julius Parker, a physician.[11] Owens's oldest sister, Evelyn Murrell Kimbro (B.A. Fisk University 1946, M.S. Atlanta University), had degrees in library science; her other sister, Willette Murrell (B.S. mathematics, Fisk University 1951), worked with the U.S. Weather Bureau on long-range weather forecasting. The majority of Owens's first cousins, the Parker children, earned degrees in the sciences and mathematics. Carolyn Parker (B.S. physics, Fisk University 1937; M.S. mathematics, University of Michigan 1941) worked on the Dayton Project of the atomic bomb during World War II at Wright Patterson Army Air Force Base; she earned a second master's degree (physics, Massachusetts Institute of Technology, circa 1950–1951); and she had completed the course work for the doctorate in physics at MIT by 1952 or 1953 but she died, apparently of job-related radiation-induced leukemia, before she could defend her dissertation.[12] Mary Parker Miller (B.S. mathematics, Talledega College 1938; M.S. mathematics, New York University 1975) and Juanita Parker Wynter (B.S. mathematics and chemistry, St. Augustine's College 1945) both taught mathematics at the high school level; Wynter later continued her education and became a guidance counselor (M.S., New York University 1957). Julia Leslie Parker majored in math (B.S., Fisk University 1947) before earning a master's degree in medical technology (Meharry Medical College circa 1950). Julius Parker Jr. earned a master's degree in chemistry at the University of Michigan and became a medical researcher. Only one Parker sibling, Martha, did not earn a degree in a natural science, choosing the social sciences instead (B.S., Tennessee State University circa 1943; M.S., Temple University).[13] Owens's other first cousin, Nathaniel Murrell, married a mathematician; their daughter, Muriel Poston, is a Howard University botanist.[14] Given the extraordinary number of family members who studied the sciences and mathematics, it is easy to understand why Owens said that she was "programmed to enter science."[15] Her late entry into the field was due largely to a lack of opportunity for Blacks and women in marine biology.[16]

In high school, Owens not only did extremely well academically, but also demonstrated a talent for art. Upon graduation from Miami's Booker T. Washington High School in 1950, she was awarded two modest scholarships: a Pepsi-Cola Company Scholarship, and a Sarah Maloney (Art) Scholarship at Fisk University. Combined, they paid only a portion of her college tuition; her father paid the major portion.[17] At Fisk, however, Owens was confronted with the fact that marine science was not then a viable career option.[18] She briefly considered a psychology major—but Fisk did not offer one. She settled on art, because she had some talent in that area and because it was not as "demanding in terms of lab requirements as chemistry."[19] For a while, however, she used her art to retain some contact with the sciences: "My art helped me to earn 'pin money,' because I did thin-section and anatomical drawings for medical students at Meharry Medical College. . . . My first job offer as graduation approached was as a surgical illustrator at the Veteran's Administration Hospital in Tuskegee, Alabama."[20]

Meanwhile, in 1953, Owens experienced her first exposure to a predominantly white social and educational environment. At that time, Fisk University had developed a kind of exchange program with a few majority white institutions, including Pomona College in California.[21] Owens and a male student from Fisk were chosen to attend Pomona during the spring 1953 semester, and to act as informal goodwill ambassadors not only for Fisk itself, but for Blacks in general. According to Owens,

> Pomona wanted to get Blacks on campus to encourage other Blacks to come. Joe [the other Fisk student] and I wanted to [go to Pomona] for the experience. We

took our courses there, and acted as ambassadors, as it were. We were the only
Blacks at Pomona, although there was another Black, from Nigeria, at the
Fairmont Business School. . . .

There were a lot of kids there who had never been around Black people
before; this was an opportunity for them to find out more about us, and they
were curious about the littlest things—what did I do with my hair? And they
were surprised to find that we do the same thing with it as whites—you wash it,
it dries, and you try to make it look nice. I took swimming and one day I was in
the shower and this girl noticed the strap marks from my bathing suit and was
surprised because she didn't think I could tan.[22]

Owens said she also discovered many of the little things that college girls had in
common, among them an attraction to "uncommonly handsome" college boys:

Richard Chamberlin [now an actor] was a student at Pomona when I was there.
Each dorm had its own cafeteria, and one night a week you could eat in any
cafeteria you wanted to. [Chamberlin worked his way through college] waiting
on tables, and on the nights we could eat anywhere we wanted to, all us girls
would try to eat where he was waiting table, because he was so handsome, you
know.[23]

Somewhat to her surprise, she even had a social life:

People were friendly, although the [white] fellows were shy about dating, but I
went to the Junior Prom, and I also went out with some people who are now very
well known. . . .

But this is not to say that there weren't occasionally problems. I had one
friend who invited me to go home with her and spend Easter vacation. She went
home a couple of weekends before Easter to discuss it with her parents, and
when they found out I was Black they said no. She came back to school in
tears—I wasn't surprised, but it really hurt her.[24]

Owens graduated from Fisk in 1954, with a major in art and a double minor
in mathematics and psychology. She decided to enter graduate school, although
she did not yet know how she would pay for it. As with a number of other women
profiled in this volume, the segregation rules applied in Southern states provided
a financial solution, which Owens described as "a bribe":

When I finished college, I felt strongly that my father's commitment to my
education was fulfilled. I discovered that the State of Florida would pay tuition,
books, and transportation to Black students wishing to pursue a field offered at a
white state college but *not* at a Black [college], *if* they would go to a [white]
college out of state. So I picked out such a field and Florida came up with all
the money I needed.[25]

She enrolled in graduate school at the University of Michigan in the fall semester
of 1954. Within a few weeks of entering, however, she changed her major from art
to guidance counseling, and earned the M.S. degree, with a specialty in reading
therapy, in 1956.[26]

After a brief stint as a reading therapist at the University of Michigan's Children's Psychiatric Hospital (1955–1957), Owens moved to Howard University as
a reading clinician in the English Department (1957–1964). She left to take a

position in Massachusetts with the Program for Pre-College Centers of the Institute for Services to Education (ISE), where she developed new teaching procedures and curriculum materials geared toward educationally disadvantaged high school students. The program she developed was later selected as a model for the Upward Bound programs. Owens moved to the position of program associate at the Institute, where by 1970, the thrust of the program had broadened beyond its original scope of developing materials for the educationally disadvantaged. As Owens recalled,

> I had been gradually pushed more and more into working on teaching materials for English teachers in the Curriculum Development Program. Although English wasn't my field, my training and earlier experience as a Reading Clinician in the English Department at Howard University allowed me to be effective in my work while working with the Upward Bound program and with college freshmen in our 13-colleges program. However, after a while, we began working on materials in literature that were geared for advanced students. I foresaw that I was rapidly reaching a point where I would need to return to school to continue to be effective in my work. At this point, I decided if I must return to school, it ought to be to pursue something I really wanted, rather than something expedient for my job.[27]

At about this time, Owens discovered she was being paid less than her colleagues, which added impetus to her decision to leave the field.[28] Until this point, there hadn't been any particular mentor who played a crucial role in her life, but that was about to change.

When Owens began to seriously consider changing careers, she spoke to a close friend, physicist Phil Morrison of MIT, who was also an advisor and consultant at the Institute for Services to Education:

> Morrison was the first person I spoke to [and] he applauded my decision. He felt strongly that I had the ability, but he also gave me a good dose of reality by pointing out some of the problems and sacrifices I would inevitably encounter. He also encouraged my change of career by challenging me to succeed where many others he knew had failed.[29]

She accepted a transfer to the Institute's Washington, D.C., offices. After a bit of research on the best way to approach her new studies, Owens decided to start over from the beginning, at the undergraduate level.

In the fall of 1970, at age thirty-seven, Owens enrolled as a part-time undergraduate at George Washington University. She wanted to pursue her old dream of becoming a marine biologist, but George Washington University did not offer a major in that field:

> No school in the D.C. area at that time offered marine biology. My professor suggested I look into attending the State University of New York at Stony Brook. But I was told—and I don't know how accurate this was—that the chairman or someone in the marine biology department said they'd love to have me because I satisfied two requirements—a Black and a woman—but they didn't have a scholarship to offer.
> I went down to the University of Miami and talked to them but by that time I had met Frank [Owens] and we got had gotten serious and I didn't want to leave.

Then a friend in the Provost's Office at George Washington told me I could get the same thing I wanted to do by studying geology. It was a small department and they were very nice and they allowed me to put together an interdisciplinary program of geology (emphasis on paleontology) and biology (emphasis on zoology) that approximated a program in marine biology and which gave me just about what I would have gotten in a marine biology department, so that's how I ended up staying at George Washington.[30]

Before the start of the spring 1971 semester, Owens had quit her job to attend school full time. "I was not married," she said, "thus had no one who would be affected by my decision."[31] The first year she lived on her savings, plus a $2,000 loan from her father. In 1972, thanks to a lab instructor friend at George Washington University, she got a job as a museum technician in the department of paleobiology at the Smithsonian Institution, where she catalogued new acquisitions and entered data on descriptions, synonymies, and localities of type specimens described in scientific journals and stored at the Smithsonian. A. G. Coates, a professor of paleontology, became her mentor and major advisor during her studies at George Washington.[32]

At that point in time, Owens said, race and gender were not the severe stumbling blocks they once were:

By the time of my "second time around" in the 1970s, my race and gender were to my advantage so long as I could "carry my own weight," meaning I was not a marginal student admitted due to affirmative action or the like. My department (geology) at GWU was very proud of me and very supportive. I had friends (both African American males) in doctoral programs in other fields—one was in mathematical statistics at George Washington and the other was in geography at Catholic University—whose departments did their darnedest to try and prevent them being granted their doctorates but failed. I did not have that type of experience.[33]

Nor was her age a disadvantage:

As an older student and one who generally maintained an A average, a number of younger students asked to study with me, but these invariably turned into my "tutoring" them. However, once I completed my course work and began research, I worked alone, but there were two or three other students doing research whom I could depend upon to help me with technical research techniques, such as making thin sections of my specimens or developing my photos of them. There was more a sense of camaraderie with a few than in a community with many.[34]

As is fairly typical of returning women students, Owens was always concerned with the quality of her work, as she explained:

I was doing an analogy of tests, called hardness tests that you do on minerals. The first thing you do is find out whether or not you can scratch the mineral with your fingernail—and my nails are extremely hard—then you work up to other more sophisticated tests. When the professor gave me back my paper, I saw that I had gotten something wrong—I had misidentified this mineral—and I went and talked to him about it.

He said, "Well, you should have known that this is not what you said it was because you couldn't scratch it with your nail."

And I said, "Yes, I could," and I took out the sample and scratched it easily.

He said, "Oh my, you're not supposed to be able to do that."

And I laughed and said, "So much for the accuracy of the hardness tests."[35]

In 1972, Owens earned the B.S. degree in geology, with a minor in zoology. In 1973, on her fortieth birthday, she married Frank Owens, a widower with two daughters, who supported her decision to continue on to graduate studies.

Owens received a Ford Foundation fellowship for graduate study at George Washington University, which in addition to paying all her tuition and fees provided her with $6,000 yearly stipend, on the understanding that she go to school full time and not work. Along with Coates, Stephen Cairns, an invertebrate zoologist at the Smithsonian, became involved in mentoring her during her dissertation research: "Coates and Cairns encouraged and supported me by making available to me whatever materials and specimens I needed for my research and always being there to help and advise me."[36] Owens held the Ford fellowship from 1973 to 1976, and earned the M.S. degree in geology, with a minor in zoology, in 1976.

At that point, she had finished her course work for the doctorate; when the department of geology and geography at Howard University offered her a position as an instructor, she accepted, but had to forfeit the fellowship:

Jobs were not that easy to get, and I decided the Howard offer was more important than the scholarship—you take a job when you can get one. By this time, I had a Master of Philosophy degree, an interim degree granted by some universities to doctoral students who have completed all their requirements except the dissertation. Once I began teaching at Howard, I paid my own tuition, which was minimal at that point, since I was not taking courses, only doing research and writing.[37]

There was, however, one other obstacle in her path. Because Owens has sickle cell anemia traits, she could not dive underwater to search for specimens. So instead, she went to the Smithsonian and did a laboratory project there.[38] Her research experiences provide an insight into the far-ranging importance of the collections at the Smithsonian Institution. She began poring over coral samples collected by a British expedition in 1880; under the direction of Stephen Cairns of the Smithsonian, her dissertation research, "Microstructural Changes in the Scleractinian Families *Micrabaciidae* and *Fungiidae* and their Taxonomic and Ecologic Implications," was based on the Smithsonian samples.[39] She completed her Ph.D. at George Washington University in 1984. In 1986, she was advanced to the rank of associate professor in the department of geology and geography at Howard University, where she taught courses in historical geology, paleontology, oceanography, stratigraphy, and geological investigations.

Her research on the coral collections at the Smithsonian did not end, however. As Owens moved into the professional arena, Steven Cairns provided assistance and advice as she began her professional writing on the results of her research, and he continued to make available to her samples from the various collections:

Certainly [Cairns's] advice and assistance in reviewing my professional papers
before I submitted them helped to get them accepted for publication. Once I got
the knack of it, the "formula," I did not need to depend on his advice and
assistance as much.[40]

She next began work on the corals in the Squires Collection at the Smithsonian.[41]
Although Owens's degrees are in geology, she stresses that geology is more than
the study of rocks and minerals:

If you look up a definition of geology you will find it is the study of the earth and
its inhabitants; it's much more encompassing than the study of rocks, especially
when it's combined with paleontology. What you're doing is combining the
living and dead world—bones, teeth and shells, the hard remnants of life, the
things that preserve well over time; these remains can tell you so much about an
organism. Then, if you have a living representative, you can make comparisons.
I study corals, and there are corals today that have similar structures to ancient
corals, but sometimes you have a specimen that does not have a modern
equivalent—then you use analogy. I see a particular structure and I know I've
never seen it before, but by the shape of it I can recall that I've seen something
like it. That's how my science builds.[42]

Her specialty is solitary corals, which, unlike reef corals, have endoskeletons. The
corals she worked with all belong to the same family, Micrabaciidae, but as she
explained:

When we examine them closely we see differences between ones in the same
family. There are different types of solitary corals; the ones I specialized in were
mobile—I knew they could move because they had growth lines going both up
and down.
    Discovering a new genus means finding more profound differences than
discovering a new species. I have found one new genus, with two distinct species
in it, and two other new species.[43]

She named one species *Letepsammia franki*, after her husband, Frank Owens.[44]
To support her work on the microstructural and microarchitectural changes in
deep sea solitary corals, she received Howard University faculty research grants
in 1986 and 1987, and a Mobil Oil Company grant in 1988. In addition to her
coral research, from 1990 to 1992 Owens worked with Bruce Dahlin, an anthropologist who was trying to explain the disappearance of the Mayans from certain
sections of Mexico.[45]

    Regardless of her professional publications as a researcher, Owens said she
was "first and foremost" a science educator:

I *loved* teaching and the intellectual stimulation in the exchange and challenge
of ideas with my students. Precious moments were when a student asked a
question that required more than a ready answer but a stretch of my own mind
and knowledge to satisfy his or her curiosity.
    I enjoyed my research and remain fascinated with my little "critters," but it
was not as stimulating as working with students seeking knowledge.[46]

As a scientist teaching in a major university—and as the only female faculty
member in the geology department—Owens impacted many lives, both male

and female. Many of her male students from Africa, the Middle East, and the Caribbean had never had a woman teacher before.[47] A number of the students she mentored during her nineteen years at Howard University merit mention here: Estella Mkwate-Atekwana is a tenured professor of geophysics at Western Michigan University; Margaret Kasim (Ph.D. hydrology, Kent State University) is with a waste management firm in Maryland; Andrea Foster is a doctoral candidate in geology at Stanford University; and Donna Montague is a geologist living in Brooklyn, New York.[48]

Owens does not believe her heavy teaching load at Howard had a negative impact on her research:

> My teaching and my research occupied different time frames. I taught from mid-August to mid-May, and during that time, teaching was my main focus and my main joy. Most of my lab work for my research was done during the summer months when I was not teaching. According to need, my research was done at home, where I had a microscope, in my lab at Howard, and at the Smithsonian, where the bulk of my specimens were housed, and at George Washington University where I had access to equipment otherwise not available to me.[49]

For more than a decade (1983–1994), Owens was involved in the Science Discovery Day programs, sponsored by the Minority Women in Science division of the American Association for the Advancement of Science, programs designed to expose minority women students to possible careers in the sciences.[50] She also served as a member of the board of reviewers of the National Science Foundation (1987–1989).[51]

In 1992, Howard University announced that the department of geology and geography would be phased out.[52] Owens was appointed special advisor to oversee the matriculation of the remaining undergraduate geology majors. Because nearly half the people on her dissertation committee had been in the biological sciences, she said, she was able to move to the biology department at Howard in 1993. Although she was finally able to teach marine biology, her "first love," the move was not without obstacles because the biology department was "heavily weighted toward the medical school," and it doubted her relevance to its program.[53]

Owens's research, which focused primarily on deep-sea corals, was supported by grants from petroleum companies and faculty grants from Howard University. Although her work redefined the classification of button corals, she described the niche her research has occupied in her field as "esoteric in the extreme":

> But [it's] important because of its contributions to our understanding of the life of the sea and some of the ways in which ecology and evolution interact. Also, it is important to others working with deep-water organisms because of some of the relationships I think I indicated—if not proved—between water depth and availability of calcium carbonate (for shells) that influenced the physical evolution of some organisms.[54]

Joan Murrell Owens retired from Howard University in 1995. She and her husband, Frank Owens, had planned a research trip to Hawaii, when he became ill; he died in May of 1996.[55]

NOTES

1. Joan Murrell Owens, interview by author at her home in Washington, D.C., 21 March 1996. Owens, author questionnaire; curriculum vitae; assorted personal papers and letters, 15 January 1996; 18 January 1996; 26 May 1996. Owens, telephone conversations with author, January–June 1996. See also "Joan Murrell Owens," biographical sketch in *Blacks in Science* calendar (Oakland, Calif.: African Scientific Institute, 1996). James H. Kessler, J. S. Kidd, Renee A. Kidd, and Katherine A. Morin, *Distinguished African American Scientists of the Twentieth Century* (Phoenix, Ariz.: Oryx Press, 1996), pp. 272–275. *Who's Who among Black Americans* (Northbrook, Ill.: Who's Who among Black Americans Publishing, 1976). Owens is also featured in the 1988 "Black Achievers in Science" exhibit at the Chicago Museum of Science and Industry; the exhibit was updated in 1992.

2. Owens, questionnaire, interview.

3. Owens, questionnaire. Booth Tarkington, *Image of Josephine* (Garden City, N.Y.: Doubleday, 1945).

4. Jacques Y. Cousteau, *The Silent World* (New York: Perennial Library, 1953); Eugenie Clark, *Lady with a Spear* (New York: Harper, 1953).

5. Owens, questionnaire.

6. Ibid.

7. Ibid.

8. Ibid.

9. Ibid.

10. Owens, interview and questionnaire. After two years at Spelman, her oldest sister transferred to Fisk.

11. Muriel Poston, interview with author at her office at Howard University, Washington, D.C., 23 March 1996. According to Poston, her father was the second Black to earn a doctorate in business administration at Harvard and was a student of John Kenneth Galbraith.

12. Juanita Parker Wynter, telephone interview with author, 26 June 1996; and Owens, interview and telephone conversation, 29 June 1996. Since all the information on Carolyn Parker is oral—and not confirmed by additional sources—I have not included her in this volume. Wynter said that Parker's work was "so secret she couldn't discuss it, even with us, her family."

13. Wynter, telephone interview; Owens, interview. It is worth noting that at least three of these individuals attended Meharry Medical College, and five attended Fisk University, although none lived in Tennessee.

14. Owens, interview. See entry on Muriel E. Poston in this volume.

15. Owens, interview and questionnaire.

16. Owens, questionnaire.

17. Ibid. Owens graduated second in her class. The Pepsi-Cola scholarship was based on a national essay contest, and the Sarah Maloney scholarship at Fisk was awarded for artistic talent.

18. Not only was marine science not an option for Black women, it was not an option for Blacks in general. As a result, Fisk did not offer a major or even a concentration in the marine sciences, which is quite understandable. The Marine Biological Laboratory at Woods Hole was still racially segregated in the 1950s, as were other marine sciences centers. And shipboard research continued to be fairly restricted for women, regardless of race, through the early 1980s. Of her later career as a paleontologist/marine biologist, Owens said, "Had I wished to be part of an oceanographic study, both my race and gen-

der would have become an issue. Even now, women—regardless of color—are often only grudgingly welcome on ships or field studies that require some sort of isolation over a prolonged period" (ibid.).

19. Owens, questionnaire.

20. Ibid. By the time Owens actually became a scientist, her talent for art, which in earlier years might have been helpful, was no longer useful because by then light microscopy and scanning electron microscopy (SEM) were widely used.

21. Whittier, Antioch, and Oberlin colleges also participated in the exchange program (ibid.).

22. Owens, questionnaire and interview.

23. Ibid.

24. Ibid.

25. Owens, questionnaire.

26. During her time at the University of Michigan, Owens was mentored by Leonard Spearman, who was later president of Texas Southern University and is a U.S. ambassador to an African country (Owens, questionnaire).

27. Ibid.

28. Owens, interview (follow-up discussion to answers supplied on the questionnaire).

29. Owens, questionnaire. According to Owens, Morrison is a noted physicist, and was a member of the Manhattan Project who lobbied against use of the atomic bomb. He was a physics professor at MIT at the time they met (Owens, interview, questionnaire).

30. Owens, interview; also Owens, questionnaire.

31. Owens, interview.

32. Ibid. A. G. Coates is now at the Smithsonian Institution's Panama Station (ibid.).

33. Ibid. However, Owens said, "When a position for a paleontologist later opened at George Washington, I was not considered. Ostensibly, the reason was that they did not want to rob a Black university (Howard) of a good teacher! This may have been their true reason because later when they needed a hydrologist and wanted a Black one, they talked to me about the 'ethics' of recruiting a Black from a Black school. My response was that they should make him the offer and let him decide" (ibid.).

34. Ibid.

35. Ibid.

36. Ibid.

37. Ibid.

38. Ibid.

39. Owens, questionnaire.

40. Ibid.

41. According to Owens, D. F. Squires had inexplicably stopped work on the collection he had amassed and left the Smithsonian (ibid.).

42. Ibid.

43. Ibid. See the appendix.

44. See the appendix.

45. Owens's part in the project was to take and analyze soil samples of cenotes to determine if there had been a change in the environment. "Project to Reconstruct Halocene Environments on a Karstic Plain, Yucatan, Mexico," Bruce H. Dahlin, Department of Sociology/Anthropology, principal investigator. Funded by NASA, November 1989; grant renewed December 1991 (Owens, curriculm vitae, ibid.).

46. Owens, questionnaire.

47. Ibid.

48. Owens, letter to author, 26 May 1996, with a postscript dated 27 May 1996. Among

the male students Owens mentioned are Allan Smith, a science teacher in the Houston (Texas) Public Schools (1990–1996) who has been accepted into the doctoral program in Educational Administration at Texas Southern University; Derrick Givens, currently with Amoco Oil Company in New Orleans; Donald Jackson, the founder of a consulting waste management firm in Maryland; and Terah Isiahiliza, who is currently with BFI (Browing Ferris Industries), a waste management and environmental affairs company that does groundwater monitoring and research into harnessing usable gases from burning trash, in Austin, Texas, where he works in the quality assurance and quality control section. A significant number of the geology department's students during Owens's tenure were from Nigeria; others came from North African countries, Trinidad, Jamaica, and Suriname. According to Owens, of the foreign students who returned to Nigeria, some are working either with the country's national oil company or in programs studying global warming at the Oceanographic Institute of Nigeria. Those graduates working in the U.S., whether native-born or permanent aliens, tend to work either in state, county, or private waste management departments, or in consulting firms. Most of the foreign students from Trinidad, Jamaica, Suriname, and North Africa—where the oil industries are no longer flourishing—have remained in the U.S.

49. Owens, questionnaire.
50. Owens, curriculum vitae.
51. Ibid.
52. Owens, interview; also Owens, curriculum vitae. According to Owens, this decision was made despite substantial outside funding, principally from major oil companies.
53. Owens, questionnaire.
54. Ibid.; also, Owens, interview.
55. Owens, letter to author, 26 May 1996.

LYNETTE PADMORE

Excellence in Teaching

Once a student of zoologist Margaret S. Collins, Florida A&M University research geneticist and teacher Lynette Padmore was named Florida's top minority educator of 1996 by the Florida Department of Education. Born in Guyana, South America, Padmore was also the 1996 recipient of the Ida S. Baker Distinguished Minority Education Award,[1] sponsored by the Department of Education and Lehman Brothers. According to Florida A&M President Fredrick S. Humphries,

> [Padmore's] record on behalf of minorities seeking careers in the sciences has been one of the more impressive educational achievements made by any faculty member in the state university system over the past decade. . . .
> She is an indefatigable first-rate scholar who demands excellence of herself and her students. Many students cite [her] as the faculty member who has had the greatest influence in their lives.[2]

Between 1983 and 1988, Padmore received more than $200,000 in grant funds to devise programs to develop scientific excellence in high school teachers and students. During a newspaper interview, she said,

I think the award will inspire my students to look around and see the rewards of hard work. I hope it will inspire them to be more persistent and determined and in the end they will succeed. When one looks at the negative manner in which education is viewed by some segments of our population, the award becomes a positive element. The public, in general, is beginning to recognize that we, as educators, do influence minds.[3]

Padmore's own teacher, Margaret Collins, was indeed proud of her influence on one of her prized pupils. Padmore, now a full professor, is currently director of BIONR, a program funded jointly by the Biology Department at Florida A&M University (BIO) and the Office of Naval Research (ONR).[4] Established with initial funding in 1989, the BIONR program is designed to increase the number of students opting for graduate and professional graduate careers in the biological, chemical, and related sciences.

NOTES

1. "Padmore Top Minority Educator," *Tallahassee Democrat*, 6 March 1996, p. 3B. The article mentions that Ida S. Baker was a deputy commissioner for Human Resource Development for the Florida Department of Education (1987–1991).

2. Ibid.

3. Ibid.

4. BIONR Home Page [web page]; http://www.tamu.edu /bionr/ [accessed 20 January 1999]. The BIONR program draws participants from both the high school and undergraduate levels; selected students receive advisement, counseling, curriculum enhancement, and peer mentoring. The program also promotes activities designed to nurture students' leadership abilities—tutoring high school and lower level undergraduate students, judging elementary and high school science fairs, and fostering community-based projects that impact on science learning for middle and high school students (ibid.).

JENNIE R. PATRICK
Chemical Engineer Working to Effect Social Change

African Americans in pursuit of achievement must contend not only with the challenge of the achievement, but also with the stress, strain, and unfairness that accompany racism. One cannot measure success strictly by the accomplishment alone. One has to consider the importance of maintaining one's integrity, dignity, and self-esteem.

—Jennie R. Patrick[1]

The fourth of five children, chemical engineer Jennie R. Patrick remembers her small-town Gasden, Alabama, childhood and close-knit family:

We lived a very simple life. . . . The fact that my parents had very little formal education did not hinder them from encouraging all their children to pursue a formal education. My parents were disciplined and willing to make necessary sacrifices to assure a better life for their children. . . .

> The warmth, love, and encouragement from my parents provided me with a
> foundation of high self-esteem, confidence, and emotional and psychological
> strength.[2]

Due to their own lack of education, according to Patrick, her parents were unable to assist her with even her elementary-school work. As a result, she said she developed a "very strong sense of independence" quite early in life. Driven by an innate sense of curiosity that she has said "often stretched the tolerance level" of those around her, she found pleasure in intellectually challenging pursuits limited only by her imagination and ability to read.[3]

In 1964, she took her place among that first generation of Black schoolchildren who integrated Southern schools. As Patrick recalled,

> The initial months were a living nightmare. The emotional, psychological,
> mental and physical violence against us was difficult to comprehend. This
> experience opened up a whole new world for me. Surviving at Gasden High, the
> previously all white school, became my greatest challenge. Not only did I have
> to survive academically, but also emotionally, psychologically and physically.
> This was the beginning of a world about which I would become more and more
> aware—a world filled with hate, abuse, unfairness, and discrimination.
> . . . I made a commitment to succeed. Perhaps even more important was my
> commitment to myself, my forefathers, and African-American people.[4]

Patrick recalls being discouraged by her white high school teachers and having counselors routinely try to sway her from taking a technical college preparatory curriculum. Despite the difficulties, she graduated from Gasden High with honors.[5]

Interested generally in science and technology, Patrick decided to major in chemistry at Tuskegee Institute. At the end of her freshman year, Tuskegee began offering a new major—chemical engineering—which Patrick quickly opted for. The program just as quickly collapsed, however, and she withdrew from the school. Accepted at the University of California at Berkeley, which did not then offer scholarships to transfer students, she soon found herself in a real scramble to support her studies. After taking a year off to earn tuition money, she completed the B.S. degree in chemical engineering in 1973, and entered graduate school at the Massachusetts Institute of Technology, where she received a series of research associate grants (1973–1979). She recalled that her days as a graduate student were "intense":

> From the beginning of my tenure, determination and commitment were
> essential if I were going to be successful. MIT is known for its challenging and
> rigorous academic program. For an African-American, the challenge was even
> greater. Not only did I have to conquer the academic challenge, but I also had to
> be emotionally and psychologically strong enough to overcome the racism.[6]

She completed her Ph.D. in 1979; her dissertation, "Superheat Limit Temperatures for Nonideal-Liquid Mixtures and Pure Components," explored high pressure separation and purification technology, and techniques of distillation and liquid-liquid extraction.[7]

Patrick soon secured a position as a staff engineer at General Electric Research in Schenectady, New York, which she described as an "exciting" technical job:

> General Electric Research, during those days, emphasized research and employed a high quality staff. Thus, the three and one half years I spent there were intellectually challenging. I initiated and developed a research chemical engineering program in the area of supercritical fluid extraction technology. . . .
>
> Because supercritical fluid technology was in its earlier stages of development in the United States, it was a particularly exciting time to head such a research effort. Since only a few people nationwide had technical backgrounds in the area, the opportunities to advance in this area [were] plentiful.[8]

In 1983, Patrick moved on to the Philip Morris Research Facility in Richmond, Virginia, where she headed the company's supercritical extraction technology program, and oversaw the design of a state-of-the-art research pilot plant.[9] She joined the Rohm and Haas Company Research Laboratories in Bristol, Pennsylvania, as a research section manager in 1985, where, in addition to her supercritical fluid extraction work, she became involved in polymer science and emulsion technology; she was later appointed a research manager for the company's first fundamental research engineering group.

During the period between 1980 and 1987, Patrick also taught as an adjunct at two universities, Rensselear Polytechnic Institute (1980–1983) and the Georgia Institute of Technology (1983–1987). Although she has described her career as a chemical engineer as "interesting and challenging," over the years Patrick says she has become "more fascinated with societal and humanistic issues":

> It amazes me to observe actions and reactions in this society which emphasize greed, lust, and youth. Solving a complex engineering problem can appear simple in comparison to the complicated and varied responses of the human mind.[10]

Patrick traces her increasing involvement in the struggle to improve conditions and educational opportunities for Blacks to the many honors and the public attention she began to receive when she completed her doctorate in 1979:

> I have appreciated and respected these honors, although receiving such honors burdened me, in a sense. A responsibility has been placed in my hands.
>
> The first six years after my graduation from MIT, I spent many hours traveling across the United States speaking to predominantly Black audiences of college and high school students. My intent was to inspire them and expose them to the truth. The response from the audience was generally emotional and deeply touching. I realized that many of the audiences had never confronted truth in an unadulterated form as it related to their lives and future in this society. . . .
>
> My presentations frequently emphasized a sense of history, self-worth, personal values, independence, duty, and self-love. History has been unkind and unfair to Black people all over the world. To reverse this appears, at times, insurmountable. Yet, if a critical mass of devoted individuals attempts to make a difference, progress is a certainty.[11]

By using some of her own personal experiences, Patrick hopes to provide examples of the kinds of survival tools that could prove particularly useful for young
Black people in predominantly white educational institutions, where, as she
points out, "the attrition rate for Blacks is often high":

> My belief is that the high attrition rate is more a reflection of the [minority]
> individual's inability to cope with the hostile environment rather than an
> inability to fulfill academic requirements. . . . Academic capabilities alone may
> not be sufficient for success in various situations. Sharing [success] tactics . . .
> could prove to be invaluable.[12]

Patrick also believes that providing young Black students with the tools they need
to cope in the workplace once they have completed their education is a vital priority:

> If a national study were conducted on the emotional, mental and physical
> health of bright, capable and ambitious Blacks, the results would be somewhat
> dismal. . . .
>   The fact of the matter is that corporate America minimizes job responsibility
> and financial reward for African Americans in general. The intent of corporate
> America is to maintain the status quo. For the most part, the activities surround-
> ing affirmative action are only tactics of extending the time requirements for real
> change. It is important that Blacks understand such tactics and not be victim-
> ized and stressed as a result of them.[13]

Patrick appears to have a good chance of putting her ideas — both on chemical
engineering and on human/social issues — into practice.

NOTES

1. Jennie R. Patrick, "Trials, Tribulations, Triumphs," *Sage* 6 (fall 1989): 51–53.
2. Ibid. Patrick was born in 1949. Vivian Ovelton Sammons, *Blacks in Science and Medicine* (New York: Hemisphere, 1990), p. 185.
3. Patrick, "Trials."
4. Ibid.
5. Ibid. Patrick graduated around 1966 or 1967.
6. Ibid.
7. Sammons, *Blacks in Science and Medicine*, p. 185.
8. Ibid.
9. Ibid.
10. Ibid.
11. Ibid.
12. Ibid.
13. Ibid.

AMANDA PEELE*

Biologist Amanda Peele was born in Jackson, North Carolina, in 1908. She
earned the baccalaureate at Virginia's Hampton Institute in 1930. Immediately

after graduation she obtained a position in Hampton's biology department, and studied for the master's degree during the summers at Cornell University, where she earned the M.S. degree in 1934.

Peele remained at Hampton for her entire career, and was mainly engaged in teaching. In May 1939, she became the first Black female to present a research paper at the Virginia Academy of Science; as was recounted in *Who's Who in Colored America*, she was "the only Colored person on the program of this particular meeting of the body. Her serious Work in her chosen field gives promise of distinction in a Virgin area for Colored women—the field of Science."[1]

NOTE

1. Vivian Ovelton Sammons, *Blacks in Science and Medicine* (New York: Hemisphere, 1990), p. 187. See also *Who's Who in Colored America: A Biographical Dictionary of Notable Living Persons of Negro Descent in America*, 1941–1944 ed. (New York: Who's Who in Colored America, 1927–), p. 400.

LINDA PHAIRE-WASHINGTON

Understanding Macrophage Cells and Monoclonal Antibodies

Immunochemist Linda Phaire-Washington has cited a lack of meaningful exposure to science and a lack of equal incentives in opportunity and pay as reasons for the low numbers of minority women involved in scientific research.[1] Nevertheless, Phaire-Washington is herself director of the Natural Sciences Research Division of the Carver Research Foundation at Tuskegee University.[2]

Born in New York City in 1948, Phaire-Washington recalled that she "always had a curiosity about the workings of the cell," and was always "excited about looking at cells under microscopes."[3] As she grew older her curiosity continued to grow. She earned a bachelor of science degree in biology at Boston University in 1970, and immediately began graduate work in immunochemistry and cellular immunology at the City University of New York/Mount Sinai Medical School (CUNY), where, as she recalled,

> I was introduced to two very stimulating immunologists [doctors Kehoe and Capra]. After taking classes with [them], I became overwhelmed by . . . how the body defends itself against disease, and the remarkable way the body can heal itself.[4]

She completed her Ph.D. in 1975, and followed it up with a postdoctoral research fellowship at Rockefeller University (1976–1977). After two years in the Department of Anatomy at Howard University's College of Medicine (1977–1979), she became an associate professor of immunology and cell biology at Tuskegee University.[5]

Phaire-Washington's research has focused on ultrastructural studies of macrophage cells and the role of the cytoskeleton in cell membrane function. Her work has been supported by grants from the National Institutes of Health and the Department of Energy; the Agency for International Development supported her investigation of monoclonal antibodies in the development of an immunodiagnostic test for malaria.[6]

In addition to her research and teaching, Phaire-Washington has been active in attempts to increase the participation of minority women in science. She believes minority women's incentives to enter science are lowered by a lack of opportunity, training, and pay.

> The women who serve as role models [for younger minority women] are often burnt out and not enough models are available . . . to stimulate [the scientific interests of] younger women. We are entering the 21st century. According to current trends, the numbers of minorities entering fields of science, biotechnology, and engineering are decreasing rapidly. It will be important as a nation for many scientists (Black, white or any ethnic origin) who have interests in the training and growth of young minds to make commitments to turn the situation around.[7]

Phaire-Washington is married and the mother of two children.[8]

NOTES

1. Rosalyn Mitchell Patterson, "Black Women in the Biological Sciences," *Sage* 6 (fall 1989): 12–13. See also *Who's Who among Black Americans*, 4th ed. (Northbrook, Ill.: Who's Who among Black Americans Publishing, 1985), p. 868.

2. She is also an associate director of the Carver Foundation.

3. Phaire-Washington, quoted in Patterson, "Black Women in the Biological Sciences," p. 12.

4. Ibid.

5. Ibid. Phaire-Washington advanced to full professor status in 1981.

6. Ibid.

7. Ibid, pp. 12–13.

8. *Who's Who among Black Americans*, p. 868.

MILDRED E. PHILLIPS*

Pathologist Mildred E. Phillips's research has focused on tissue and organ transplantation. Born in New York City in 1928, she received her baccalaureate degree from Hunter College in 1946, and her M.D. degree from Howard University in 1950. She became the first Black to intern at King's County Hospital in Brooklyn, New York. She completed her residency at Mt. Sinai Hospital in New York, and then became a surgical assistant at Presbyterian Hospital (1954–1955).[1]

Phillips began her career as an instructor of pathology at the State University of New York's Downstate Medical Center in 1955. She had moved up to the rank of assistant professor by 1968, when she took a position as an associate professor of pathology at New York University Medical Center. She moved on to a professorship at the State University of New York at Stony Brook in the 1980s.[2]

NOTES

1. Vivian Ovelton Sammons, *Blacks in Science and Medicine* (New York: Hemisphere, 1990), p. 190. Marianna W. Davis, ed., *Contributions of Black Women to America*, vol. 2 (Columbia, S.C.: Kenday Press, 1982), p. 385. *American Men and Women of Science: The Physical and Biological Sciences*, 16th ed. (New York: Bowker, 1971–), p. 966.
   2. *Blacks in Science and Medicine*, p. 190.

BARBARA WRIGHT PIERCE

Cancer Research

Physician Barbara Wright Pierce's career has been almost completely overshadowed by that of her older sister, Jane Cooke Wright. The little that has been written about Pierce is peripheral—usually in the form of brief mentions in articles about her sister. She nevertheless deserves note as a physician who was a product of the Black elite—the younger daughter of a famous Black physician and the younger sister of another famous Black physician and researcher.

Pierce was born in New York City in 1920, one year after her sister. She too attended private schools in Manhattan, and then went on to Mount Holyoke College, where she earned the baccalaureate degree in 1943 or 1944. She received her M.D. degree from Columbia University College of Physicians and Surgeons in 1946 or 1947. Like her sister, she interned at Bellevue Hospital and completed her residency at Harlem Hospital, with a specialty in industrial medicine.

For a short time—how long is not clear—Pierce was engaged in research at the Cancer Research Center at Harlem Hospital with her father (the Center's founder) Louis Tompkins Wright, and her sister. She then became a member of the medical staff of the Metropolitan Life Insurance Company and a member of the medical staff of Union Health Center, which was operated by the International Ladies Garment Workers Union.[1] Her husband was the former New York Court of General Sessions Judge S. V. Pierce.[2]

NOTES

1. From an article on Jane Cooke Wright, "Outstanding Women Doctors: They Make Their Mark In Medicine," *Ebony*, May 1964, p. 76. See also the article on Jane Cooke Wright in George L. Jackson, *Black Women Makers of History: A Portrait* (Sacramento, Calif.: Fong and Fong, 1977).
   2. "Outstanding Women Doctors."

GWENDOLYN WASHINGTON PLA
Changing Fields within Science

Gwendolyn Washington Pla

Nutritionist Gwendolyn Washington Pla is the quintessential superwoman. An associate professor of nutrition at Howard University's Colleges of Dentistry and Allied Health Sciences and a nutrition consultant for eating disorders in the Department of Psychiatry at George Washington University, Pla also maintains a private practice, is taking graduate psychology courses at Howard University in preparation to enter the Baltimore-Washington Psychoanalytic Institute for training as an analyst, and is the divorced mother of two grown sons (Raymond, who is just completing his residency in anesthesiology, and Richard, who is just completing law school).[1]

Born in North Little Rock, Arkansas, in 1939, Pla grew up in the city proper, in a sprawling home occupied by a close-knit family: her mother, a teacher; her father, a railroad man; and a grandmother, two uncles, two aunts, and "many, many" cousins—all mostly older than she. As Pla recalled,

> I grew up in a very proud family that stressed the value of education. I was told how my mother's first cousin had been supported through Meharry Medical College by his sister, a school teacher, and his brother, a railroad man. My

father talked about how proud it would make him if I were to become a "lady doctor," but the very idea of blood repelled me.[2]

Pla's mother, who had always wanted to play the piano but had never had the opportunity to learn, insisted that she have piano lessons. Pla recalls fondly that her mother "always got season's tickets to concerts and took me to every event that was, in her words, 'uplifting.'"[3] She also described books and music as her "constant childhood companions. . . . My mother and one of my cousins, who was older, inspired my interest in books. I particularly enjoyed reading poetry, [and] also tried my hand at writing it."[4]

A gifted student, Pla attended segregated public schools; by the time she was about ten years old, it had become clear that she had a talent for science, but music was still her first love:

I wanted to be a singer and had voice lessons. But while I suppose my voice was pleasant enough, I wasn't talented enough for that goal, but I enjoyed all my music lessons, and even played trumpet in the high school band.[5]

In high school, Pla encountered a teacher who fostered her interest in science:

Mr. Dyer was my high school chemistry teacher and he was wonderful. He really stimulated my interest in chemistry and pushed me to go further than the usual. He gave me more challenging assignments and encouraged me to think and solve problems independently.[6]

Pla did very well in high school, as a result, she said, "of my mother, father, cousins, and teachers all encouraging me to make the honor roll and to be the best."[7] She graduated valedictorian of her 1957 class at James High School.

Like so many of the other women in this volume, Pla said there was never any doubt that she would attend college; "so in that sense it wasn't a conscious decision—I just knew I was expected to go."[8] Although she had been offered a scholarship to Howard University, her mother wanted her to stay closer to home, so her parents paid for her to attend Fisk University, in Nashville, Tennessee.

Despite her other academic talents and gift for science, when Pla went off to Fisk she had intended to major in music. However, "I made my decision when I got there. Once I heard my classmates sing and play the piano, I realized that I didn't have that kind of talent for music. It was very clear to me that my talents were in another area—I knew I could do chemistry."[9] Pla's parents' decision to send her to Fisk and her own decision to major in chemistry rather than music were good ones, she recalled. "Fisk, at that time, had an incredibly gifted science faculty, including Prince Rivers (chemistry), Marian Williams (botany), and Mary McKelvey (genetics)."[10] She graduated in 1961 with a B.S. degree in chemistry.[11]

For the next year (1961–1962), Pla worked as a chemistry instructor at Shorter College in her hometown of North Little Rock, Arkansas, but her professional science career began in earnest when she secured a job as a research chemist at the National Institutes of Health (1962–1963), where she participated in research on amino acids. In 1963, she moved to a position at the Walter Reed Army Institute of Research, where "I received my first exposure to nutrition [research]. I worked with a man who was studying the effects of chromium supplementation of Type II diabetics."[12]

During this time, Pla started graduate school, with her tuition paid by a federal job-related training program; she continued her research on the relationship between chromium status and glucose tolerance at Walter Reed, and attended school part time. In 1966, she took at position as a research chemist at the Food and Drug Administration's Division of Nutrition, and in 1968 she completed the M.S. degree in biochemistry at George Washington University.[13]

Pla was unsure about continuing on with doctoral study when she encountered a special mentor:

> After a few years I started working for James C. Fritz, [who was] perhaps an unusual mentor. He was not at graduate school but at the Food and Drug Administration. . . . He encouraged me, pushed me along.
>
> [Fritz] gave me the opportunity to present papers at meetings—I worked for him in the lab, and he'd say, "I think you ought to present this work at the next meeting. I've presented enough in my time." Or he'd say, "Why don't you write this up?"
>
> He was a very special man, a white man . . . and really encouraged me to begin work on my doctorate, although he himself did not have one. At his encouragement, I enrolled in the doctoral program at the University of Maryland in nutritional sciences.[14]

Pla continued working full time, with her graduate study again supported by a job-related federal training program. She completed the Ph.D. in nutritional sciences at the University of Maryland in 1978.[15]

The following year, 1979, Pla was promoted to the position of supervisory chemist at the Food and Drug Administration's Atlanta Center for Nutrition, where she did analyses of nutrients in foods and food supplements. She returned to Washington, D.C., in 1980 with a position as a supervisory food science and nutrition specialist at the U.S. Department of Agriculture's Food and Nutrition Service. From 1981 to 1983, she was a public health policy analyst with the Executive Office of the President in the Office of Management and Budget, where she made recommendations on national public health policy.[16]

In 1982, Pla began her private practice, which has focused on the management of diabetes, hyperlipidemias, obesity, renal failure, and eating disorders. After brief stints at the University of Maryland and the University of the District of Columbia, she became an assistant professor of nutrition at the Howard University Colleges of Dentistry and Allied Health Sciences. At Howard her responsibilities included developing the nutrition component in the curriculum for dental students and dental hygiene students; organizing and presenting faculty development in nutrition in dental education workshops; teaching nutrition courses for dental, dental hygiene, and nursing students; and compiling statistics on experimental design for pediatric dentistry and orthodontic residents.[17]

A closer examination of Pla's duties at Howard and the level of her professional status there echoes the lament of other Black women scientists, and provides part of the reason why Pla began a career change in 1994 and embarked on graduate studies in the psychology program at Howard University: "I spent nine years in the Colleges of Dentistry and Allied Health Sciences at Howard without tenure, and eleven years there before receiving a promotion from assistant professor to associate professor."[18] Another reason for her change of field, Pla said, was the fact that the patients she sees in her private practice "present with such complex issues and problems":

> Many have such a difficult time changing [habits or outlooks] and this is very destructive in some cases.
>
> I've been frustrated. . . . I am recognizing the issues but I can't treat them. [When] I've tried to refer them to therapy, sometimes they feel that the mere fact of my having made the referral makes them believe that I think they are mentally ill, which is not the issue. . . . So I would prefer to be in a position to treat such patients myself.[19]

Becoming an analyst allows Pla to do just that. Pla earned her M.S.W. (master of social work) degree at Howard University in 1998.[20] In the meantime, she continues her superwoman's juggling act, apparently loving every minute of it.

NOTES

1. Gwendolyn Washington Pla, interview with author at her home in Washington, D.C., 20 March 1996. Pla, personal communications including résumé, 20 March 1996; questionnaire, 20 March 1996; personal account, 20 March 1996; telephone interviews, 16 February 1996, 21 February 1996.
2. Pla, interview, questionnaire.
3. Pla, questionnaire.
4. Ibid.
5. Ibid.
6. Ibid.
7. Ibid.
8. Ibid.
9. Ibid.
10. Ibid.
11. Pla, interview.
12. Ibid.
13. Pla, personal account, résumé.
14. Pla, questionnaire, personal account.
15. Pla, résumé.
16. Ibid.
17. Ibid.
18. Pla, interview.
19. Ibid.
20. Pla, personal communication, 21 March 1999.

MURIEL E. POSTON

Juggling Tropical Botany Research, Family Life,
Teaching, Service, and Law Studies

Botanist Muriel E. Poston, an associate professor in the Department of Biology at Howard University and a research collaborator at the Smithsonian Institution's

National Museum of Natural History, is an expert on the flora of Central and South America, particularly Panama, Nicaragua, and Ecuador. Active in her field, both as a teacher of undergraduate and graduate students and as a researcher who has done a great deal of fieldwork, she has been a frequent presenter at professional meetings. Poston has also participated vigorously in faculty issues at Howard University, most recently serving a two-year term as president of the university's faculty senate.

Yet, frustrated with the pace of her career in botany, as well botany's steadily weakening position in the academic biological sciences, Poston has said that she is ready to give up science for a career in law: "It is difficult for female scientists to advance at Howard; in fact, it is difficult for female scientists at a majority of HBCU's [historically Black colleges and universities]."[1] Poston also indicated that the diminishing place occupied by academic botany at Howard is similar to that at other large universities:

> Botany departments are dying across the country, being swallowed up by other departments with their own agendas. When there was a Botany Department here at Howard, plant science was important; now that botany is just a part of Biology, plant science has been devalued, and this at a time when we should be increasing the number of trained botanists. . . . There's not much chance to do research here—there's no office, no equipment, no institutional support, and no infrastructure in place.[2]

Increasingly, she said, universities often see botany and zoology as introductory-level courses in pre-medical or pre-dental programs, and not as intrinsically important in their own right. Finally, she pointed out that another reason for her move away from science is that her duties in faculty concerns have presented a "competing interest" with her research, because Howard grants no release time for such activities. "I still have my regular teaching load; I still have my regular departmental duties."[3]

Born in Detroit, Michigan, in 1950, Muriel Poston grew up in a family of aunts, uncles, and cousins with a strong tradition for higher education.[4] Both sets of grandparents had completed high school—an unusual accomplishment in their day—and both parents were college graduates:

> My mother had gone to Talladega College and my father did his undergraduate work at the University of California at Berkeley; his two brothers also graduated from Berkeley. They were one of the first cohorts to move through Berkeley back in the 1930s and 1940s.[5]

Her father, Nathaniel Murrell, a student of John Kenneth Galbraith at Harvard University, was the second Black to receive a doctorate in business administration from that school, according to Poston. He completed the Ph.D., worked a year as a faculty member at Northeastern University, then died of cancer when his daughter was only ten years old.[6]

Her mother, a high school math teacher, moved the family, including two younger sons, west to Danville, California. Teaching full time, she raised her children with help from her own mother, from an extended family of aunts, uncles, and cousins, from other nearby families, and from her husband's brother, Marvin Murrell—a divorced, single parent, who was raising his own daughter, Marlene. Murrell acted as a surrogate father to his brother's children, and Poston's mother reciprocated by acting as a surrogate mother to his daughter.[7]

When Murrell, who was an optometrist with a practice in Oakland, moved to Danville, a white, bedroom community of about 10,000 outside Oakland, his was the first Black family in the town. When Poston's mother decided to buy there as well, it was a tricky endeavor, according to Poston:

> The way my mother ended up buying her house was to get a white buyer to agree to do it for her—a traditional way for African Americans to get a house was to have a white person buy the house and then buy it from them. When we moved in the neighbors came and said our family was lowering the property values and asked us to please move out. . . . Obviously, since we all had the same last name, townsfolk thought we were this huge Black family [trying to take over] the town. . . . It stayed that way pretty much until I graduated from high school. Neither my mother, who taught 30 miles away at the fully integrated Berkeley High School, nor my Uncle Marvin ever worked in the town.[8]

Poston and her younger brother were the only Black children in the school. Each was tested during the fifth and sixth grades, and both were placed on the honors gifted-and-talented track and entered special classes. "We were the only Black kids in the school," Poston recalled, "but we were smart enough that we were integrated into the curriculum."[9] Poston recalls that is was difficult for teachers to tell her apart from her cousin:

> Marlene had gone to junior high and high school [about four or five years] before me, but when I entered high school just after she had graduated a lot of teachers thought I was Marlene—still there for another four years. They couldn't differentiate us, apparently.[10]

Poston went to San Ramon High School—where she was the only Black female teenager in school:

> It was interesting in some respects because, obviously, since we were the only Black family there was no social life that we could be associated with. This was the early 1960s so there was no [interracial] dating. The student government would have their balls and they'd hold them at the local country club and I couldn't go, not only because I didn't have a date but because they didn't let Negroes into the country club. . . . So even if some young white boy had wanted to invite me, I wouldn't have been allowed in. There were all those issues, but I was active in student government and worked on the newspaper and did all that kind of stuff.[11]

At times, she encountered racism in class:

> Once when I came home and complained that my Spanish teacher made me pay every time I used an English word in class but didn't make any of the other students pay, my mother and my uncle went down and confronted [the teacher]. They said what the teacher was doing was inappropriate and, "You are not going to do this to Muriel, and we want it stopped now! If you don't we will go to the

principal's office and the superintendent's office, and you will be sorry you ever did this." The teacher refused, and so my mother and uncle went over the teacher's head and it was finally stopped.[12]

Other teachers, however, were supportive and encouraging:

> I took a biology course in high school where my teacher, Mr. Bruce, recruited a bunch of us students and took us out to do a botanical treatment where we collected plants in the local state park. He never attempted to depress my ambition, and was the first one who took me out to collect plants, so I would have to credit Mr. Bruce for my becoming a scientist.[13]

She won a National Merit Scholarship, a National Achievement Scholarship, and a Daughters of the American Revolution (DAR) scholarship. Poston commented that "the DAR had a fit when they found out I was Black, but they didn't take the scholarship back."[14] She also said that because of local newspaper publicity regarding her scholarships, her mother received quite a bit of hate mail.

There was no question that Poston would go to college. The family attitude was that all the children in the family would all go to college and they would go to the *best* college possible. The scholarships she received would pay for college completely, and there were some funds left from her father's social security and life insurance. Because her father had gone to graduate school at Harvard, her mother wanted her to attend Radcliffe. But the year she applied for admission, Poston said, "there was already a junior from my high school accepted; they were only admitting one from my high school that year and she got it. My second choice was Stanford, but my mother is bitter about it to this day."[15]

Poston majored in biology at Stanford University, where she recalled encountering few race or gender issues. "Medicine was my first choice," she said, "and any scientist who tells you it wasn't is probably lying."[16] That was when her high school botany project influenced her career path at Stanford:

> Stanford had a study abroad program and I was fortunate enough to go for six months to Austria. This was in the spring of 1970 when we bombed Cambodia, and everything fell apart. I had to try to register, and I couldn't get into embryology, which I would have had to take the next semester if I were going to continue with pre-med, so I looked for something to fill the time slot. I had enjoyed my high school science field project so much that I decided to sign up for the anatomy and morphology of vascular plants.[17]

She then took courses in evolutionary biology, taught by Peter Raven, and taxonomy courses, taught by John Thomas, and said she never looked back:

> I worked with Raven, one of the preeminent botanists in the world, and after working with him, I thought, "I don't need to be in premed courses with all those crazies. . . . People do this [botany] for a living and they get paid and this is fun, exciting and interesting." I thought I could stay outside and watch insects pollinate plants. . . . I could travel . . . I could go to tropical places. . . . I decided I wasn't going to go to medical school, I was going to go to graduate school.[18]

Poston worked more with Thomas than anyone else, particularly in the herbarium, which was essentially his laboratory. He would later serve on her dissertation committee as an external member:

If there's any sort of continuing theme of a person who mentored my science career it would be Thomas. He was also on my graduate thesis committee and has since mentored my science career.[19]

In the meantime, Poston's brother had been accepted to Harvard University.[20] Poston learned that because of her summer semester overseas she could double up on some courses and graduate early:

I realized my mother didn't need to have two of us in college at the same time, so I graduated early. But then it was like, what am I going to do with myself? I talked to my advisors, and they said, "Well you should go to graduate school," and I asked where I ought to go, and they advised me to apply to a couple of places that were doing really hot systematic biology at the time—Harvard, Michigan, Oregon State, the University of California at Los Angeles, and the co-terminal program at Stanford. . . . Just the fact that I was gifted and interested was a big plus; I could go to my professors with problems and they were receptive.[21]

Poston was accepted both to UCLA with four years of support, as well as to Stanford's co-terminal program. She was unsure which to choose. Her mother and uncle convinced her to go to UCLA:

My Uncle Marvin said, "You have the momentum, don't drop the ball. You must keep moving ahead." In retrospect, if I had to do it over again, I think it might have been useful to take that extra year at Stanford and get a master's and then go to grad school. Sometimes, I feel like I've been running ever since.[22]

Another Stanford professor, Hal Mooney, sponsored her for an undergraduate research project, and she received one of the first undergraduate research grants awarded by the National Science Foundation (NSF).[23] She completed her B.S. in 1971, and continued to work in Mooney's lab during the summer after her graduation.[24]

Loaded with scholarships, Poston had just turned twenty-one when she entered graduate school at UCLA. She had a four-year graduate fellowship from UCLA's Minority Affairs Office, a Department of Biology teaching assistantship, and a systematic botany fellowship; later, she also received an NSF dissertation grant.[25] As it turned out, Poston found that UCLA at that time was a good place to be for a woman, a minority, and a "hot" biology student:

The thing that was interesting about UCLA's Ecology and Evolutionary Biology Group was that there was a fairly large cohort of us that came in at the same time. The year I was admitted was the last year of the Botany Department, immediately after that they merged Botany and Zoology, so my advisor was the first chairman of the Biology Department. It also meant that where there had once been a smaller cohort there was now this larger mass, and there were several minority students who were admitted at the same time because UCLA was making this concerted effort to create diversity in its graduate programs.[26]

Although she had no strong mentor at UCLA, she did work well with Harry Thompson, who was associated with her former professors; she continued to rely on him by dint of his own graduate studies at Stanford. She worked well with her peers and made one "truly good friend," Barbara Pleasant, with whom she did fieldwork:

Barbara's an ornithologist, a Sephardic Jew, so she has very curly hair, and we'd always go to the same place to get our hair cut—kind of like soul sisters, this Jewish lady and this African-American lady. We worked well together, and formed a strong sense of community, and she's truly my friend. . . . She's in the Zoology Department at Iowa State now, and we still keep in touch—the four of us [Poston, Pleasant, and their husbands] are long-standing friends.[27]

Poston completed her M.S. in botany in 1973, married, and soon became pregnant. As luck would have it, Elaine Tobin had just been hired by the biology department, as Poston recalled:

One day she asked me if I had done any planning about my child care—and of course I hadn't. She's the one who pointed me in the right direction and she said, "As a female graduate student in the sciences you need to have all this stuff lined up or you are going to drown." If she hadn't intervened I wouldn't have known the questions to ask.[28]

Interestingly, Poston said the people in her department—particularly the professors—did not much change their attitudes toward her when she got pregnant.[29]

Poston's first child, a son, was born in 1975, and she recalled that trying to combine the responsibilities of a young mother and a graduate student was difficult:

I must admit my family planning wasn't the best, but in retrospect I'm sure glad I did it my way. I had my son in the middle of graduate school because I thought he would suck his fingers and toes while I wrote my dissertation. But babies don't do that, as I quickly found out.[30]

She completed her Ph.D. in biology in 1979, quickly secured a postdoctoral position at the Missouri Botanical Garden in St. Louis, and soon got promoted to the position of administrative curator. In 1980, Poston accepted an assistant professorship in what was then the Department of Botany at Howard University in Washington, D.C.; she was advanced to the associate professor level in 1988.[31]

In terms of her career path and her research, Poston, a specialist in tropical botany, said the only way she can publish is to "get away," yet the only sabbatical she had was during the 1990–1991 academic year: "I have four classes to teach and I teach every day, so my treatments [on new species] are sitting there, waiting."[32] She is convinced that if she had been a white male, or even a male of any race, she would have produced more research:

It would have been different—my research would have been easier to do if I had been male, and there would have been more of it. A white male friend who I collaborated with in grad school, and who was not mentored any more than I was, has made more substantive contributions [than I], not because he is brighter or harder working necessarily, but because he is a male and at a college where the teaching load is not what it is here.[33]

Gender and traditional expectations for homemakers and mothers have an impact on women across the board, Poston says: "There's no question that a woman's home life is an issue. I was going up for tenure the same time as I had kids in diapers, and that takes it out of you."[34] Yet she has managed to carve out a place to do botany at Howard, although it has not been easy.

Pointing out that Howard lacked any substantive botany facilities, Poston eventually convinced the administration to give up a men's bathroom in the Ernest Everett Just Building, which she turned into an herbarium. Proud of her small victory, she explained to a visitor how each plant specimen in the collection is dried and fixed to an herbarium sheet label, which tells what it is, and where, when, and by whom it was collected. "It's mostly a teaching collection," Poston said, adding,

A systematic collection is valuable when you can't actually be in the field. Many collections [at other universities and botanical gardens] are already computerized, and so it's fairly easy to see what the holdings of each collection are, but it's hard to see the detail in the specimen. They are scanned into the computer, but this is not as good as seeing something first hand.[35]

Securing an herbarium was just one of the competing interests that have taken time away from her research; another was her involvement in faculty issues:

We had some substantive issues facing the Faculty Senate during my tenure. The University Handbook presented a real problem, as it involved tenure; they [the administrators] inserted a clause that said tenure notwithstanding, if we want to fire an individual we can; which in essence abrogated all tenure rights, so we have been struggling with them on that issue, although I think it's going to be ultimately resolved.[36]

Poston stressed the importance of such service, but acknowledged that it was not reflected on her curriculum vitae: "I have spent fifteen years working on my study of the flora of Ecuador, which, if there had not been other issues draining my time, could have been done in two years."[37]

Married to a physician, Poston says that her children have been a big part of her life and have impacted her productivity, especially when they were small and she went home every day at five o'clock. She is well aware that male scientists are not overwhelmed by such considerations. With her children now grown, Poston appears determined to move into the field of law, which would represent a gain for the legal community but a loss for the scientific community.

NOTES

1. Muriel Poston, telephone interview with author, 1 February 1996. Poston, e-mail to author, 28 February 1996, 5 March 1996, 29 March 1996. Poston, interview with author at her office at Howard University, Washington, D.C., 21 March 1996. Poston, curriculum vitae and author questionnaire, 21 March 1996. This quote was taken from the telephone interview.

2. Poston, interview. Poston's comment, "and this at a time when we should be increasing the number of trained botanists," was partially in reference to the threat to the Amazon rain forest and the expected loss of many valuable plants, many of which might have tremendous potential use as medicines.

3. Ibid.

4. See the entry in this volume on Joan Murrell Owens for more information on this remarkable family.

5. Poston, questionnaire and interview.

6. Poston, questionnaire.

7. Poston said there was also an aunt who lived sixty miles away in Sacramento, and many other family connections (ibid.).

8. Ibid.

9. Ibid.

10. Ibid.

11. Ibid.

12. Ibid. Since I have not conducted an interview with this teacher, I have deleted the teacher's name in the interest of fairness.

13. Ibid.

14. Ibid.

15. Ibid.

16. Poston, questionnaire and interview.

17. Poston, questionnaire.

18. According to Poston, the courses she took with Raven were seminars he gave for his graduate students (Poston, interview). Peter Raven is now director of the Missouri Botanical Garden; John Hunter Thomas is professor emeritus at Stanford; and Harold Mooney is still teaching at Stanford.

19. Poston, interview.

20. Her brother graduated from Harvard and went to MIT for graduate studies; her youngest brother attended Fisk for two years before joining the Navy, where he made a career.

21. Ibid.

22. Ibid.

23. This particular type of NSF grant is now called Research Experiences for Undergraduates (ibid.).

24. Poston, questionnaire.

25. Poston, questionnaire and curriculum vitae. Poston held a teaching fellowship in biology at UCLA from 1971–1977; during the 1974–1975 school year, she was a National Science Foundation project teacher at Washington High School in Los Angeles, California; and during the 1977–1978 school year she was an instructor (Botany I) at Santa Monica College in California.

26. Poston, interview.

27. Ibid.

28. Ibid.

29. Ibid. Poston does, however, believe that her colleagues at UCLA "always thought I was basically 'out to lunch.'" She also recalls one instance when she was working in the herbarium and felt morning sickness coming on: "I just knew I was going to lose a bad ham sandwich and one of the people on my committee was there and I said, 'I think I'm going to be sick.' If you've ever been in a herbarium, it's like a library with old musty books, and so I'm coming down the stairs, and Frank is getting ready to come up and suddenly you never saw anybody run so fast in your life" (ibid.).

30. Ibid.

31. Poston, curriculum vitae.

32. Poston, interview.

33. Ibid.

34. Ibid.

35. Ibid.

36. Ibid.

37. Ibid.

JESSIE ISABELLE PRICE
Groundbreaking Work in Veterinary Microbiology

Jessie Isabelle Price
Photo by James Runningen

Veterinary microbiologist Jessie Isabelle Price has spent her entire career in major research laboratories, where she has developed vaccines against a variety of organisms that once killed ducks and other waterfowl. Born in Montrose, Pennsylvania, in 1930, Price was the only child in a single-parent home.[1] Raised by her mother, Teresa Price, she attended the public schools in the predominantly white town of Montrose, Pennsylvania. "I was the only Black in my class," Price said. "There was a Black male student three years ahead of me, and a Black female three years behind me, and that was it."[2] Although the family was poor, Price's success in school was strongly encouraged by her mother, as well as by her teachers; she recalls three teachers had a particularly strong impact and supported her desire for a college education.[3]

Price graduated from Montrose High School and was accepted to Cornell University, but feared she might not have a strong enough academic background to succeed at Cornell. As a result, she and her mother moved to Ithaca, New York, where she attended high school for an additional year, taking advanced math and English courses, along with French and German.[4]

Because she had spent a year living in Ithaca and attending high school, Price had established New York state residency; although she did not have an official scholarship to attend Cornell University, her tuition was waived because of her resident status and her high school grades. Other than living expenses, she only had to cover the cost of books and a few nonrefundable fees.[5] She wanted to become a physician, but financially it was not an option. During her first semester at Cornell, Price developed a friendship with a microbiologist, who, she said, "Told me what it entailed—it sounded so interesting I decided to major in veterinary microbiology."[6] In 1953, she earned the B.S. degree in microbiology at the College of Agriculture of Cornell University.[7]

As was the case for so many of the women in this study, a mentor proved crucial to Price:

> Dorsey Bruner was my major professor, my thesis advisor and my dissertation advisor. When I graduated, he suggested that I apply to grad school. At that point, I thought I'd had enough of school, but he encouraged me to think about it and I did. Still, money was a problem.[8]

Price got a job as a laboratory technician at the Poultry Disease Research Farm of the New York State Veterinary College at Cornell University, where she worked from 1953 to 1956.[9] During this period, she continued to receive encouragement from Bruner, until finally she decided to attend graduate school: "I told [Bruner] that I'd only go for a master's degree, but all along he kept encouraging me, asking me why I didn't stay on for the doctorate, and so I did."[10]

Owing in large measure to Bruner's support, Price received research assistant-ships for graduate studies (1956–1959), and was awarded the M.S. degree in veterinary bacteriology, pathology, and parasitology by the New York State Veterinary College at Cornell University in 1958; her master's thesis, "Morphological and Cultural Studies of Pleuropneumonia-like Organisms and Their Variants Isolated from Chickens," was done under Bruner's direction.

Price earned her Ph.D. in 1959. She did her dissertation research, "Studies on *Pasteurella anatipestifer* Infection in White Pekin Ducklings," under Bruner's direction; her study was later published in the *Journal of Avian Diseases*.[11] She was on her way to becoming a recognized authority on avian diseases, particularly those that afflict waterfowl.

For the next eighteen years, from 1959 to 1977, Price worked as a research specialist at the Cornell University Duck Research Laboratory in Eastport, Long Island, where she conducted research on the identification and control of bacterial diseases in commercial white Pekin ducklings. When she began her investigations, eastern Long Island was the major producer of the edible waterfowl, and farmers were losing approximately 30 percent of their animals to bacterial diseases during their eight-week growth period.[12] Price discovered a method for isolating and reproducing the *Pasteurella anatipestifer* organism that causes respiratory disease in ducklings. She described her experience as similar to that of the eponymous "father" of the various forms of *Pasteurella*, Louis Pasteur, who had one day accidentally used a weakened strain of the microbe that causes chicken cholera: "It was purely an accident. I simply happened to use the right medium and through pure chance used certain temperatures. Only good luck made me stop the process at the right point."[13] She then developed and tested the first injectable vaccine, initially under controlled laboratory conditions and then under field conditions. She employed flocks of 3,000 to 4,000 ducklings for the studies, intermingling equal numbers of vaccinated and unvaccinated ducklings in each trial. A deft veterinary surgeon, she conducted daily autopsies and collected cultures throughout the ducks' eight-week growth period. Her intensive study of the ducks' mortality revealed multiple causes of death; "the birds were dying from *Pasteurella multocida*, *Escherichia coli*, and duck hepatitis in many instances."[14] In 1966, Price received a National Science Foundation travel grant under the auspices of the American Society for Microbiology to attend the Ninth International Congress for Microbiology in Moscow and present her research findings.

By 1974, Price had developed and tested dual *anatipestifer-coli* vaccines that were injectable, and she then began preliminary field studies testing a *Pasteurella anatipestifer* vaccine that could be delivered orally.[15] The *Pasteurella multocida* and *Pasteurella anatipestifer* vaccines she developed were used commercially by the Pekin duck farmers on Long Island and by other farmers throughout the Midwest and Canada. Her *Pasteurella multocida* vaccine was also used by commercial turkey growers, and her *Salmonella typhimurium* vaccine was used by a commercial pigeon farmer.

During her tenure at the Duck Disease Research Laboratory, Price also taught at two local colleges. Between 1963 and 1969, she taught microbiology at the Mitchel College Branch of Long Island University in Westhampton Beach, where she moved from the rank of adjunct lecturer to adjunct assistant professor. Between 1969 and 1976, she taught earth sciences at Southampton College of Long Island University in Southampton, New York, where she moved from the rank of adjunct associate to adjunct professor.[16] Price has not only taught others, but throughout her career has supplemented her own education with continuous training in the technical advances in her field.[17]

In 1977, Price left the Duck Disease Research Laboratory and took a position as a research microbiologist with the National Wildlife Health Center of the National Biological Service in Madison, Wisconsin.[18] Price's current research includes studies on the interactions of environmental contaminants and diseases among wildlife in general, and waterfowl in particular. She develops new microbial detection and characterization techniques, identifies and characterizes newly discovered microbial diseases of wildlife, and develops approaches for reducing wildlife mortality and morbidity occurring as a result of microbial infections.

Since joining the National Wildlife Health Center, Price has undertaken a number of research projects that have made substantive contributions to the field. To enhance the survival rate of captive-reared Aleutian Canada geese when they are released into the wild, she developed and tested a *Pasteurella multocida* serotype-1 vaccine to prevent avian cholera.[19] Her vaccine was tested for efficacy in a number of other bird species, including mallards, Pekin ducks, redhead ducks, snow geese, and Sandhill cranes; the results suggest that her vaccine may be useful in some species of endangered birds being raised for propagation and also in display flocks when there is risk of exposure to avian cholera.[20] Another of Price's studies—conducted in cooperation with the National Biological Service, the U.S. Fish and Wildlife Service, the states of California, Oregon, and Washington, the Canadian Wildlife Service, and Russian scientists—was designed to discover a relationship between snow geese and outbreaks of avian cholera; Price found the first tangible evidence that snow geese may be carriers of *Pasteurella multocida*.[21] She has also conducted intensive investigations on the survival of *Pasteurella multocida* organisms in pond water, which suggested that high concentrations of calcium ions, and a synergistic response between calcium and magnesium ions, enhanced the survival of *Pasteurella multocida* organisms. Her findings provided the first step toward understanding why some wetlands have a higher frequency of avian cholera outbreaks than other nearby wetlands, and were valuable in establishing criteria for determining wetland management strategies to control avian cholera.

Price has also conducted studies on *Mycobacterium avium* in Sandhill cranes and snow geese in an effort to understand why avian tuberculosis occurs at such a high rate in whooping cranes, an endangered species. She now provides laboratory assistance to the Whooping Crane Recovery Team as part of a health monitoring program that is done prior to the relocation and release of whooping cranes.[22] Price's microbiological analysis of Hawaiian brown tree snakes sought to determine their role as a possible disease vector that limits the population of Hawaiian forest birds; she recovered *Pasteurella multocida*, the cause of avian cholera, for the first time in wild Hawaiian birds. She has also conducted a microbial analysis of the brown tree snake on Guam to determine its normal parasites and microbes with an eye toward its role in decreasing the bird populations on the island; for this study Price developed two new microbial culture protocols for isolating microorganisms.[23] Among the new isolates she reported in Guam birds were *Listeria gravi/murrayi* and *Bordetella bronchiseptica.*[24]

Price's most recent work has involved two studies on the development of mutant and subcellular *Pasteurella multocida* vaccines. She is seeking new methods for providing birds with protective immunity against *Pasteurella multocida*, and has shown that oral immunization with a subcellular vaccine can effectively protect ducks from the organism. According to Price, a single outer membrane protein can elicit protective immunity, which makes a recombinant live vaccine for use on free flying birds possible.[25]

In addition to her own research at the National Wildlife Health Center, Price continues to help other researchers. She served as a research advisor and graduate committee member for three students enrolled in the United States Forest and Wildlife Service/National Biological Service Cooperative Education Program,[26] and hosted Dr. Atef Kamel, a visiting scientist from the Veterinary College of Egypt, who spent two weeks in Price's lab gaining experience working with *Pasteurella multocida* serotyping and ELISA techniques.[27]

Price is currently a member of the University of Wisconsin's Institutional Biosafety Committee, and she participates in a number of scientific organizations, including the American Association for the Advancement of Science, the Association for Women in Science, and the American Institute of Biological Sciences.[28]

Price has been a very active member of the American Society for Microbiology, and has been the chair of its Predoctoral Minority Fellowship Ad Hoc Review Committee and its Summer Research Fellowship and Travel Award Program since 1982. She is also a member of the organization's Committee on the Status of Minority Microbiologists (1980 to present), and has participated in the Committee on the Status of Women Microbiologists (1976 to 1979, chair from 1978 to 1979). Price has been very active in the honorary Sigma Delta Epsilon — Graduate Women in Science, where she has served as national president (1974–1975), national second vice-president (1972–1973), and a member of the national board of directors (1976–1980).

NOTES

1. Jessie Isabelle Price, personal communications to author, including a curriculum vitae with summary of accomplishments, 4 September 1995; and assorted letters during 1995 and 1996. Price, telephone interview by author, 27 June 1995, 25 June 1996.

2. Price, interview, 27 June 1995.

3. The teachers were Mr. Birchard, her history teacher; Mrs. Highhouse, her math teacher; and the school's principal, Mr. O'Brien (Price, interview, 25 June 1996).

4. Ibid.

5. Ibid.

6. Ibid.

7. Price, curriculum vitae. See also *American Men and Women of Science: The Physical and Biological Sciences*, 14th ed. (New York: Bowker, 1971–), p. 4036. "Doctor to Long Island Ducks: N.Y. Woman Scientist Fights Disease That Causes $250,000 Damage Yearly," *Ebony*, September 1964, pp. 76–78, 80, 82. "'Salute to Women' Fete: Governor Is Speaker," *New York Amsterdam News*, 20 October 1962, p. 12. "Veterinary Microbiologist," in *Contributions of Black Women to America*, ed. Marianna W. Davis, vol. 2 (Columbia, S.C.: Kenday Press, 1982), p. 432–433.

8. Price, interview, 25 June 1996.

9. Price, curriculum vitae.

10. Price, interview, 25 June 1996.

11. J. I. Price, "Studies on *Pasteurella anatipestifer* Infection in White Pekin Ducklings," Ph.D. dissertation, Cornell University, 1959; published in *Avian Diseases* 3, no. 4 (1959): 486–487.

12. White Pekin ducks are also grown commercially in Indiana, Wisconsin, and Canada (ibid.).

13. "Doctor to Long Island Ducks," p. 76. On Pasteur, see Paul DeKruif, *The Microbe Hunters* (New York: Harcourt, Brace, 1926), pp. 140–141. According to DeKruif, Pasteur used microbes from a bottle that was several weeks old, which were thus weakened enough to work as a vaccine (p. 141).

14. Price, "Studies on *Pasteurella anatipestifer*."

15. Ibid. J. I. Price, "Immunization Trials Using a Low-Virulent *Pasteurella anatipestifer* Culture in White Pekin Ducklings," in *1974 Duck Research Progress Report* (Cornell University Duck Research Laboratory, 1974), pp. 77–105; Price, "Immunization of Ducklings with Low-Virulent *Pasteurella anatipestifer*," in *1975 Duck Research Progress Report* (Cornell University Duck Research Laboratory, 1975), pp. 95–123; Price "*Pasteurella multocida* Studies," in *1975 Duck Research Progress Report* (Cornell University Duck Research Laboratory, 1975), pp. 124–132. See the appendix for more of Price's publications.

16. Price, curriculum vitae.

17. The additional training courses Price has taken are pathology of diseases of laboratory animals (Armed Forces Institute of Pathology, Washington, D.C., 1963), American Society for Microbiology anaerobic traveling workshop (New York Public Health Laboratories, New York, 1971), *Pasteurella* serotyping training course (USDA—National Animal Disease Laboratory, Ames, Iowa, 1971), microbiology workshops in Enterobacteriaceae, parasitology, syphilis serology, immunofluorescent techniques and mycobacteriology (New York State Department of Health at C. W. Post College, New York, 1971), clinical cellular immunology workshop (American Society for Microbiology, Los Angeles, California, 1979), application of statistics to microbiological examination of pharmaceuticals, food and water (American Society of Microbiologists, Dallas, Texas, 1981), latest methods of detection of *Clostridium botulinum* and its toxins (ASM, Dallas, Texas, 1981), new developments in drinking water microbiology (ASM, Las Vegas, Nevada,

1985), advanced anaerobic bacteriology for the clinical laboratory (ASM, Washington, D.C., 1986), nucleic acid probe technology (ASM, Atlanta, Georgia, 1987), nonmedical applications of nucleic acid probes–gene probes as a tool in environmental monitoring (ASM, Atlanta, Georgia, 1987), laboratory and clinical aspects of nontuberculous mycobacteria (ASM Traveling Workshop, State University of New York, Farmingdale, NY 1987), mycobacteriology bench training (USDA-National Veterinary Services Laboratory, Ames, Iowa, 1987), clinical mycobacteriology—update and rapid methods workshop (ASM, Miami Beach, Florida, 1988), rapid detection of microorganisms by using nucleic acid probes and other related techniques (ASM, New Orleans, LA, 1989), theoretical and hands-on introduction to frequently used techniques in biotechnology (ASM, Anaheim, California, 1990), recombinant DNA technology (Madison Area Technical College, Madison, Wisconsin, 1991), detection, recovery, and amplification technique for plasmids in the environment (ASM, New Orleans, LA, 1992), and use of gene probes in clinical microbiology laboratory (ASM, New Orleans, Louisiana, 1992) (Price, curriculum vitae).

18. Ibid. According to Price, "The National Wildlife Health Center provides agencies within the Department of Interior with the capability to inventory, monitor, prevent, and control disease problems in wildlife species under Department of Interior stewardship. The Center houses the only program in the federal government with a basic mission of combating disease in free-ranging wildlife populations for the specific benefit and protection of those species; pursues an integrated program of research, inventory, monitoring, field response, diagnostic services, and training in the identification and control of wildlife diseases; and provides a wide range of technical assistance to Department of Interior agencies, particularly the Fish and Wildlife Service and the National Park Service, as well as other federal agencies, state conservation departments, and the private sector. Price plans and conducts research on the diagnosis, pathogenesis, and epizootiology of bacterial diseases that can be applied in the development of control strategies" (Price, curriculum vitae).

19. "At the request of the Sacramento-Wilcox Wildlife Management Area, an experimental use permit was obtained from the USDA and sufficient quantities of vaccine were produced to immunize the giant Canada goose propagation flock at the Sacramento-Wilcox Wildlife Management Area. The vaccine eliminated avian cholera as a cause of mortality in that flock and was successfully used for five years due to collaborative efforts by Price. The vaccine was found to be effective and safe for both captive and wild-trapped Canada geese" (Price, curriculum vitae).

20. "There was great variation in the degree of efficacy for each species. One hundred percent protection was achieved in Pekin and mallard ducks, while protection of the redhead ducks, Sandhill cranes, and snow geese was 25%, 67% and 42% respectively. Significantly higher serum antibody levels measured by ELISA were noted in the vaccinates that survived challenge compared to those that died. A live oral vaccine and vaccines prepared with each of two cellular fractions of *P. multocida* organisms were tested for efficacy in Pekin ducks by challenge with virulent organisms. One or two doses of the live avirulent vaccine protected 100% of the ducks when given subcutaneously, and two doses given orally protected 100% of the ducks. A vaccine prepared with the purified capsular polysaccharide fraction of the cells protected 60% of the challenged vaccinates and a vaccine prepared with the outer membrane proteins protected 98% of the ducks when challenged with virulent organisms" (Price, curriculum vitae).

21. "For this continuing study, lesser snow geese were captured on Wrangel Island, Russia in 1993 and 1994 and Banks Island, Canada in 1994" (Price, curriculum vitae).

22. "The study suggested that an extremely low prevalence of tuberculosis existed in

free-flying cranes and snow geese. This was in contrast to the extremely high rate of infection in the whooping cranes, suggesting that whooping cranes are extremely susceptible to this disease or are becoming infected by some other means. Special procedures were developed for culturing tissues to ensure recovery of the slowly growing, fastidious *M. avium* organisms" (Price, curriculum vitae).

23. Ibid.

24. Price obtained and examined tracheal or pharyngeal swabs from twenty-five snakes. Sixty-seven isolations were made from these samples, of which only eight had been reported previously from snakes (ibid.).

25. "Components of *P. multocida* cells were prepared and used to develop a subcellular vaccine. The bacterial cells were grown in large volume and fractionated to prepare purified capsular polysaccharide (PCP) and outer membrane proteins (OMP). The PCP was found to be a poorer immunogen that protected 60% of the vaccinates while a dose of 1.0 mg OMP was highly efficacious and protected 98% of the challenged vaccinates. A strong serum antibody response developed which protected ducks when challenged with virulent organisms" (Price, curriculum vitae).

26. Hildy Reiser (1978–1981) received an M.S. degree in wildlife ecology; Amna El Tayeb of the Sudan (1989–1993) received her Ph.D. from the Department of Animal Health and Biomedical Sciences; Susan Clark completed the M.S. degree in August 1995 and is continuing her studies for a Ph.D in the Department of Animal Health and Biomedical Sciences. Each student project was partially funded through a research work order and Price served as project officer and consultant for each study. Price also provided guidance and information and laboratory training in morphological, biochemical, and serological identification of *P. multocida* organisms to Lidija Cicnjak-Chubbs (1991–1994), a graduate student and M.S. candidate at the Cooperative Wildlife Research Laboratory, Southern Illinois University at Carbondale (Price, curriculum vitae).

27. The enzyme-linked immunoadsorbent assay (ELISA) is a standard laboratory serological test.

28. She was appointed to the committee in 1986 (ibid.).

INEZ BEVERLY PROSSER

Educational psychologist Inez Beverly Prosser graduated from Prairie View Normal College with a degree in teacher training around 1916. She taught for a while in segregated schools in Yoakum, Texas, before moving to Austin. She then taught in Austin's segregated public schools and attended Samuel Houston College, where she earned a B.A. in education around 1924.[1] After her marriage, she attended the University of Colorado, where she earned a master's degree in educational psychology in 1927.

Prosser returned to Austin and taught education courses at Tillotson College, where she was also a dean and the registrar. In 1930, she became a professor and administrator at Tougaloo College in Mississippi.

Prosser was awarded a Rockefeller Foundation general education board fellowship in 1931, which allowed her to pursue her doctorate at the University of Cincinnati; she received her Ph.D. in educational psychology in 1933. Her dissertation, which studied the non-academic development of Black children in mixed and segregated elementary schools, was one of the earliest investigations into the relationship between school and the socialization of children. One year later, in 1934, she was killed in an automobile accident.[2]

NOTES

1. The exact date of Prosser's B.A. degree is not available. Beth Jungwirth of the registrar's office at the University of Colorado indicated "for informational purposes only" that Prosser's bachelor's degree was awarded by Samuel Houston College (telephone conversation, 29 April 1996). Samuel Houston College is now Houston Tillotson College; according to the registrar, no records are extant under the maiden name Beverly.

2. Robert V. Guthrie, *Even the Rat Was White: A Historical View of Psychology* (New York: Harper & Row, 1976), pp. 134–135.

JOHNNIE HINES WATTS PROTHRO

Amino Acid Research

Although chemist Johnnie Hines Watts Prothro and Cecile Edwards were both at Tuskegee Institute and the Carver Foundation in the mid-1950s, I have not been able to find any evidence that they ever collaborated, although it is likely that they did associate with one another. Born in Atlanta, Georgia, in 1922, Prothro earned the baccalaureate at Spelman College in 1941, and immediately obtained a position as an instructor of foods and nutrition at Atlanta's Washington High School, where she remained until 1945.

Although the records are not clear, she most likely pursued summer-semester studies at Columbia University, which awarded her an M.S. degree in chemistry in 1946.[1] She earned her doctorate at the University of Chicago in 1952; her dissertation was titled "The Relation of the Rates of Inactivation of Peroxidase, Catecholase, and Ascorbase to the Oxidation of Ascorbic Acid in Vegetables."[2]

Immediately after obtaining her Ph.D., Prothro became a professor of foods and nutrition at Tuskegee Institute in Alabama, where she was also a research associate at the Institute's Carver Research Foundation.[3] In 1954, she was awarded a three-year grant ($47,414) by the U.S. Department of Health, Education and Welfare to study amino acid nutrition problems in adults. From 1958 to 1959 she had a National Institutes of Health fellowship to the University of California at Los Angeles for the study of public health.[4]

Nutritional problems related to the availability of amino acids in foods were prevalent in young adults during the mid-1950s — particularly among poor Blacks living in the South. Prothro's research focused on, among other areas, comparisons of essential amino acid patterns and how levels of nitrogen might affect the biological availability of amino acids in human subjects.[5] She published more than twenty papers based on her research.[6]

NOTES

1. *Spelman College Catalogue, 1941–2* (Spelman College, Atlanta, Ga., April 1942). Vivian Ovelton Sammons, *Blacks in Science and Medicine* (New York: Hemisphere,

1990), p. 244. "Physical Chemists," in *Contributions of Black Women to America,* ed. Marianna W. Davis, vol. 2 (Columbia, S.C.: Kenday Press, 1982), pp. 445–446. See also *American Men of Science,* 10th ed. (New York: Bowker, 1967), p. 4308. Julius H. Taylor, Clyde R. Dillard, and Nathaniel K. Proctor, eds., *The Negro in Science* (Baltimore: Morgan State College Press, 1955), p. 186.

2. Sammons, *Blacks in Science and Medicine,* p. 244.

3. Prothro was also named chair of the Department of Home Economics and Food Administration at Tuskegee sometime in the mid-1950s ("Physical Chemists," p. 445).

4. Speaking of People, *Ebony,* October 1960, p. 7. See also Hattie Carwell, *Blacks in Science: Astrophysicist to Zoologist* (Hicksville, N.Y.: Exposition Press, 1977), p. 28.

5. *Spelman Messenger,* February 1972, p. 32.

6. Herman A. Young, and Barbara H. Young, *Scientists in the Black Perspective* (Louisville, Ky.: Lincoln Foundation, 1974).

MARY LOGAN REDDICK

Neuroembryologist, Harvard/Radcliffe Ph.D.

Neuroembryologist Mary Logan Reddick was born in Atlanta, Georgia, in 1914. She attended the Laboratory High School of Spelman College, graduated in 1929 at age fifteen, then entered the college to major in biology. In 1929, Spelman College, an all-female school, became affiliated with Atlanta University and Morehouse College (then an all-male institution) and, as a result, faculty at one college in the consortium often taught courses at another.[1] At Spelman, Reddick studied biology under two white senior faculty members, Helen Tucker Albro and Anna Grace Newell.[2] Reddick also took cross-listed courses in biology at Morehouse College and at Atlanta University with the Black biologist Samuel Milton Nabrit.[3] Reddick earned her bachelor's degree in biology at Spelman College in 1935. She then received a Rockefeller Foundation General Education Board fellowship for further study at Atlanta University, where she earned the M.S. degree in embryology in 1937.[4]

Immediately after securing her master's degree, Reddick began her professional career as an assistant in biology at Spelman College; she was promoted to the rank of instructor the following year. In 1939, she became the first female biology instructor at Morehouse College. In 1942, she received a second General Education Board fellowship, and was granted a two-year leave from Morehouse to attend Radcliffe College of Harvard University, where she concentrated on studying techniques for tissue transplantation and conducted research into factors affecting the formation and development of living organisms. She earned a second master's degree in 1943 in biology. She earned a Ph.D. in neuroembryology in 1944; her dissertation was "The Differentiation of Embryonic Chick Medulla in Chorioallantoic Grafts." While at Radcliffe, Reddick was elected to the Phi Beta Kappa and Sigma Xi honor societies, and was a classmate of neuroembryologist Geraldine Pittman Woods.[5]

Reddick returned to Morehouse College, where she became only the tenth member of the faculty to hold an earned doctorate. She was named acting chair of the biology department—the first and only female to hold this position—and received a series of promotions, culminating in a full professorship. In 1952, she received a Ford Foundation fellowship to study embryology in England at Cambridge University's School of Anatomy.[6]

Upon her return to the United States in 1953, Reddick left Morehouse College and joined the faculty at Atlanta University, where she was again awarded the rank of full professor and named chair of the biology department.[7] During her tenure as chair, Reddick became engaged in facility building; she lobbied for and then oversaw the construction of a biology research building, the first separate science facility built by Atlanta University.[8] In 1956 she received a neuroembryology research grant from the National Science Foundation under the Faculty Research Program.[9] During the 1950s and 1960s, Reddick directed the research of more than twenty students who earned their master's degrees in the biological sciences.[10] Although the bulk of her career focused on teaching and administration, Ford and National Science Foundation grants allowed her to engage in a certain amount of research; she published articles in a number of journals, among them the *Journal of Comparative Neurology* and the *Anatomical Record*.[11] Mary Logan Reddick died in 1966.[12]

NOTES

1. In their junior and senior years, students from Morehouse and Spelman could enroll in these cross-listed courses.

2. Both Helen Tucker Albro (A.B., Brown University, 1919; A.M., 1923; Ph.D., 1927) and Anna Grace Newell (A.B., Smith College, 1900; A.M., 1908; Ph.D., University of Illinois, 1916) joined the Spelman faculty in 1931. See the *Catalogue of Spelman College*, especially for the years 1933–1934, 1934–1935, 1941–1942, 1943–1944 (Atlanta University Archives, Archives and Special Collections Division, Atlanta, Ga.). Clearly, judging from the course offerings in the biological sciences, the two senior faculty members, even with the help of assistants, could not have taught all the offerings; students would have had to have taken some courses at Morehouse or Atlanta. Among the courses offered were general biology (required for all freshmen); comparative anatomy and general botany (each of which was offered in both semesters); and introductory botany, vertebrate embryology, introductory entomology, household bacteriology, histology, and cytology (each of which was offered in one semester per year). General botany, introductory entomology, household bacteriology (required for all home economics majors), histology, and cytology were all upper division courses (see the *Spelman College Catalogue*).

3. Samuel Milton Nabrit (B.S., Morehouse College, 1925; M.S., Brown University, 1928; Ph.D., 1932) also did research at the Marine Biological Laboratory at Woods Hole, Massachusetts, in the summers of 1934, 1936, 1938, and 1942 (*Spelman College Catalogue, 1944–5*, p. 24).

4. *New York Age*, 1 July 1944 (Atlanta University Archives). Nabrit had moved from Morehouse to Atlanta University in 1931, so it is probable that she again studied with him.

5. "News of the College," *The Morehouse Alumnus*, July 1944, p. 17 (Atlanta University Archives). Reddick was one of twenty Radcliffe College graduates profiled in a study on the Radcliffe Ph.D. that was published by the Radcliffe Committee on Graduate Education for Women. She appears in a section devoted to the fields of work undertaken by Radcliffe graduates. See the entry for Geraldine Pittman Woods in this volume.

6. "Two Atlanta Educators Win Fellowships to Study Abroad," *Atlanta Daily World*, 2 April 1952. According to some accounts, she may have been the first Black female scien-

tist to receive a Ford science fellowship to study abroad. See "Meet the New Members of the Faculty," *Atlanta University Bulletin*, no. 84 (December 1953): 15 (Atlanta University Archives). See also Rosalyn Mitchell Patterson, "Black Women in the Biological Sciences," *Sage* 6 (fall 1989): 8–14.

7. "Meet the New Members of the Faculty," p. 15.

8. "Faculty Items," *Atlanta University Bulletin*, no. 95 (July 1956): 19 (Atlanta University Archives).

9. "Dr. Mary Logan Reddick of Atlanta U. Dies," *Atlanta Daily World*, 11 October 1966. See also the *Spelman Messenger*, November 1966, p. 38 (Atlanta University Archives).

10. Patterson, "Black Women in the Biological Sciences."

11. "Embryologist," in *Contributions of Black Women to America*, ed. Marianna W. Davis, vol. 2 (Columbia, S.C.: Kenday Press, 1982), p. 433.

12. *Spelman Messenger*, November 1966, p. 38; "Dr. Mary Logan Reddick of Atlanta U. Dies," *Atlanta Daily World*, 1 October 1966; "Mary Logan Reddick," *Jet*, 3 November 1966, p. 17.

YVONNE A. REID

American Type Culture Collection Researcher

Yvonne A. Reid

Margaret Collins once described microbiologist Yvonne A. Reid as "one of the few people able to work at the American Type Culture Collection, which keeps

pure strains of bacteria and cultures and performs important, fundamental research."[1] The American Type Culture Collection, recently relocated to Manas-
sas, Virginia, is a private, nonprofit organization that maintains collections of
living microorganisms, viruses, DNA materials, plant tissues, and metazoan cells;
its primary responsibility is to authenticate and preserve its unique specimens.[2]

Born and raised in Jamaica, British West Indies, Reid entered Howard University in 1969, and soon came in contact with Collins, the zoologist who became
both her role model and mentor.[3] In 1973, Reid completed the B.S. degree in
microbiology at Howard. From 1978 to 1980, she worked as a research associate at Meloy Laboratories, a private company in Springfield, Virginia. She then
returned to Howard, where she earned the Ph.D. in zoology in 1986 and was
elected to Sigma Xi, the scientific honor society.

Reid's relationship with Collins continued until the latter woman's death in
1996, as Reid recalled:

> In 1993 or 1994, we went on a field trip to Dominica, in the eastern Caribbean.
> It still has a natural tropical rain forest, basically untouched, and many of the
> flora and fauna are not catalogued, so that's why we went. Maggie was so
> knowledgeable, so terrific in the field—she would go anywhere, do anything,
> eat anything.[4]

Collins, for her part, never tired of keeping track of Reid's accomplishments and
publications, as she did with many of her former students.[5]

Reid began her career at the American Type Culture Collection as a research
associate in 1980; she held this position, concurrent with her graduate studies,
until 1986, when she became a research immunologist. She has earned a steady
series of promotions: she was an assistant staff scientist from 1988 to 1991, an associate staff scientist from 1992 to 1995; and a collection scientist from 1995 to the
present.[6]

Over the years, the American Type Culture Collection has amassed more than
80,000 items, which are cataloged and available for use by the national and international scientific communities. In collaboration with the Institute for Genomic
Research and the I.M.A.G.E. consortium, ATCC can make available to researchers over 300,000 human cDNA clones[7] representing the majority of genes in the
human genome. Because researchers all over the world must work with standard-
ized stocks, among the organization's research priorities are cryopreservation to
insure against contamination or degradation; developing taxonomic guidelines
for identification and characterization; and the cultivation of unadulterated organisms and cells from its own stocks for use by outside researchers.[8]

Reid is an expert in recombinant DNA techniques and applications, cell and
tissue culture, and specific cryopreservation techniques. Her research interests
have focused on using DNA hypervariable probes in cell line characterization,
identifying hypervariable regions by polymerase chain reaction and its application to DNA profiling of human cell lines, identifying species-specific probes,
and identifying and characterizing specific proteins involved in leukocyte adhesion and differentiation.[9] Her work is useful in fingerprinting human cell and
human tumor genetic lines, in human and animal cancer research, and even in

studies aimed at determining the genetic ancestors of humans. She has received considerable funding from the National Cancer Institute (1994–1998), the National Institutes of Allergy and Infectious Disease, and the National Institute of Dental Research. Reid has consulted on research as diverse as genome research in African American pedigrees for Howard University, and standardizing procedures for the isolation of leukocytes from whole blood for the National Cancer Institute.

Reid is a member of the Society for In Vitro Biology, the American Association for the Advancement of Science, and the American Society for Cell Biology. She has been active on committees for the National Capital Area Branch of the Tissue Culture Association, and has served as the organization's secretary and vice-president (1993–1994 and 1994–1995, respectively). She has also served as chair of the cell culture standardization committee of the Society for In Vitro Biology.[10]

NOTES

1. Margaret S. Collins, telephone conversation with author, 21 August 1995.
2. "ATCC: Research and Infrastructure," a historical overview of the organization (provided by Yvonne A. Reid). Information was also provided in Reid, telephone interview by author, 30 April 1996.
3. See the entry on Margaret Strickland Collins elsewhere in this volume.
4. Reid, telephone interview.
5. Collins, personal communications, 12–13 September 1995.
6. Yvonne A. Reid, résumé, 28 August 1996.
7. cDNA is complementary DNA. The organization provides clones of genes, not clones of entire human beings.
8. "ATCC: Research and Infrastructure."
9. Ibid.
10. Reid, résumé.

GLADYS W. ROYAL

Researching Radiation's Effects

Chemist Gladys W. Royal, a specialist in human nutrition, did extensive work during the Cold War period on the effects of radiation on human tissue. Born in Dallas, Texas, most likely in the late 1920s or early 1930s, Royal earned her baccalaureate degree at Dillard University (exact date unknown),[1] and her M.S. degree in organic chemistry at Tuskegee Institute in 1954.[2] After doing pre-doctoral work at the University of Wisconsin, she completed her Ph.D. at Ohio State sometime around 1957 to 1958.[3]

During the 1950s and 1960s, Royal and her husband, George C. Royal, were faculty at North Carolina A & T College in Greensboro. George Royal, a professor of bacteriology, joined the faculty in 1952 and Gladys Royal, a professor of chemistry, in 1953.[4] At North Carolina A&T, the Royals' joint research project (with George Royal as project director and Gladys as associate director), "Biochemical and Immunological Comparisons of Irradiated Mice and Rats Treated with Bone Marrow Transplants," was funded by the Atomic Energy Commission.[5] The project developed data on the effects of radiation overdoses in mice, and the efficacy of using bone marrow transplants to treat the irradiated mice. The Royals worked from two basic inferences: that bone marrow could be typed in a way similar to blood typing, and that selected bone marrow transplants could be used to protect animals exposed to injurious doses of radiation.[6] Among the project's goals were determining and quantifying the differences and similarities of bone marrow material, and discovering ways that donor (foreign) bone marrow might be made compatible to a recipient. They also sought ways to lessen the deleterious effects on bone marrow of the high doses of radiation used in cancer therapy, and worked to develop new treatments for animals and humans exposed to high levels of radiation from other sources.[7] Their research findings were published in a number of articles and at the Fifth International Congress on Nutrition in Washington, D.C., in 1960, and at the International Congress of Histochemistry and Cytochemistry in Paris that same year.[8]

By 1982, Gladys Royal had become a biochemist with the U.S. Department of Agriculture's Food Service Program, Cooperative State Research Services, where she evaluated federal projects in human nutrition.[9]

Royal is a strong proponent of providing undergraduates with research experiences, as well as a forum in which they might report their findings. Royal was active in the Beta Kappa Chi scientific society, and served as national president of the organization in the 1970s.[10] In a 1982 interview, Royal discussed her own route to success:

> The extent to which I have achieved . . . has been inspired from a fortunate mesh between "wanting to do" and "having to do." I have been inspired by many individuals at home, at school, at church, at work, and some I only read about, who collectively taught or reinforced the truths that constructive work can be fun. Earnest effort is its own reward.[11]

The Royals had three children: the first born in 1949, most likely while Gladys Royal was still an undergraduate; the second born in 1957, around the time she completed her doctorate; and the third born in 1959, while she was teaching and working on the radiation study.[12]

NOTES

1. The Registrar's Office at Dillard University was not able to help because I have been unable to obtain Royal's maiden name. See "Husband-Wife Team's Project Draws International Interest," *Afro-American*, 10 September 1960, p. 10; "Physical Chemists," in *Contributions of Black Women to America*, ed. Marianna W. Davis, vol. 2 (Columbia, S.C.: Kenday Press, 1982), pp. 446–447.

2. "Husband-Wife Team," p. 10 and "Physical Chemists," pp. 446–447.

3. The entry "Physical Chemists," in the Davis volume states that Royal earned the Ph.D. in 1954 at Ohio State (p. 447). This must be in error, as she had only earned the M.S. in 1954; I estimate that the date for the doctorate must be somewhere between 1956 and 1958. There is no doubt that she had the degree by 1960, when the *Afro-American* article was published. That article also listed her as earning the doctorate at Ohio State University but did not mention a date.

4. George Royal, a native of Wilmington, South Carolina, earned the B.S. at Tuske-

gee Institute, the M.S. at the University of Wisconsin, did additional pre-doctoral work at Ohio State University, and completed the Ph.D. at University of Pennsylvania as a U.S. Public Health Service Research Fellow ("Husband-Wife Team," p. 10).

5. Ibid.

6. Ibid.

7. Ibid. It is important to note that this research was funded during the Cold War period, when the threat of nuclear war was high.

8. Ibid. Wording in the *Afro-American* article also points to the Cold War impact on their research: "[The project has] excited national and international attention. . . . Requests for copies of the published articles . . . have come from as far away as Australia and New Zealand, from behind the Iron and Bamboo Curtains, practically every country in Europe and nearly every state in this country" (p. 10).

9. "Physical Chemists," p. 446.

10. Ibid., p. 447.

11. Ibid., p. 447.

12. This chronology was pieced together from information in the *Afro-American* article, which states, "The Royals have three children: George III, 11; Geraldine, 3, and Guerick, 1," so it was a matter of counting back from the article's 1960 publication date ("Husband-Wife Team," p. 10).

ALBERTA JONES SEATON*

Embryologist Alberta Jones Seaton is the only woman in this volume who went abroad for doctoral study; she was awarded the Sc.D. degree from the University of Brussels, Belgium, in 1949.[1] Born in Houston, Texas, in 1924, she earned her bachelor's degree at Howard University in 1946, and her master of science degree in 1947.

Seaton began her career as chair of the biology department from 1950 to 1951 at Wiley College, a historically Black college in Marshal, Texas. She then moved to Makerere College in Uganda, Africa, where she was a visiting professor of biology and embryology from 1952 to 1953. When she returned to the United States, Seaton obtained a position as assistant professor of biology at Spelman College from 1952 to 1954, before becoming an associate professor at Texas Southern University, a position she held until 1961. Seaton was awarded a National Science Foundation faculty research fellowship and returned to the University of Brussels for the 1965–1966 academic year. She became director of freshman studies at Texas Southern University in 1970.[2]

Notes

1. Julius H. Taylor, Clyde R. Dillard, and Nathaniel K. Proctor, eds., *The Negro in Science* (Baltimore: Morgan State College Press, 1955), p. 180. See also Vivian Ovelton Sammons, *Blacks in Science and Medicine* (New York: Hemisphere, 1990), p. 211. *American Men and Women of Science: The Physical and Biological Sciences,* 12th ed. (New York: Bowker, 1971–), p. 5666. *Who's Who of American Women* (Chicago: Marquis Who's Who, 1974), p. 862. *Who's Who in the South and Southwest,* 13th ed. (Chicago: Marquis Who's Who in America, 1973–1974).

2. Sammons, *Blacks in Science and Medicine*, p. 211. See also *Who's Who in the South and Southwest.*

MARIANNA BECK SEWELL*

Nutritionist Marianna Beck Sewell earned the baccalaureate degree in home economics at Howard University in 1937.[1] She was engaged in research during some portion of the next ten years, and was part of a group that published a paper on ascorbic acid metabolism in 1949.[2] She then studied for her master's degree in nutrition at Howard, and was awarded the degree in 1950; her thesis was "A Study of Changes of Rutin and Ascorbic Acid in the Blood of Normal Subjects and Surgical Patients."[3]

Immediately after earning her master's degree, Sewell became an instructor in the Howard University home economics department.[4] In 1951 she was again part of a research group on ascorbic acid metabolism.[5] It appears that Sewell remained at Howard until sometime around 1960, although what her position was in the department is not clear.

Sometime in the early 1960s, Sewell enrolled in New York University, where she earned the Ed.D. in 1965 and was elected to the Sigma Xi honorary society. Her focus continued to be on nutrition, as reflected by her dissertation, "An Evaluation of Food Habits of Nigerian Students in American Colleges and Universities."[6]

Sewell returned to Howard, and was named lecturer in the department of home economics, a position she held until at least 1971. Sewell was not as heavily engaged in research as the other faculty and most likely spent the major portion of her career in teaching.[7]

NOTES

1. The only information I found on nutritionist Mariana Beck Sewell was contained in the Flemmie Kittrell Collection at the Moorland-Spingarn Research Center at Howard University. See Flemmie P. Kittrell, chairman and principal author, "Proposal for a Program Leading to the Doctor of Philosophy Degree in Nutrition in the Department of Home Economics, Howard University," 6 May 1971, p. 31 (Kittrell Collection, Moorland-Spingarn Research Center, Howard University).

2. Ibid. See R. L. Hayes, W. Booker, and M. B. Sewell, "Ascorbic Acid Metabolism in Patients with Certain Gingival Disturbances," *Federation Proceedings* 8 (March 1949): 70. It is not entirely clear where this research was conducted, but, judging from a 1951 publication involving Sewell and two of the other authors of this paper, it was probably Howard University.

3. "Proposal for a Program" (Kittrell Collection).

4. See Hettie Pearson, "Scrapbook 2: Howard University Home Economics Department: The Open House and Career Conference, April 13–14, 1951 (Senior Day)" (Kittrell Collection). The scrapbook lists Sewell as "Instructor." Mrs. Lydia J. Rogers is listed as acting head of the department—Flemmie Kittrell was in India at the time.

5. "Proposal for a Program" (Kittrell Collection). The paper is W. M. Booker, R. L. Hayes, M. B. Sewell, and F. M. Dent, "Experimental Studies on Ascorbic Acid Metabolism," *American Journal of Physiology*, August 1951, p. 374.

6. New York University, 1965. See "Proposal for a Program" (Kittrell Collection).

7. I made this judgment based on the number of publications by faculty members listed in the 1971 "Proposal for a Program Leading to the Doctorate of Philosophy Degree in Nutrition." Cecile H. Edwards, for example, earned the baccalaureate degree ten years later than Sewell, yet she had more than fifty publications through 1971 (see the entry for Edwards in this volume). Other faculty members also had a high number of publications; for example, Clarence F. Winchester (B.S., 1924, M.S., 1935, University of California, Berkeley; Ph.D., 1939, University of Missouri) was also a home economics lecturer in nutrition, and had a (partial) listing of 23 publications through 1971.

DOLORES COOPER SHOCKLEY

Pharmacy Professor, Researcher

Pharmacologist Dolores Cooper Shockley was the first Black female to receive a doctorate in a science from Purdue University, and the first Black female to earn a doctorate in pharmacology at a United States university.[1] But she struggled against the sexism of a predominantly male field. "Some men thought that I was just working temporarily and that I wouldn't stay long enough to contribute anything," she said. "I was greeted with a lot of suspicion until they realized that I was there to do my job."[2]

Born in Clarksdale, Mississippi, in 1930, Shockley attended segregated schools. Because there were no Black pharmacists in Clarksdale, her original ambition was to become a licensed pharmacist and open a drugstore in her home town.[3] She earned the B.S. degree in pharmacy at Louisiana State University, then a historically Black college, in 1951.[4] During her undergraduate training, however, Shockley developed an interest in research; even though she was now a licensed pharmacist, she decided to continue on to graduate school.[5] She was awarded a two-year assistant instructorship (1951–1953) at Purdue University in West Lafayette, Indiana, where she earned the M.S. and Ph.D. degrees in pharmacology in 1953 and 1955, respectively.[6]

Immediately after earning her doctorate, Shockley became an assistant professor of pharmacology at Meharry Medical College in Nashville, Tennessee. Before she began her duties at Meharry, she was granted a leave to accept a Fulbright Fellowship (1955–1956) for postdoctoral study at the University of Copenhagen in Denmark. Upon her return to Meharry, Shockley, like most faculty members at historically Black institutions, encountered a heavy teaching load; nevertheless, she managed to continue with her research, which centered around the measurement of non-narcotic analgesics, the effects of drugs on stress conditions, and the effect of hormones on connective tissue.[7]

In 1959, Shockley was again granted a leave from Meharry to accept a visiting professorship at the Einstein College of Medicine in New York City, where she remained until 1962. Between 1963 and 1966, she held a Lederle Pharmaceutical Company Faculty Research Award at Meharry and was promoted to the rank of associate professor in 1967.[8] In the early 1970s, she was named chair of the pharmacology department, became the school's foreign student advisor, and liaison officer for international activities to the Association of American Medical Colleges.[9] Married to a scientist, Shockley is the mother of four children.

NOTES

1. Vivian Ovelton Sammons, *Blacks in Science and Medicine* (New York: Hemisphere, 1990), p. 213. See also *The Crisis*, March 1955, p. 184.

2. Shawn D. Lewis, "A Veteran at Meharry Medical College," *Ebony*, August 1977, p. 116.

3. Ibid.

4. Sammons correctly lists Shockley as having earned the B.S. degree at Louisiana State University (Sammons, *Blacks in Science and Medicine*, p. 213); but another source incorrectly lists her as having earned her B.S. at Xavier University (Lewis, "Veteran at Meharry," p. 116). Both sources agree on the date. The confusion may originate in both Louisiana State and Xavier being historically Black colleges/universities; Xavier is a Catholic institution. See also *American Men and Women of Science: The Physical and Biological Sciences*, 14th ed. (New York: Bowker, 1971–), p. 4641 (also see the 18th edition). *National Medical Association Journal* 55 (May 1963): 246–247.

5. Lewis, "Veteran at Meharry," p. 116.

6. Sammons, *Blacks in Science and Medicine*, p. 213; Lewis, "Veteran at Meharry," p. 116. See also *The Crisis*, March 1955, p. 184.

7. Sammons, *Blacks in Science and Medicine*, p. 213; Lewis, "Veteran at Meharry," p. 116. See the appendix for several of Shockley's publications.

8. Sammons, *Blacks in Science and Medicine*, p. 213.

9. Lewis, "Veteran at Meharry," p. 116.

GEORGIA CALDWELL SMITH*

Like Angie Turner King, mathematician Georgia Caldwell Smith experienced a long delay in obtaining her doctorate. Born in 1909 in Atchison, Kansas, Smith attended segregated public schools, and enrolled at the University of Kansas, where she earned the bachelor's degree in 1928. She earned her master's degree in 1929, and was elected to the Phi Beta Kappa honorary society. Smith then became an assistant professor at Spelman College in Atlanta, where she taught mathematics for the next ten years. From 1939 to 1943, she taught at Lincoln University of Missouri, then moved on to Alabama State College in the fall of 1943 before returning to Spelman in 1945.

Smith was almost fifty before she had the opportunity to continue her studies for the doctorate, which was awarded by the University of Pittsburgh in 1960. Her dissertation, "Some Results of the Anticenter of a Group," was done under the direction of Norman Levine. She died in January 1961, shortly after her defense of the dissertation.[1]

NOTE

1. Patricia C. Kenschaft, "Black Women in Mathematics in the United States," *Journal of African Civilizations* 4 (April 1982): 71. Kenschaft, "Black Women in Mathematics," *Association for Women in Mathematics Newsletter* 8 (September 1978): 6. "Mathematicians," in *Contributions of Black Women to America,* ed. Marianna W. Davis, vol. 2 (Columbia, S.C.: Kenday Press, 1982), p. 458.

DOLORES MARGARET RICHARD SPIKES

Mathematician, University System President

Mathematician Dolores Margaret Richard Spikes served as president of the Southern University and A&M College System in Louisiana from 1991 to 1997. Born in Baton Rouge in 1936, Spikes attended parochial (Catholic) and public schools in that city. She earned the baccalaureate degree, summa cum laude, in mathematics, with a minor in the sciences, at Southern University in 1957. She was awarded a University of Illinois fellowship to study for the master's degree, which she completed in 1958.

From 1960 to 1964, Spikes taught science at the secondary school level at Mossville School in Calcasieu Parish, Louisiana, before leaving to accept a position as an assistant professor of mathematics at Southern University. By 1965 Spikes, then married and the mother of one child, had begun to consider the possibility of graduate school.[1]

Spikes received a National Science Foundation fellowship for further study in 1966, and took a leave from Southern to attend Louisiana State University (LSU). She returned to her position, and then in 1968 received a three-year Ford Foundation fellowship for doctoral studies at LSU, again on official leave from Southern. She was awarded the Ph.D. in 1971, becoming the first Black, male or female, to receive a doctorate in mathematics from LSU. Her degree was in pure mathematics, with a specialty in commutative ring theory; her dissertation, directed by Jack Ohm, was "Semi-valuations and Groups of Divisibility."[2]

Immediately after completing the doctorate, Spikes was advanced to the level of associate professor of mathematics at Southern University, and in 1975, was promoted to the rank of full professor. By 1981, in addition to her faculty position, Spikes was named part-time assistant to the chancellor of Southern University, and coordinator of the Mathematics Development Program. A year later, in 1982, she became full-time assistant to the chancellor, a position she held until 1985, when she was made executive vice-chancellor and vice-chancellor for academic affairs at Southern University, Baton Rouge.

Spikes attended the Institute for Educational Management at Harvard University (1986–1987), and then served as chancellor of Southern University at

New Orleans (1987–1988). She was named chancellor of Southern University at Baton Rouge in 1988.[3] In 1991, she became president of the Southern University and A&M College System, "the first female in the United States to head a university system, and the first female to lead a Louisiana public college or university."[4] In 1997 she became president of the University of Maryland Eastern Shore (another historically Black institution), a position she currently holds.

Throughout her career, Spikes has been active in efforts to increase the participation of minority students in the sciences and mathematics. She has said that her own expectations with regard to doctoral studies in mathematics were limited by a lack of available role models:

> When I finished my Bachelor's degree at Southern University in 1957, I journeyed to the University of Illinois where I studied for the Master of Science degree in mathematics. That was to be the end of my training because I had never known a Black Ph.D. in mathematics. I did a few years later, but not at that time. Certainly I had known no Black women mathematicians.[5]
>
> I honestly believe that the question of role models weighed heavily here. Of course, this was prior to graduate education's being (comparatively) more open to Blacks in southern institutions of higher learning.[6]

Spikes believes that, due to the legacy of racism, majority white institutions in the South have failed to meet the needs of Black mathematicians:

> Southern University . . . is the largest predominantly Black university in the country, and . . . it sits in a city with another institution of higher education [Louisiana State University] that does offer a Ph.D. degree. But through the years that institution has failed to respond to the needs of Black mathematicians.
>
> In 1971 when I received a doctorate in mathematics from that institution [LSU], I was the first Black to do so and even the first graduate of Southern University to do so. I am not proud of that fact. I regard it as a shame—a blight on the state of Louisiana and on education in general.[7]

The high financial cost of extending a college education also tends to keep women from pursuing graduate studies, Spikes acknowledges.

> Another problem in the recruitment of Black women mathematicians is often job offers. About 80 to 90% of our students are on financial aid. That means they come to us with the expectation that in a few years they will be able to earn a little money and indeed live a little more decently than they have in the past. So when the lure of a job comes to them upon receipt of their B.S. degree, they are sometimes hesitant even to think about going to graduate school.[8]

And she finds that Black women with advanced degrees in mathematics face special problems:

> Black women mathematicians generally work in institutions which have high teaching loads, thus limiting their ability to create mathematics, and to become, what in the minds of many . . . are "real mathematicians." Somehow some people don't regard persons who have had training in mathematics but who somehow can't get around to doing research as "real mathematicians." . . .

Another problem is that because of race and other societal influences, Black women mathematicians usually must engage themselves in activities outside mathematics that take up much of their time so they have a limited amount of time to devote to mathematics and mathematical professional organizations.[9]

Spikes has presented papers and given lectures at a number of symposia with regard to increasing minority women's involvement in science and mathematics.[10]

During her tenures as vice-chancellor for academic affairs and later as president of the Southern University College System, Spikes was instrumental in establishing the Center for Energy and Environmental Studies at the university. She later worked to establish a series of academic programs that would focus on issues of hazardous wastes and energy sources by establishing a consortium involving the Center for Energy and Environmental Studies, the campuses of the Southern University System, California's Lawrence Livermore Laboratories, and the Historically Black Colleges and Universities/Minority Institutions Consortium. She also established the Office of Planning, Assessment and Institutional Research at Southern University at Baton Rouge; and a mentoring program for Southern's Baton Rouge students that involved faculty and staff, retired Southern University employees, and other professionals. She was an active participant in discussions and litigation involving the desegregation of public higher education in Louisiana.[11] Finally, she has served as a consultant and proposal reviewer on various National Science Foundation evaluation committee panels, and as codirector of the Ford Foundation–funded consortium to improve mathematics teaching in public schools.[12]

NOTES

1. Dolores Margaret Richard Spikes, curriculum vitae, 24 June 1996. See also Patricia Clark Kenschaft, "Delores Margaret Richard Spikes," in *Black Women in America: An Historical Encyclopedia*, ed. Darlene Clark Hine, Elsa Barkley Brown, and Rosalyn Terborg-Penn, vol. 2 (New York: Carlson, 1993; Bloomington: Indiana University Press, 1994), pp. 1096–1097. Dolores M. Spikes, personal account, in "Excerpts from the Association for Women in Mathematics Panel on Black Women in Mathematics" (Atlanta, 7 January 1978), ed. Patricia C. Kenschaft, *Newsletter of the Association for Women in Mathematics* 8 (September 1978): 8–11; and 10 (May–June 1980): 5–8. Kenschaft, "Black Women in Mathematics in the United States," *Journal of African Civilizations* 4 (April 1982): 75–76.

2. Spikes, curriculum vitae.

3. Spikes, curriculum vitae.

4. Kenschaft, "Delores Margaret Richard Spikes," p. 1096.

5. Spikes, personal account, in "Excerpts," pp. 8–11.

6. Kenschaft, "Black Women in Mathematics," 1982, pp. 75–76.

7. Ibid.

8. Spikes, personal account, in "Excerpts," pp. 8–11.

9. Ibid.

10. Spikes presented the following papers at symposia: "Science Education: A Tool for Unifying America's Communities," National Conference, National Organization for the Professional Advancement of Black Chemists and Chemical Engineers (1995); "Sav-

ing the At-Risk Female," AACTE Women's Breakfast, Chicago (1990); "Improving the
Campus Climate for Women," Conference for Women in Higher Education Adminis-
tration, University of Northern Iowa (1989). In 1995, she presented testimony before the
U.S. Congress Committee on Energy and Natural Resources, "Testimony on the Depart-
ment of Energy National Competitiveness Technology Partnership Act of 1993" (1995)
(Spikes, curriculum vitae).

11. A partial negotiated settlement was arrived at, but the outstanding issue of Land
Grant status is still under discussion. In 1990, Spikes presented a paper, "Achieving a
Global University—A Land Grant Imperative," at the National Association of Land Grant
Colleges and Universities' Division of Agriculture in Kansas City (Spikes, curriculum
vitae). In 1991, she presented a paper, "The Reorganization of the USDA and Its Impact
on Rural Development and Land Grant Status," at Tuskegee University. She is an active
participant in the National Association of State Universities and Land-Grant Colleges.

12. Spikes, curriculum vitae. The consortium consists of the schools in Southern University system, plus Xavier College of New Orleans, Dillard University, and Grambling
State University (all in Louisiana).

JEANNE SPURLOCK
Psychiatrist, Journal Editor, Researcher

Psychiatrist Jeanne Spurlock's research interests have covered a wide range: the
effects of racism on children, the self-concept of Black children, and therapeutic
psychiatric intervention for minorities. Born in Sandusky, Ohio, in 1921, she was
the oldest of Frank and Godene (Anthony) Spurlock's seven children. The family
moved to Detroit shortly after Spurlock's birth, and was later especially hard hit
by the Depression. Spurlock graduated from a segregated Detroit high school in
1940, and entered Spelman College on a tuition scholarship. Despite the scholarship, she had to work almost full-time to meet her other expenses and was forced
to quit Spelman in 1942. She then attended Roosevelt University in Chicago for
almost a year.[1]

America's entry into World War II created a tremendous need for trained
physicians, and Howard University, along with a number of U.S. university medical schools, instituted an accelerated medical degree program. Although she
had not earned a baccalaureate degree, Spurlock was accepted with funding into
Howard's accelerated program in March of 1943; she was awarded the M.D. in
1947. She completed her internship at Chicago's Provident Hospital in 1948, and
then performed a two-year residency in general psychiatry at Cook County Hospital. From 1950 to 1951, she studied child psychiatry in the fellowship program
at the Institute for Juvenile Research in Chicago.[2]

In her work with adult patients, Spurlock noted that a sizeable number described troubled childhoods; as a result, she decided to focus her career on early-
stage intervention. She continued as a staff psychiatrist at the Institute until 1953,
served concurrently as a staff psychiatrist at the Mental Hygiene Clinic at Wom-

en's and Children's Hospital, and consulted at the Illinois School for the Deaf in Jackson, Illinois (1951–1952). In 1953, she began advanced psychoanalytical training on a part-time basis at the Chicago Institute for Psychoanalysis. While attending the Institute, she served as the director of the Children's Psychosomatic Unit at Chicago's Neuro-Psychiatric Institute (1953–1959), and was an assistant professor of psychiatry at the University of Illinois College of Medicine (1953–1959). In 1960, she became an attending psychiatrist at the Michael Reese Hospital in Chicago, and finally completed her psychoanalytic training in 1962. She then became chief of the Child Psychiatry Clinic at Michael Reese Hospital, where she remained until 1968.[3]

Spurlock moved to Meharry Medical College in Nashville, Tennessee, where she was named chair of the Department of Psychiatry in 1968. She remained in that position until 1973, when she accepted a post as a visiting scientist at the National Institute of Mental Health's Division of Special Mental Health Programs (1973–1974). In 1974, she became deputy medical director of the American Psychiatric Association. During the 1970s and 1980s, Spurlock also held clinical professorships at George Washington University College of Medicine and Howard University College of Medicine, both in Washington, D.C.[4]

Spurlock has written on sexism in medicine and psychiatry, and minority issues with regard to training physicians in child and adolescent psychiatry.[5] She has also served on the editorial boards of the *Journal of the American Academy of Child Psychiatry* (1969–1974), the *Journal of Medical Aspects of Human Sexuality* (1972–1987), *Integrated Education* (1974–1980), and *The Journal of Psychiatric Education* (1975–present). She served on the board of the Carnegie Corporation (1973–1980), the Hillcrest Children's Center, Washington, D.C. (1977–1979), and the Green Door, a halfway house program for former psychiatric patients in Washington, D.C. (1978–1984).[6]

NOTES

1. Vivian Ovelton Sammons, *Blacks in Science and Medicine* (New York: Hemisphere, 1990), p. 219. See also Susan Brown Wallace, "Jeanne Spurlock," in *Notable Black American Women*, ed. Jessie Carney Smith, vol. 2 (Detroit: Gale Research, 1992), pp. 1072–1074. *Who's Who among Black Americans*, 4th ed. (Northbrook, Ill.: Who's Who among Black Americans Publishing, 1985), p. 790 (also see the 1990–1991 edition). "Outstanding Women Doctors: They Make Their Mark in Medicine," *Ebony*, May 1964, p. 74. *Ebony Success Library*, vol. 1 (Chicago: Johnson Publishing, 1973), p. 291. *ABMS Compendium of Medical Specialists*, vol. 5 (Chicago: American Board of Medical Specialists, 1986), p. 747.

2. Sammons, *Blacks in Science and Medicine*, p. 219; Wallace, "Jeanne Spurlock," pp. 1072–1074.

3. Sammons, *Blacks in Science and Medicine*, p. 219; Wallace, "Jeanne Spurlock," pp. 1072–1074.

4. *Ebony Success Library*, p. 291.

5. See the appendix for a partial list of Spurlock's publications.

6. Wallace, "Jeanne Spurlock," pp. 1072–1074.

NETTIE S. STRANGE*

As of 1979, Nettie S. Strange, a geologist with a specialty in paleontology, was working in the petroleum exploration industry and was studying the mechanisms affecting the deposition of sandstone and shale and the micro-fossils contained in oil well samples. Since 1974 she had worked for the Atlantic Richfield Company of Houston, Texas. In her work, Strange used the micro-fossils as "environmental indicators, age determinants, and indicators of favorable zones of hydrocarbons."[1]

Born in Roanoke, Virginia, Strange earned her bachelor of science degree in geology at Virginia State College in 1972, and her master of science degree in geology and paleontology at the University of Texas at Austin in 1975.[2] She was a senior geological associate with Atlantic Richfield by 1982.[3]

NOTES

1. "Oil Firm Paleontologist," *Ebony*, March 1979, p. 4.
2. Ibid. Nettie S. Strange's master's thesis was "A Census Study of Foraminifera in the Lower Taylor Group, Upper Cretaceous, Travis and Williamson Counties, Texas." See also "Geologist," in *Contributions of Black Women to America*, ed. Marianna W. Davis, vol. 2 (Columbia, S.C.: Kenday Press, 1982), p. 449.
3. "Geologist," p. 449. The material for the profile in the Davis book was obtained from a personal description provided to the Black Executive Exchange Program, National Urban League.

MARIE CLARK TAYLOR

Botanist, Mentor

Marie Clark Taylor was, according to her former colleague Margaret Collins, "a powerhouse, who for almost three decades trained most of the botanists who came out of Howard University, and worked tirelessly to improve teacher training in the sciences."[1] Taylor was also very effective in using a series of National Science Foundation teaching grants to improve the science education of Black students and Black educators at Howard University.

Taylor was born in 1911 in Sharpsburg, Pennsylvania, to William S. Clark and Mattie (Reed) Clark. Her mother died when Taylor was quite young, and her father moved the family to Washington, D.C., so that his sisters could help raise his five children.[2] Taylor attended the segregated Dunbar High School in Washington, graduated with honors in 1929, and entered Howard University, where she earned a B.S. (magna cum laude) in 1933. She then entered the master's program in botany at Howard with a research fellowship, and was awarded the degree in botany, magna cum laude, in 1935.

For the next six years Taylor taught biology at Washington's historically Black Cardozo High School.[3] She concurrently entered the doctoral studies program at

Fordham University in New York, where she earned her Ph.D., cum laude, on June 11, 1941, almost six months before the start of World War II.[4] She was the first woman, of any race, to earn a science doctorate at Fordham University, and she was elected to Sigma Xi, the scientific honor society. In September 1941, Taylor, doctorate in hand, returned to teach at Cardozo High School.

Early in 1942, Taylor interrupted her teaching career to join the Army Red Cross. She served in the Pacific,[5] and while she was stationed in New Guinea, she met Richard Taylor, also from Washington, D.C., who was an officer in the 93rd Airborne Infantry Division, an all-Black unit.[6] In 1945, soon after she had returned from duty, Taylor joined the botany department at Howard University, first as an instructor and then as an assistant professor. In 1947 she succeeded Dr. Charles Parker as department chair. She and Richard Taylor were married on New Year's Day, 1948.[7] In 1950, the same year she achieved the rank of associate professor, she bore a son at Howard University's Freedman's Hospital. Marie Taylor reportedly enjoyed the campus joke that "because of her professorial dedication, she had deliberately planned her child's delivery after final exams."[8]

During the 1950s and 1960s Taylor headed a series of summer science institutes for high school teachers of biology, under the auspices of the National Science Foundation.[9] She was a strong proponent of using botanical materials in high school biology courses, because plant materials were generally easy to afford and offered both students and teachers an opportunity to observe firsthand a variety of biological processes such as light-induced seed germination, hereditary factors, and gas exchange.

Taylor used the summer institutes to put forward her teaching methods. She stressed the usefulness of light-microscopes to study the physiology of living cell processes—mitosis, meiosis, cell replication, enlargement, elongation, and cytokineses; she pointed out that plant specimens were particularly good for illustrating these processes, because their three-dimensional quality enhanced viewing and their cells lived longer than most isolated metazoan cells under simple mounting techniques.[10] Also, the greater simplicity of plant material "facilitates investigations into the manifestations of life. [P]lants are the equal of animals in the study of the various kinds of utilization of nutrients, the biotic conditions, and the ability to produce new cells and individuals of the same heredity."[11] No other source material could duplicate the ease of plants to demonstrate these processes, she said, adding that "since the organs of plants are generally external and easily identified, observations and measurements of organogenesis and organ responses to stimuli may be obtained with little difficulty."[12] She decried the fact that eighty to eighty-five percent of high school biology teachers were unacquainted with the versatility that plant materials offered for dynamic instruction in the cycles of living organisms.[13]

In the mid-1960s, President Lyndon B. Johnson requested Taylor's services for both domestic and overseas consulting. As a result, the scope of her NSF summer institutes was expanded to include international teachers.

Like many of the women in this study, Taylor was involved in facility building. She was instrumental in the design, development, and construction of Howard

University's biology building, which was named after Ernest Everett Just, and she oversaw the construction of a greenhouse laboratory atop the building.[14] She chaired the botany department for twenty-eight years, until her retirement in 1976. Taylor died in December 1990, and in November 1991 an auditorium in the Ernest E. Just Hall was named in her honor.[15]

NOTES

1. Margaret Strickland Collins, interview by author, Washington, D.C., October 1995. One of Taylor's former students, Dr. Geraldine Twitty, is now a senior faculty member in Howard University's biology department.

2. Marie Clark Taylor's birthplace is listed as Sharpsville, Pennsylvania, in "Service for the Life of Marie C. Taylor, 1911–1990," Wednesday, 2 January 1991, printed by the McGuire Funeral Service, 7499 Georgia Avenue, NW., Washington, D.C. (provided by Dean Harris, archival technician, Moorland-Spingarn Research Center, Howard University, Washington, D.C., December 1995.) However, it is listed as Pittsburgh in "Marie Clark Taylor, Botanist," Howard University in-house biography (provided by Dean Harris, Archival Technician, Moorland-Spingarn Research Center, Howard University, December 1995). Also see "Botanist," in *Contributions of Black Women to America*, ed. Marianna W. Davis, vol. 2 (Columbia, S.C.: Kenday Press, 1982), pp. 433–444; this profile lists Sharpsburg, Pennsylvania, as Taylor's birthplace.

3. "Marie Clark Taylor, Botanist."

4. Date verified by Anna Maria Conte, assistant registrar at Fordham University, letter of certification, 1 March 1996. According to Howard University in-house biographies, Taylor "interrupted her teaching career and became an officer in the Army Red Cross, serving at various posts in the Pacific, including New Guinea. . . . At the end of the war she returned to the United States and entered the doctoral studies program at Fordham University, in New York." A portion of the chronology of this statement is in error. Taylor did teach at Cardozo High school for a few years beginning in 1935. Around 1938, however, she took a leave of absence and enrolled in the doctoral program at Fordham University. This error in chronology has persisted in a number of Howard "in-house" publications on Taylor. See also the pamphlets "Dedication, Marie Clark Taylor Auditorium in Ernest E. Just Hall," 12 November 1991; "Memorial Service for the Life of Marie C. Taylor: 1911–1990," Andrew Rankin Memorial Chapel, Howard University, 14 January 1991; and "Marie C. Taylor Auditorium Dedicated," *The Capstone* (Howard University) 12, no. 23 (1991), p. 1. (Photocopies of these documents were provided by Dr. Marjay Anderson, Department of Biology, Howard University, Washington, D.C.)

5. "Marie Clark Taylor, Botanist." See also "Service for the Life."

6. Charlie J. Black, "Upward Mobility," *New Observer*, January 1991.

7. "Service for the Life."

8. Ibid.

9. "Dedication, Marie Clark Taylor Auditorium."

10. Marie C. Taylor, "Live Specimens," *The American Biology Teacher* 27 (February 1965): 116–117.

11. Ibid.

12. Ibid., p. 116.

13. Ibid.

14. "Dedication, Marie Clark Taylor Auditorium."

15. Ibid.

MARGARET ELLEN MAYO TOLBERT
Biochemist and Administrator

Margaret Ellen Mayo Tolbert
(*seated, second from left*), March 1999

Pharmaceutical chemist and science administrator Margaret Ellen Mayo Tolbert is currently the laboratory director at the U.S. Department of Energy's New Brunswick Laboratory, in Argonne, Illinois, a position she has held since 1996.[1] Immediately prior to accepting this position, Tolbert was director of the Division of Educational Programs at Argonne National Laboratory, also in Argonne.[2] Born in 1943, Tolbert grew up in Suffolk, Virginia. Both her parents had died before she entered her teens, so Tolbert was raised by her grandmother and a group of close neighbors, who functioned as an extended family.[3] She graduated from high school and entered Tuskegee Institute, where she earned the bachelor of science degree in chemistry in 1967; she immediately enrolled at Wayne State University, where she earned the master's degree in analytical chemistry in 1968. With research focused on regulation of sugar production in rat hepatic cells, she earned her Ph.D. in biochemistry at Brown University, in Providence, Rhode Island, in 1974.[4]

Tolbert began her career as an instructor of mathematics at Tuskegee (1969–1970), became an assistant professor of chemistry (1973–1976), and then moved on to the School of Pharmacy at Florida A&M University where she was an associate professor of pharmaceutical chemistry and an associate dean (1977–1978). She returned to Tuskegee as a full professor (chemistry) and was named director of the Carver Research Foundation at Tuskegee Institute in 1979. Her duties at the foundation included supervision of research projects, fund-raising, and the oversight and approval of foundation-employed researchers' requests for patents on their work. Concurrently, she was Tuskegee's associate provost for research and development.[5] During this period, her own research focused on the effects of various drugs on the human liver.[6] Early in 1980, she also spent six months at the International Institute of Cellular and Molecular Pathology in Brussels, Belgium.

In 1987, Tolbert took a leave from her position at the Carver Research Foundation to take a position with BP America at the company's research center in Warrensville Heights, Ohio; she then formally left Carver and remained with BP America as a senior planner and senior budgets and control analyst until 1990. From 1990 to 1993, Tolbert was director of the RIMI Program and a staff associate at the National Science Foundation in Arlington, Virginia. In 1994, she became a consulting scientist at Howard Hughes Medical Institute in Chevy Chase, Maryland, before moving on to the positions at Argonne and New Brunswick.

The New Brunswick Laboratory is government owned and operates under the auspices of the Department of Energy's Chicago Operations Office. Included in Tolbert's responsibilities as director of the lab are the development, evaluation, certification, and dissemination of nuclear materials and material technology safeguards related to U.S. nuclear energy programs. Throughout her career, she has maintained strong professional interests in science education; she has participated in programs to enhance the participation of minorities and women in science, and has worked with both national and international science and government agencies.[7] Married and twice divorced, Tolbert is the mother of one son.

NOTES

1. Margaret Ellen Mayo Tolbert, e-mail communication, 10 March 1999.

2. Margaret Ellen Mayo Tolbert, interview by author at the Graduate Women in Science and Engineering Conference, National Academy of Science, Washington, D.C., October 1995. Tolbert, correspondence to author, 6 November 1995.

3. Brenda Flanagan, "Essence Women," *Essence*, August 1980, pp. 37–38.

4. Vivian Ovelton Sammons lists Tolbert as earning the Ph.D. at Wayne State University; Sammons, *Blacks in Science and Medicine* (New York: Hemisphere, 1990), p. 233. Note that Flanagan lists the school as Brown University ("Essence Women"). Tolbert's dissertation was titled "Studies on the Regulation of Gluconeogenesis in Isolated Rat Hepatic Parenchymal Cells" (Sammons, *Blacks in Science and Medicine*, p. 233). See also *American Men and Women of Science: The Physical and Biological Sciences*, 16th ed. (New York: Bowker, 1971–), p. 151. *Who's Who among Black Americans*, 4th ed. (Northbrook, Ill.: Who's Who among Black Americans Publishing, 1985), p. 151.

5. Tolbert, e-mail, 10 March 1999.

6. Flanagan, "Essence Women."

7. Tolbert, e-mail, 10 March 1999.

ALBERTA BANNER TURNER
Psychologist

Psychologist Alberta Banner Turner was born in Chicago, Illinois, in 1909. She was raised and attended public schools in Columbus, Ohio, graduating from East High School in 1925 at age sixteen. She earned the bachelor's degree at Ohio State University in home economics in 1929; she had planned to enter graduate school immediately, but the Depression intervened. At that point, Turner took the first in a series of temporary positions.

Turner taught home economics and was department head at Wilberforce University (now Central State College) of Xenia, Ohio. She attended Ohio State during the summers and earned an M.S. in education, with a specialty in child development, in 1931. As her education major indicates, Turner was already moving away from home economics; her concentration in child development shows her movement toward psychology. In 1935, she completed the doctorate in psychology at Ohio State University but was unable to obtain a position in the field.

During the next ten years, Turner held a series of posts, the majority of which were in home economics, her baccalaureate major. She was head of the home economics department at Winston-Salem College in North Carolina from 1935 to 1936; professor of psychology and head of the home economics department at Lincoln University in Missouri from 1936 to 1937; head of the home economics department at Southern University in Louisiana from 1938 to 1939; and head of the home economics department at Bennett College for Women in North Carolina from 1939 to 1942. She then took a maternity leave and returned to Columbus, Ohio.

In 1944, Turner was offered her first full-time position in psychology, as a clinician at the Ohio Bureau of Juvenile Research, a division of the Ohio Youth Commission. She remained with the Bureau of Juvenile Research for the next twenty-seven years, and was concurrently an adjunct professor of psychology at Ohio State University and a psychologist at the Ohio Reformatory for Women. During her tenure at the Bureau of Juvenile Research she was promoted from clinical psychologist to supervising psychologist to chief psychologist; in 1963 she moved up to the Central Administrative Office of the Ohio Youth Commission, where she remained until her retirement in 1971.[1]

NOTE

1. Robert V. Guthrie, *Even the Rat Was White: An Historical View of Psychology* (New York: Harper & Row, 1976), pp. 146–148. See also Vivian Ovelton Sammons, *Blacks in Science and Medicine* (New York: Hemisphere, 1990), p. 239.

OLA B. WATFORD*

As of 1982, Ola B. Watford was chief of the Space Management and Design Branch of the National Oceanic and Atmospheric Administration (NOAA) in

Rockford, Maryland.[1] Born in Kingston, North Carolina, Watford earned her baccalaureate degree in mathematics at Johnson C. Smith University and her master's degree in physical science at the University of Northern Colorado at Greely. She became a statistical clerk at the Naval Oceanographic Office in 1951, then she joined the Coast and Geodetic Survey, a division of NOAA, in 1953.[2]

In 1973, when Watford was the only female staff geophysicist employed by the U.S. Department of Commerce, she said that she believed "the scientific field is wide open for all minorities, including women."[3] By the time she was chief of NOAA's Space Management and Design Branch, Watford was the "mother of six children and married to an economist," and she "successfully managed a career, motherhood, and the role of wife."[4]

NOTES

1. Marianna W. Davis, ed., *Contributions of Black Women to America*, vol. 2 (Columbia, S.C.: Kenday Press, 1982), pp. 450–451. A call to NOAA's Public Affairs Office (18 June 1996) yielded no information, other than that there are "not too many people in the Rockville location anymore," and no listing for Ola B. Watford. It was, however, suggested that I try Carol Watford, who works at the U.S. Census Bureau. I spoke with Carol Watford, who said she was not related to Ola B. Watford and was not familiar with the name (telephone conversation, 26 June 1998).

2. Ibid.

3. Ibid, p. 451. The Davis book is quoting from the September 8, 1973, issue of the *Kingston [North Carolina] Daily Press*.

4. Ibid, p. 451. The Davis book is quoting from the 1973 issue of the *Boulder [Colorado] Daily Camera* (this citation was already a decade old when used in the Davis book).

ROSA CLARK WEBSTER*

Born and raised in Virginia Beach, Virginia, physicist Rosa Clark Webster worked at the Langley Research Center of the National Aeronautics and Space Administration (NASA) during the 1970s and 1980s. She earned the bachelor of science degree in physics at Virginia's Norfolk State College in 1969, and the master's degree at Old Dominion University, before continuing to study for her doctorate.[1]

Webster's research centered on optimizing the optical properties of ion implanted gallium-arsenic-iridium light emitting diodes (LEDs), the photovoltaic properties of ion implanted gallium-arsenic solar cells, and the electro-optical properties of p-n junction devices. This type of research required both strong theoretical knowledge and specialized laboratory skills and techniques.[2]

NOTES

1. "Physicists," in *Contributions of Black Women to America*, ed. Marianna W. Davis, vol. 2 (Columbia, S.C.: Kenday Press, 1982), pp. 448–449.

2. Ibid.

MYRA WILLARD*

In 1961, Myra Willard was a "space age research chemist" employed at the Hughes Aircraft Company's Aerospace Engineering Division. Willard, a cum laude graduate in chemistry from Immaculate Heart College in Los Angeles, California, was involved in testing materials related to the space program for "Project Surveyor, a $50 million program under which Hughes will design and build space vehicles to land on the moon."[1]

NOTES

1. "Space Age Research Chemist," *Ebony*, April 1961, p. 6. My only information on Willard was provided by this article.

MARGUERITE THOMAS WILLIAMS
Geologist

Geologist Marguerite Thomas Williams provides a more positive example of Ernest Everett Just's legacy at Howard University than does Roger Arliner Young.[1] Like Young, Williams earned the baccalaureate at Howard University in 1923; Williams was, in fact, a science major, while Young was not. Oddly, Just chose Young for the assistant professorship at Howard, although Williams was clearly more qualified.[2] Immediately after graduation, Williams secured a position as assistant professor at Miner Teachers College, which is now part of the University of the District of Columbia. She was granted a leave from Miner to attend Columbia University, where she earned the master's degree in geology in 1930.[3]

Williams completed her doctorate in geology at Catholic University in 1942, nineteen years after she had earned the baccalaureate, and became the first Black woman to earn a doctorate in that field.[4] Her dissertation was "The Study of the History of Erosion in the Anacostia Drainage Basin."

Although Williams's advanced degrees were in geology, her career was spent primarily in teaching geography and the social sciences. She was chair of Miner Teachers College's Division of Geography from 1923 to 1933, and moved from the rank of assistant professor to full professor of social sciences there between 1943 and 1955.[5] She also taught night classes at Howard University during the 1940s.[6]

NOTES

1. Also see the entries on zoologist and endocrinologist Lillian Burwell Lewis and on Roger Arliner Young in this volume.

2. The 1923 edition of the *Bison* is dedicated to E. E. Just: "To E. E. Just, whose exemplary scholarship has been a source of inspiration to all of us, we, the Class of Nineteen Hundred and Twenty-three, dedicate The Bison."

3. Williams was born in Washington, D.C., in 1895. See Vivian Ovelton Sammons, *Blacks in Science and Medicine* (New York: Hemisphere, 1990), p. 253. See also *How-*

ard *University Directory of Graduates, 1870–1980* (Bernard C. Harris Publishing, White Plains, N.Y., 1982), p. 322. Harry W. Greene, *Holders of Doctorates among American Negroes: An Educational and Social Study of Negroes Who Have Earned Doctoral Degrees in Course, 1876–1943* (Boston: Meador, 1946), pp. 159–160.

4. Verified by telephone call to Maria Consolacion, Registrar's Record Office, Catholic University, Washington, D.C., 23 February 1996. Williams is not among the graduates listed in the 1923 edition of *The Bison*, nor in the 1922 or 1924 editions. There is, however, an entry on another illustrious 1923 Howard graduate, Zora Neale Hurston. The yearbook states, "Zora's greatest ambition is to establish herself in Greenwich Village where she may write stories and poems and live an unrestrained Bohemian" (*The Bison*, 1923, no pagination).

5. Sammons, *Blacks in Science and Medicine*, p. 253; Greene, *Holders of Doctorates*, pp. 159–160.

6. Sammons, *Blacks in Science and Medicine*, p. 322; Greene, *Holders of Doctorates*, pp. 159–160.

DONELLA JOYCE WILSON
Sickle Cell Anemia Researcher

Molecular geneticist Donella Joyce Wilson's research in cloning has focused on developing techniques that may someday be used in gene-based therapies to correct genetic blood diseases such as sickle cell anemia.[1] Born in Milwaukee, Wisconsin, in 1952, Wilson earned the baccalaureate degree at Johnston College of the University of Redlands in 1973. She completed two master's degrees—one in immunogenetics at Texas Southern University in 1977, and a second in microbiology at Purdue University in 1979—before earning her Ph.D. in molecular genetics at Purdue in 1981.[2] Wilson spent a brief postdoctoral fellowship working on recombinant DNA research in *Salmonella typhimurium* bacteria at Washington University in St. Louis, Missouri.

Wilson became a research associate in the Department of Oral Biology and Pathophysiology at Harvard Medical School, where between 1981 and 1983 her research involved the cloning and genetic engineering of the prolactin gene in pituitary gland tumors. Between 1983 and 1985, she had a postdoctoral fellowship in molecular biology at the Massachusetts Institute of Technology, and in 1984 she was named a Mary Ingraham Bunting Fellow at Radcliffe College and a science scholar at the Whitehead Institute of MIT. During this time, she conducted research on gene expression during the development of murine erythroleukemic Friend cells.

In 1985, Wilson secured an assistant professorship at Meharry Medical College in Nashville, Tennessee, and began conducting research on the expression of the mouse glycophorin gene to discover an identifying marker for red blood cell differentiation. An identifying marker in glycophorin, a red blood cell membrane protein, might prove useful in the study of certain types of genetically-caused blood diseases.[3] Soon after Wilson's arrival at Meharry, the National Institutes of Health awarded a joint grant to her and to the college's Sickle Cell Center for a study on the role of glycophorin in malaria.

Wilson served on National Science Foundation cell biology grant review panels, consulted on site visits for the National Institutes of Health, and received a 1988 National Science Foundation Research Opportunity Award.[4] In 1995, Wilson was reportedly on temporary leave from Meharry, conducting research with the American Cancer Society in Atlanta, Georgia.[5]

NOTES

1. Rosalyn Mitchell Patterson, "Black Women in the Biological Sciences," *Sage* 6 (fall 1989): 11–12.

2. Ibid. A 1985 *Ebony* magazine article stated that Wilson had earned her Ph.D. at Princeton University, but this is incorrect ("The Future-Makers: They Clone Cells and Smash Atoms in Search of Tomorrow's World Today," *Ebony*, August 1985, p. 66). Sammons apparently picked up the error; see Vivian Ovelton Sammons, *Blacks in Science and Medicine* (New York: Hemisphere, 1990), p. 255. Wilson's dissertation was titled "An Analysis of the Hisu Regulatory Mutation in *Salmonella typhimurium:* Altered Stable RNA Synthesis and the Pleiotropic Effects on ILV Regulation" (Sammons, ibid.).

3. Patterson, "Black Women in the Biological Sciences," p. 11. See also "The Future-Makers."

4. Ibid.

5. Isabella Finklestein, interview at her office at Atlanta University, 12 November 1995.

GERALDINE PITTMAN WOODS
Creating Opportunities for Minorities in Science

Dr. Geraldine Pittman Woods

Of all the women profiled in this volume, neuroembryologist Geraldine Pittman Woods has had the most unusual career in science. Shortly after earning her doctorate, she married, moved west, raised a family, and took a twenty-five-year leave from science. She never again entered a laboratory as a researcher or a teacher, but instead used her educational background to become an innovative science facilitator. Woods was blessed with strong family support, particularly from her mother and aunt, and clearly this was crucial to her early success.

The only child of poorly educated but economically successful parents, Woods was born in West Palm Beach, Florida, in 1921. She attended a segregated private Episcopal grammar school for a time, and then transferred to the public school system (also segregated). Although Woods excelled in science courses during high school, she was a rather indifferent student in her other courses.[1] She graduated from Industrial High School in West Palm Beach in 1938, and enrolled in Talladega College in Alabama. Unlike many of the women scientists of her day, her quest for a science education was entirely funded by her family. Although neither parent had advanced beyond an eighth grade education, her father, Oscar Pittman, had established himself in a number of business enterprises in West Palm Beach, Florida. He died when Woods was still in high school, but her mother, Susie (King) Pittman, continued to maintain the family businesses despite a two-year illness when Pittman Woods was in college. According to Woods, her mother was especially determined that she get an education, so she set a strong pattern of support:

> My mother couldn't help me, even in grade school, as she was not fortunate
> enough to have had an education. But if I needed help in a subject—like I did
> in math—my mother would find someone to help me, and then she would be so
> pleased when I did well and pleasing her made me happy.[2]

Her mother continued to support Woods throughout her college and graduate school years—to the extent that she never had to work to afford school and was able to move forward from the baccalaureate level though graduate school without interruption.[3]

Woods attended Talladega College in Alabama for two years, from 1938 to 1940, then due to her mother's illness was forced to withdraw at the end of her sophomore year. Mother and daughter traveled to Baltimore, Maryland, and stayed with friends, the Coasey family, while Mrs. Pittman underwent treatment at Johns Hopkins University Medical Center. Woods might have dropped out of school at that point, but others recognized her abilities and provided encouragement. Both Coasey, a physician, and his wife were Howard University alumni; in addition, they had maintained a close friendship with Dr. Charles Wesley, then dean of the graduate school at Howard. Within this atmosphere of strong support and encouragement, Woods transferred to Howard in 1940 to major in biology.[4]

At Howard, Woods was mentored by Louis Hansborough, a Harvard University Ph.D. who had earlier been a student of Ernest Everett Just and Roger Arliner Young. According to Woods, Hansborough "encouraged me and gave me books to read which we would then discuss."[5] By her senior year, he had begun to talk seriously with her about going on to graduate school. By the time Woods graduated from Howard in 1942, her mother had recovered her health. Hansborough had suggested she apply to Harvard University or travel to England for advanced training, but Woods was reluctant to move so far away from home. With her mother's encouragement, however, she chose Radcliffe College, in Cambridge, Massachusetts.[6]

In 1942 women were not permitted to enroll at Harvard University, but there were close ties between Harvard and Radcliffe; women graduate students enrolled at Radcliffe generally took the majority of their courses at Harvard.[7] Woods took all her graduate courses at Harvard, and despite the high cost of tuition and other expenses, never applied for funding:

> My mother paid for everything. She had a tremendous commitment to education in general and my education in particular. She just paid the bills when they were sent to her each semester. She had property—rental property—and my college bills were her first priority. She probably sacrificed in some areas, but we didn't suffer at all. I wouldn't say we were well off financially, but we were comfortable, we didn't want for anything. Since she was always aware that she had never had the opportunity to get an education, she wanted me, her only child, to have all the education I desired. In spite of her lack of education, she was an excellent businesswoman.[8]

Woods said her mother believed that scholarships were better left to those in greater need.[9]

Woods recalled that during her time at Radcliffe/Harvard there were only two Blacks in a doctoral science program—herself and Mary Logan Reddick—and they were both in neuroembryology. Reddick was older and had taught for some years with her master's degree before continuing on for the doctorate.[10] As a result, Woods and Reddick had the advantage of being able to study together and bounce ideas off one another: "Mary and I had our research laboratory together, so we spent hours there; we studied together many times."[11] There was also a group of Black female students at Boston University and a few Black males at Harvard (although none were in the sciences) with whom Woods and Reddick were friends. Woods said she also "had friends in the Science Building who I had lunch with practically every day. We were of different ethnic groups, [they were] mostly Caucasian, but that was never a problem."[12]

During her first year at Radcliffe, Woods had to play "catch up," taking courses that other graduate students had generally taken in college. There were two reasons for this—the lack of scientific equipment at segregated high schools and historically Black colleges, and the fact that Woods had changed colleges as an undergraduate:

> In my high school teachers motivated students, although we didn't have the finances for equipment and supplies for each student to have a *hands on* experience. I attended a segregated high school and the teachers inspired and taught us in the best way they could. . . .
> In college, particularly at Talladega, the professors usually performed an experiment and the class stood around and watched.[13]

Other women in this volume have mentioned the lack of proper scientific equipment at segregated high schools and historically Black colleges; however, like Woods, they also spoke of the diligence with which their teachers and professors tried to make up for the lack of equipment. When Woods walked into her first laboratory session at Harvard, students were using apparatus that she had never seen before: "I wasn't as familiar with as much of the scientific equipment as I should have been, *at first.* So I had to work harder, but after becoming adjusted, I worked about as hard as the other students did."[14] She was mentored by Leigh Hoadley, a Harvard embryologist, who served as her graduate advisor and guided her through her research on the early development and subsequent specialization of nerves in the spinal cord.[15] She earned the master's degree in embryology in 1943. In 1945 she was elected to the Phi Beta Kappa honorary society and received her Ph.D. in neuroembryology from Radcliffe/Harvard.[16] Her dissertation was "The Development of the Spinal Cord of the Chick Embryo in Chorioallantoic Grafts."[17]

Immediately after earning her doctorate, Woods took a position as a biology instructor at Howard University.[18] She was already engaged to Robert Woods, who had interrupted his dental studies at Meharry Medical College to serve in World War II.[19] He was mustered out in 1945 and resumed his studies; they were married at the end of the fall semester, although she continued to teach at Howard until the end of the 1946 spring semester. When her husband had completed his studies, they moved to California where he set up a practice in Los Angeles.[20] Woods settled down to raise a family and took a twenty-five-year hiatus from science. For the first fifteen years, Woods was a housewife and mother; when her children reached their teens, however, she turned her energies to volunteer work.[21] Her activities started at the local level, but gradually broadened to the state and finally the national levels.

By the early 1960s, Woods had entered a world in which she was selected for a series of upwardly mobile positions. In 1963, she began a four-year term on the personnel board of the California Department of Employment. Then in 1964, at the suggestion of Vice-President Hubert H. Humphrey, the United States Surgeon General appointed her to a four-year term as a member of the National Advisory Council of the National Institute of General Medical Sciences, part of the National Institutes of Health.[22] In 1965, as a result of her activities as national president of Delta Sigma Theta sorority—where Woods directed projects involving mental health, social action, community service, and child development programs—the First Lady, Mrs. Lyndon Baines Johnson, invited her to the White House to participate in the launch of Project Head Start. In 1967, Woods began a three-year term as a member of the Foreign Service Officers Selection Board for the United States State Department. In 1968, President Lyndon Johnson appointed her chairman of the Defense Advisory Committee on Women in the Services.[23]

With the exception of her membership on the National Institute of General Medical Sciences Advisory Council (1964–1968), Woods had not yet returned to science. But it was while Woods was serving on the NIGMS Advisory Council that she first became involved in an effort to increase support for science and science training for Black students and Black scientists at historically Black colleges. She said, "I noticed that [NIH and NIGMS] received thousands and thousands of grant applications, but so few of them were from Black colleges. And

of those few that did come from Black colleges, even fewer were awarded funding."[24] President Lyndon Johnson's vision of a "Great Society," coupled with the
continued Cold War demand for trained personnel in the sciences and related
fields, provided a framework within which Woods could use the power of the
federal government to improve science opportunities for minorities. In 1969, she
was appointed as first Black female special consultant to NIGMS. Woods's initial mandate was to advise its director and staff on possible ways to improve the
level of biomedical science research and training at predominantly Black colleges
and universities.[25] To accomplish this, she first had to analyze the reasons why
Black scientists had difficulty competing for research grants from government
agencies and philanthropic organizations. Were scientists at historically Black institutions presenting topics in their grant proposals that were sufficiently important to the scientific community? Were the laboratories at these institutions sufficiently equipped to do the proposed research? Was the science training of the
faculty adequate to the proposed tasks?

> By the time I became a *special* consultant, I had already begun to keep a record
> of professors applying for research grants and fellowships. I had been invited to
> talk to the director of the National Institutes of Health after he had heard of my
> [earlier] report to the National Institute of General Medical Science Council.
> He said that the staff and I must have been doing something right, as our per-
> centage of funding [for Blacks] at NIGMS had become higher than the average
> at NIH. He then became interested in developing programs for minorities.[26]

Woods found that, although the proposals being submitted were in general sufficiently interesting, the schools lacked the equipment to carry them out properly. Additionally, the faculty members at historically Black institutions were so
overworked by heavy teaching loads that they had difficulty advancing their educations to the doctoral levels; for the same reason, those who had doctorates had
difficulty in keeping current with their fields.

By the early 1970s, Woods had initiated and played a major role in the development of the Minority Access to Research Careers (MARC) program and the
Minority Biomedical Research Support (MBRS) program at NIGMS. However,
when the programs were formally initiated in 1972 she had a difficult time persuading administrators and staff at predominantly Black colleges to become involved.[27] "They were reluctant . . . to participate," she said. "They had been dis-
illusioned by all the years of neglect."[28] According to Woods, another obstacle was

> impressing the staff at NIGMS/NIH and the Congress with the need for money
> to implement these programs . . . and also impressing these individuals with
> what could be done in the sciences at historically Black colleges and convincing
> them of the potential for science that was being lost. . . .
>   Many did not know the true problems of the [HBCUs]. . . . I was the only
> African-American woman who was a special consultant. I was alone—people
> had talked about helping minorities but they didn't do anything. . . . So I was a
> sort of liaison person between the [Black] colleges and the NIH.[29]

She credits Charles Miller, who was then in charge of the effort, with playing a
crucial role. "[In the beginning], there wasn't much support for the programs, but

Dr. Miller was committed. [He] was also helpful in giving me a free hand to call the colleges, so I could urge them—despite their reservations—to participate."[30]

The MBRS and MARC programs have had an unprecedented impact on the development of scientific research at historically Black institutions. Prior to the 1970s, federal research funding for Black scientists was quite limited. With the establishment of the MARC and MBRS programs, significant federal funding became available for the first time for biomedical research programs at historically Black colleges, and the programs served as catalysts to broaden the scope of funding available to Black scientists in other science fields. When they were first begun, the two programs were aimed specifically at the historically Black colleges, but soon after the government changed the designation to include all colleges with significant minority, rather than Black-only, populations.[31] As a result, the programs increased federal funding for the biomedical sciences at all colleges and universities with significant minority populations. By 1980, approximately 50 MBRS students had completed their doctorates at minority institutions and were involved in postdoctoral research at various research institutions; 443 MBRS students graduated with B.S. degrees in 1980 (compared with 275 students in 1973); and student and faculty research had produced 402 publications in 1980 (compared to only 75 in 1974).[32] Woods's initiative on behalf of these colleges began a new era of scientific research that has significantly increased the contributions of these institutions to the sciences.

Despite her close ties to science, Woods no longer considers herself a scientist:

> Many people just call me a scientist, and I've been named "one of the famous Black scientists," but I don't consider myself a true scientist, just a facilitator, a science-trained academic who has worked to make science available for others by developing programs to provide better access for those who, historically, have been bypassed.[33]

She has also been active in other areas. In 1969, she was appointed to the board of trustees of Howard University; she was named chairman in 1975, and chair emerita in 1988. Between 1984 and 1988, she also served as chair of the Howard University Foundation's board of directors. Although Woods said that her contacts at the National Institutes of Health did not directly help in her work on the board of directors at Howard University, indirectly her contacts helped Howard professors receive grants.

> I would refer persons to NIH to give them suggestions. Before I was chairman, I could work with the [NIH] staff, but after becoming chairman at Howard, NIH would not allow it. I was able to meet many professors by working with NIH, though, so when I became chairman, I knew many in the sciences and this was often beneficial.[34]

In 1974, she was elected to membership to the Institute of Medicine of the National Academy of Sciences, a position she held until 1995. She was also elected to the board of trustees of Atlanta University (1974) and served as chair of the Educational Policy Committee and a member of the Executive Committee until 1986.[35]

Although she has some regrets that she never returned to scientific research and thus never published, Woods has said that "I was so busy 'driving' that I couldn't find time to research and write. But I think I have made a lasting contribution to science nevertheless."[36] A fellowship in biology at Howard University and one in chemistry at Atlanta University have been established in her name.

NOTES

1. Geraldine Pittman Woods, correspondence sent to author (curriculum vitae, author questionnaire, assorted papers), 16 May 1995, 12 June 1995, 20 June 1995. Woods, telephone interviews by author, May through July 1995. See also James H. Kessler, J. S. Kidd, Renee A. Kidd, and Katherine A. Morin, *Distinguished African American Scientists of the Twentieth Century* (Phoenix, Ariz.: Oryx Press, 1996), pp. 346–350. *American Men and Women of Science: The Physical and Biological Sciences*, 16th ed. (New York: Bowker, 1983). Tia Gindick, "Able Advocate of Higher Education: Dr. Geraldine Woods Has Words for 'Qualifying Minorities,'" *Los Angeles Times*, 18 July 1980, pp. 1, 12. *Who's Who among Black Americans* (Northbrook, Ill.: Who's Who among Black Americans Publishing, 1985), p. 928; 1988 edition, p. 775; 1990–1991 edition (Detroit: Gale Research). Jessie Carney Smith, "Geraldine 'Jerry' Pittman Woods," in *Notable Black American Women*, ed. Jessie Carney Smith, vol. 2 (Detroit: Gale Research, 1992), pp. 1278–1280. "Neuro-Embryologist," in *Contributions of Black Women to America*, ed. Marianna W. Davis, vol. 2 (Columbia, S.C.: Kenday Press, 1982), pp. 434–435. Rosalyn Mitchell Patterson, "Black Women in the Biological Sciences," *Sage* 6 (fall 1989): 3–13. William Glitz, "The Minority Biomedical Support Program," *The Black Collegian*, October–November 1981, pp. 38–40. Vivian Ovelton Sammons, *Blacks in Science and Medicine* (New York: Hemisphere, 1990), p. 257.

2. Woods, correspondence. According to Woods, her mother's parents died when Susie King Pittman was quite young; she was raised by others and had to go to work as soon as she finished eighth grade. Oscar Pittman was involved in restaurant, farming, lumber, and real estate businesses.

3. Ibid.

4. Ibid.

5. Woods, questionnaire. Hansborough's degree was in embryology.

6. Ibid.

7. Woods, interviews.

8. Ibid.

9. Ibid.

10. Woods, questionnaire.

11. Ibid.

12. Ibid.

13. Ibid.

14. Ibid.

15. According to Woods, in the early embryo stage nerve cells are barely distinguishable from other kinds of cells. As the fetus matures, they become specialized and begin transmitting messages. Her research focused on determining whether the specialization process was governed by hereditary factors or by stimulation from other nearby cells. Using chick embryos, she showed that both effects took place. Specialization was also affected by the number of muscle cells present—a larger number of muscle cells appeared to cause the development of a larger number of nerve cells, which became specialized as nerve activators (Woods, interviews).

16. Woods, curriculum vitae.

17. Ibid.

18. She was hired by Hansborough.

19. Kessler, *Distinguished African American Scientists*, p. 348. Also Woods, questionnaire.

20. Robert Woods received his D.D.S. from Meharry Medical College in 1946.

21. Woods has three children.

22. According to Woods, members of the National Advisory Council of NIGMS were elected, and worked under the auspices of the Secretary of Health and Human Services. The job of the Council was to focus on ways of facilitating scientific training and research. "To the best of my knowledge," Woods said, "I was one of the first Blacks on the Council." Although there were minorities employed as staff at NIGMS and NIH, Woods believes she was the only Black and the only female special consultant during that period (Woods, questionnaire).

23. According to Woods, "The Defense Advisory Committee on Women in the Services was a group of women who traveled to the four branches of the Armed Services— the Army, Navy, Air Force, and Marines—to see how things were going for women in the services—general treatment, opportunities for advancement, utilization of their skills, and so on. We traveled to bases and reported to the Secretary of Defense." At this time, women were becoming more important in the military, and according to Woods, "The first female general of the Army was appointed during this time. . . . I was invited to the White House when President Johnson signed the order for women to become generals" (Woods, questionnaire).

24. Ibid.

25. Ibid.

26. Ibid.

27. The MARC and MBRS programs enable professors and students at minority institutions to increase their participation in research and research training through federal funding for faculty release time, student tuition, supplies, and equipment. The two programs also provide salaries and stipends for undergraduates, pre-doctoral and post-doctoral students, as well as faculty fellowships and visiting scientist awards (Glitz, "Minority Biomedical," pp. 38–40).

> One cause of minority underrepresentation in science has been the limited involvement of minority institutions in biomedical research prior to the advent of the [two programs]. This situation resulted from several factors, including: institutional attitudes [at HBCUs] that emphasized teaching as the predominant mission; heavy teaching loads that left little or no time for faculty to conduct research; lack of funds and inadequate facilities and equipment for research; and inadequate faculty staffing for research.
> Starting with the general institutional support focus, the program evolved to encourage individual research projects involving faculty and student participants. A unique feature is the emphasis placed on undergraduate research participation which ranges from conceptualization of important research question to design of experiments, review of the literature, collection and analysis of data, and finally, interpretation and reporting of results at meetings and in scientific journals.
> In 1972, thirty-eight institutions were funded for a total of $2 million. [By 1981], the program had expanded to eighty institutions, including two-year colleges and Indian reservations through community colleges within the reservation; by 1987, the program had been expanded to include institutions with significant numbers of minority students. (Ibid.)

28. Woods, interviews and correspondence.

29. Ibid.

30. Ibid.

31. These now include, for example, colleges with Native American, Hispanic, and Asian populations (ibid.).

32. Glitz, "Minority Biomedical," p. 40. In addition, student investigators supplied with MBRS funds made 550 presentations at scientific meetings in 1980, as compared with 212 in 1974.

33. Woods, questionnaire and interviews.

34. Woods, questionnaire.

35. Woods, curriculum vitae. Other positions Woods has held include: elected member of the board of directors of the National Commission for the Certification of Physicians Assistants (1974–1981); appointed member by the California State Senate Rules Committee to a four-year term on the California Postsecondary Education Commission (1974–1978); elected vice-chairman of the California Postsecondary Education Commission (1976); appointed member of the Air Pollution Manpower Development Advisory Committee of the Federal Environmental Protection Agency (1973–1975); appointed member of the board of directors for the Robert Wood Johnson Health Policy Fellowships Committee of the Institute of Medicine of the National Academy of Sciences (1973–1978); appointed member of the Board of Graduate Studies of Meharry Medical College in Nashville, Tennessee (1972–1975); elected member of the Board of Trustees of the California Museum Foundation of the California Museum of Science and Industry (1971–1979); chair of the California Museum Foundation's Education Committee (1970–1973); member of the General Research Support Program Advisory Committee for the Division of Research Resources of the National Institutes of Health (1977–1978); elected member of the Board of Directors of the Charles R. Drew University of Medicine and Science (1991–present). Through her involvement in Delta Sigma Theta sorority, Woods was a strong supporter of Head Start programs across the country. She encouraged sorority chapters all over the country to participate. Los Angeles now has fifteen centers, and the one dedicated in October 1993 was named the Dr. Geraldine P. Woods Head Start Pre-School Center.

36. Woods, interviews.

JANE COOKE WRIGHT

Expanding a Family Tradition in Medical Research

Physician and cancer researcher Jane Cooke Wright was born into America's "Black elite." The eldest of Louis Tompkins Wright and Corinne (Cooke) Wright's two daughters, she was born in New York City in 1919.[1] Her father was one of the first Blacks to graduate from Harvard University Medical School, and his father had been one of the first graduates of Meharry Medical College.[2] Her father's stepfather, William Penn, was the first Black to graduate from Yale Medical School. In addition, one of Jane Wright's uncles, biochemist Harold Dadford West, was both a Rosenwald and a Rockefeller General Education Board Fellow and later became president of Meharry Medical College.[3]

Jane Wright's career path was influenced by her famous father's successes. After his graduation from Yale University in 1915, Louis Tompkins Wright served as an Army officer in World War I; when the war ended, he became the first Black doctor appointed to a staff position at a municipal hospital in New York City. In 1929, he also became the City's first Black police surgeon.[4] During the early portion of his career, Wright was involved in a variety of research efforts: he explored new methods for the treatment of skull fractures and a more effective treatment of hookworm.[5] He was also responsible for a number of other innovative developments: he pioneered in new methods of treating gunshot and knife wounds, invented a neck brace that allowed accident victims to be moved without risk of further spinal cord injury, developed an internal splint that permitted multiple breaks in major bones to heal without permanent impairment,[6] headed a team that was among the first to use aureomycin, and originated an intradermal method of vaccination against smallpox.[7] In the late 1930s, he founded the Cancer Research Center at Harlem Hospital, which became an important research venue for a new generation of Black doctors and scientists. In 1934 he became the first Black elected as a fellow to the American College of Surgeons. Louis T. Wright also had a long and active commitment to civil rights: he was chairman of the board of directors of the NAACP from 1932 until his death in 1952.[8]

As the daughter of Dr. Louis Wright, Jane Wright spent much of her childhood in a predominantly wealthy, largely white world. Her fellow students at the schools she attended — Ethical Culture and Fieldston — were almost entirely whites of the privileged class, along with a sprinkling of gifted scholarship students. Yet Wright appears to have had little difficulty fitting in — an excellent student, she was also captain of the swimming team.[9] In 1938, Wright entered Smith College on a four-year academic scholarship to study art. Her father, however, did not feel that art offered a proper future, so he encouraged her to switch her major to pre-med, which she did in her junior year.[10] At Smith, a majority white institution, Wright continued to fit in exceptionally well: she was literary editor of the yearbook and a member of the varsity swim team.[11] Her family was so prominent that when she graduated from Smith College in 1942, her picture appeared on the cover of *The Crisis*, the official publication of the NAACP.[12]

With the advent of World War II, many U.S. colleges accelerated their medical programs from the traditional four years to three. Wright was accepted into the accelerated medical program at New York Medical College with a full academic scholarship. Here again, she fit in exceptionally well with her predominantly white, now largely male, classmates: she was elected vice-president of her class and president of the honor society.[13] In June 1945, she graduated with honors; she was third in a class of ninety-five students, and the only female among the predominantly white males.

Wright interned at Bellevue Hospital (1945–1946), and served as an assistant resident in internal medicine from April to December of 1946.[14] She then completed a residency in internal medicine at Harlem Hospital (1947–1948), during which time she married David Jones Jr., a Harvard Law School graduate and the son of David D. Jones, the president of Bennett College.[15] In 1948, she took a six-month maternity leave for the birth of her first child; she returned to Harlem Hospital as chief resident in July 1948 to complete the last phase of her training.[16]

In January 1949, Wright took a position as a staff physician with the New York City Public Schools, and was concurrently a visiting physician at Harlem Hospital. After six months, however, she was unhappy with the way her career was progressing, and left the city schools job to become a clinician and researcher at the Cancer Research Foundation at Harlem Hospital, which had been founded and was then directed by her father, Louis Tompkins Wright.[17]

The first, faint glimmer of the idea of treating cancers with poisonous chemicals occurred during World War I, when doctors noticed that American infantrymen who had been exposed to mustard gas attack suffered from, among other things, a markedly reduced ability to produce white blood cells. Since leukemia is marked by an overproduction of white blood cells, doctors speculated that some derivative of mustard gas might prove useful in combating it.[18] Little research was done in the following two decades to follow up on the idea, perhaps because scientists had not as yet developed even the most rudimentary techniques to offer protection for noncancerous cells. Surgery remained the only option for cancer treatment until the 1930s when radiation therapy began to be used as well. Then, in 1942, a supply ship carrying a quantity of nitrogen-mustard gas was sunk in an Italian port, again subjecting troops to the chemical agent. After doctors again noticed that the exposed survivors had significantly reduced white blood cell counts, they renewed speculation that some form of the compound might be useful in treating leukemia.[19] As nitrogen mustard, an alkylating chemical, disrupted cell division, some thought it might also be useful in the treatment of other cancers.[20] By 1946, researchers had begun experimenting with nitrogen mustard in liquid form, injecting it intravenously into a small number of leukemia patients.[21] Louis Wright, then the chief of surgery at Harlem Hospital, had already directed the focus of the Cancer Research Foundation toward the study of anti-cancer chemicals. He, along with Sidney Farher in Boston, became a pioneer in treating advanced cancer patients with the newly synthesized folic acid antagonists aminopterin and amethopterin (methotrexate).[22]

When Jane Wright began her research on using chemicals to treat various types of cancer, chemotherapy was still largely experimental. Her first experiments studied the efficacy of triethylenemelamine (TEM), an alkylating agent that had already been shown effective against leukemia in mice.[23] Working both as a team and independently, in 1949 the Wrights began to use TEM on certain human leukemias and cancers of the lymphatic system. Louis Wright would do the laboratory work and Jane Wright would perform the trials on patients. In 1950, Louis Wright was joined in the laboratory by cell biologist Jewel Plummer (Cobb), then a National Cancer Institute/National Institutes of Health postdoctoral fellow at Harlem Hospital.[24] Jane Wright began by administering daily doses of TEM to eleven Hodgkin's disease patients; the daily doses lasted for a period of one week, and were followed by once-weekly doses for five more weeks. She reported that ten of the eleven patients began to experience varying degrees of remission.[25] The tests were then expanded to include other types of lymphatic cancers similar to Hodgkin's, and produced promising results. Wright's findings were corroborated by her father's and Jewel Plummer Cobb's laboratory experiments.[26] Despite TEM's promise in treating Hodgkin-type cancers, it was not effective against more severe leukemias nor on other types of cancers. The group then began experimental work on a new chemical group, antimetabolites, which were thought to block the production of certain chemicals in the nuclei of cancer cells.[27]

Louis Wright died in 1952, and Jane Wright took over his position as head of the Cancer Research Foundation. Jewel Plummer Cobb also left New York at this time, but she returned two years later; Cobb and Wright continued to collaborate during the next ten years.[28]

In the 1950s, scientists around the country were researching an ever-increasing number of chemical agents for possible use in cancer treatment. One problem, as Wright saw it, was to find a way to determine which particular chemical worked best on which specific form of cancer. Within a few years of her taking over as director of the Cancer Research Foundation, Wright, both independently and in collaboration with Cobb, began to focus her investigations on ways to *predict* the clinical response of a patient to a particular chemical agent by treating selected tissue from the patient's tumor in the laboratory.[29] They surgically removed tumor sections, divided them, then cultured them in flasks of protein-rich growth medium; once tumor cell colonies were established in the flasks, a different anti-cancer agent would be injected into each group. The agent that appeared to have the most lethal effect on a colony's cells could then be administered to the patient.[30] As a bonus, the tissue culture method provided a means to determine exactly *how* each chemical agent affected various cancers.

In 1955 Wright joined the faculty of New York University as an associate professor of surgical research and director of cancer chemotherapy research at NYU Medical Center and its affiliated Bellevue and University hospitals. The clinical and laboratory research group of Wright's Cancer Research Foundation moved to New York University's Post-Graduate Medical School as part of Fifth Division Surgery, with Wright as director.[31] By 1960, in addition to Wright's work on TEM and nitrogen mustard (in liquid form), she had conducted studies on a panoply of anti-cancer agents, including triethylenethiophosphoramide (thiotepa), CB 1348, dihydro E. 73, puromycin, corticotropin (ACTH), and folic acid antagonists.[32] She developed a database on each type of cancer cross-referenced with drug types and patient profiles to aid in predicting the efficiency of the various anti-cancer agents.[33]

Wright also embarked on a new avenue of research: finding more efficient methods for delivering an anti-cancer agent to the tumor site. The tissue culture research had indicated that administering chemicals orally or by injection was not the most effective means of transporting the substances to tumors lying deep within a patient; also exposing healthy tissues to these chemicals resulted in understandably negative side effects. The dosages could not be increased beyond a certain level because of the drugs' destructive effect on all cells, but this also meant that the amount of chemical reaching the cancer site was often greatly reduced. Researchers elsewhere had developed a perfusion technique for injecting medications more directly into a specific bodily site. Wright hit upon the idea of using the perfusion technique to inject anti-cancer chemicals into the specific blood vessel supplying a tumor.[34] The technique was especially promising in the treatment of tumors that were not amenable to surgery.[35] As an outgrowth of this research, Wright also developed a new method of temporarily re-routing the arteries that fed the site of the cancer, further restricting a drug to the cancer site.[36] By this time many cancer researchers, including Wright, began to favor the idea that most cancers were caused by viruses; based on that idea, she expanded her research to a new class of anti-cancer agents made up of compounds similar to antibiotics, such as dactinomycin, streptonigrin, mithramycin, and fluorouracil. She also began experiments using combinations of anti-cancer agents.[37]

In 1964, Lyndon B. Johnson appointed Wright to the President's Commission on Heart Disease, Cancer, and Stroke. The Commission's report led to the establishment of a national network of centers for the study and treatment of these diseases. In 1966, Wright was elected to the Sigma Xi honorary society.[38]

Wright left New York University in 1967 to take a position as professor of surgery, head of the cancer chemotherapy department, and associate dean at New York Medical College, her alma mater. With this appointment she became the highest ranked Black woman at a nationally recognized medical institution, at a time when there were only a few hundred Black women physicians in the country.[39] She also joined the staffs at New York Medical College's affiliated hospitals—Flower Fifth Avenue, Metropolitan, and Bird S. Coler Memorial—where she continued her research.[40]

In addition to pursuing her own research and administrative duties at the college, Wright developed a new comprehensive program to study cancer, heart disease, and stroke, and implemented another program to educate doctors about chemotherapy.[41] During her forty-year career, she produced more than 75 research papers on cancer chemotherapy.[42] She was a member of the American Association for Cancer Research, the American Medical Association, the National Cancer Advisory Council, the Board of Directors of the American Cancer Society, and served on the editorial board of the *Journal of the National Medical Association*.[43] Wright was the distinguished lecturer at the annual convention of the National Medical Association in 1983.[44]

Jane Cooke Wright retired in 1987. Her analyses of a wide range of anti-cancer agents, her explorations of the relationship between patient and tissue culture response, and her new techniques for administering anti-cancer agents contributed significantly to her field of science.

NOTES

1. George L. Jackson, *Black Women Makers of History: A Portrait* (Sacramento, Calif.: Fong and Fong, 1977), pp. 182–184. Jean Elder Cazort, "Jane C. Wright," in *Notable Black American Women*, ed. Jessie Carney Smith, vol. 2 (Detroit: Gale Research, 1992), pp. 1283–1285. See also Robert C. Hayden and Jacqueline Harris, *Nine Black American Doctors* (Reading, Mass.: Addison-Wesley, 1976), pp. 110–117. *Who's Who among Black Americans*, 4th ed. (Northbrook, Ill.: Who's Who among Black Americans Publishing, 1985), p. 932. *Ebony Success Library*, vol. 1 (Chicago: Johnson Publishing, 1973), p. 338. "Outstanding Women Doctors: They Make Their Mark in Medicine," *Ebony*, May 1964, pp. 68–69, 72–74, 76. "Homecoming for Jane Wright," *Ebony*, May 1968, p. 72–77. *The Crisis*, August 1942 (her picture appeared on the cover of this issue). "Dr. Jane

Wright Honored," *The Crisis*, May 1965, pp. 328. "Black Women as Medical and Nursing Administrators," in *Contributions of Black Women to America*, ed. Marianna W. Davis, vol. 2 (Columbia, S.C.: Kenday Press, 1982), p. 410.

2. Jackson, *Black Women Makers*, p. 182; and Cazort, "Jane C. Wright," p. 1283; and "Homecoming for Jane Wright," p. 72. For more information on L. T. Wright, see *Negro Year Book*, 10th ed. (Tuskegee, Ala.: Negro Year Book Publishing, 1947), p. 25. According to Sammons, L. T. Wright graduated Harvard Medical School, cum laude, in 1915 and was ranked fourth in his class; see Vivian Ovelton Sammons, *Blacks in Science and Medicine* (New York: Hemisphere, 1990), p. 259. At the 25th reunion of his graduating class at Harvard Medical School, his peers voted Louis T. Wright the man in the class who had contributed the most to medical knowledge (Jackson, *Black Women Makers*; Cazort, "Jane C. Wright"; Sammons, *Blacks in Science and Medicine*).

3. Sammons, *Blacks in Science and Medicine*, pp. 188, 246, 259. The records on Louis Wright's natural father are not clear, nor do they reveal his first name. Since divorce was not common in those days, it seems logical to assume that he died at a fairly young age and that Wright's mother then married William Penn. Sammons does not list the year in which Penn earned the M.D. at Yale. Harold West was the son-in-law of William Penn, and apparently maintained a close relationship with the Wright family (Cazort, "Jane C. Wright," p. 1293). He earned the Ph.D. from the University of Illinois in 1930 (Sammons, *Blacks in Science and Medicine*, p. 246). West was the first Black president of Meharry (Cazort, "Jane C. Wright," p. 1293).

4. Sammons, *Blacks in Science and Medicine*, p. 259. L. T. Wright was appointed to the staff at Harlem Hospital, which was then a white institution. May Edward Chinn had interned at Harlem Hospital in 1926–1928, but did not become a member of the staff there until the late 1930s. See my entry for May Chinn in this volume.

5. *Negro Yearbook*, p. 25. L. T. Wright wrote a chapter, "The Standard Treatment of Skull Fractures," in a medical text edited by Charles L. Scudder (ibid.).

6. James H. Kessler, J. S. Kidd, Renee A. Kidd, and Katherine A. Morin, *Distinguished African American Scientists of the Twentieth Century* (Phoenix, Ariz.: Oryx Press, 1996), p. 350.

7. Sammons, *Blacks in Science and Medicine*, p. 259.

8. "Dr. Jane Wright Honored," p. 328. See also *The Negro Yearbook*, p. 25. In 1938, Louis T. Wright was the recipient of the 25th annual Spingarn Award, presented to him by the NAACP for having been "a consistent and persistent foe of every form of segregation and denial of opportunity," for playing a major role in "the establishment of a yardstick of medical proficiency which has done much to change the attitude of both the white and Negro medical world and the public at large," and for distinguishing himself as a surgeon to the extent that it "led to the color line being broken in the American College of Surgeons through his election as a fellow in 1934." Ernest Everett Just had been an earlier recipient of the Spingarn Award, per Kenneth W. Manning, *Black Apollo of Science: The Life of Ernest Everett Just* (New York: Oxford University Press, 1983).

9. Cazort, "Jane C. Wright," p. 1284.

10. It appears that Wright made the change from art to medicine rather reluctantly. She was quoted as saying that as a freshman at Smith College she had convinced herself that she could never compete in medicine under the shadow of her father ("Homecoming for Jane Wright").

11. Cazort, "Jane C. Wright," p. 1284. According to Cazort, she established a number of swimming records which held for many years.

12. The caption for the photo read "Jane Cooke Wright, daughter of Dr. Louis T. Wright of New York City, chairman of the NAACP Board of Directors. Miss Wright graduated from Smith College this year with the degree of Bachelor of Arts" (*The Crisis*,

August 1942, p. 426). Her picture appeared on the cover of *The Crisis* again for the January 1953 issue to note her recognition by *Mademoiselle* magazine as one of the "outstanding young women" of 1952, for her "outstanding contribution to medical science with her evaluations of the efficacy of drugs in cancer treatment—evaluations that are now being translated, abstracted, and quoted all over the world" ("Dr. Jane Wright Honored," pp. 4–5).

13. Jackson, *Black Women Makers*, p. 183.

14. During the late 1930s and early 1940s, Bellevue, along with New York Memorial Hospital, was a leading center of radiation treatment for cancer. According to an article in *Life* magazine, in 1937 Bellevue and NYMH each had "nine grams of radium, more than any other U.S. hospitals" ("U.S. Science Wars Against an Unknown Enemy: Cancer," *Life*, 1 March 1937, pp. 11–17).

15. Cazort, "Jane C. Wright." See also "Homecoming for Jane Wright," p. 76, and Jackson, *Black Women Makers*, p. 183.

16. Jackson, *Black Women Makers*, p. 183.

17. Ibid.

18. Hayden, *Nine Black American Doctors*, pp. 111–112.

19. Kessler, *Distinguished African American Scientists*, pp. 351–352.

20. Hayden, *Nine Black American Doctors*, pp. 111–112.

21. Ibid.

22. Jewel Plummer Cobb, "A Life in Science: Research and Service," *Sage* 6 (fall 1989): 39–43.

23. Jackson, *Black Women Makers*, p. 183; Cazort, "Jane C. Wright."

24. Jewel Plummer Cobb, curriculum vitae, 19 February 1996. Also Cobb, written communications with author, 15 March 1996 and April 1996. For a brief time during this period, Wright also collaborated with her sister, Barbara Wright Pierce (see below).

25. Hayden, *Nine Black American Doctors*, pp. 111–112.

26. G. Antikajian, L. T. Wright, J. Plummer, and S. Weintraub, "The Effect of Triethylene Melamine, Aureomycin, and Some 4–Amino Derivatives of Folic Acid on Tissues in Vitro," *Journal of the National Cancer Institute* 12 (1951): 269–274; J. Plummer, L. T. Wright, G. Antkajian and S. Weintraub, "Triethylene Melamine in Vitro Studies. I. Mitotic Alterations Produced in Chick Fibroblast Tissue Cultures," *Cancer Research* 12 (1952): 796–800.

27. J. C. Wright, J.I. Plummer, R. S. Coidan, and L. T. Wright, "The in Vivo and in Vitro Effects of Chemotherapeutic Agents on Human Neoplastic Diseases," *The Harlem Hospital Bulletin* 6 (1953): 58–63.

28. See the appendix.

29. Hayden, *Nine Black American Doctors*, pp. 111–112; Kessler, *Distinguished African American Scientists*.

30. The primary paper resulting from this research was J. C. Wright, J. P. Cobb, S. L. Gumport, F. M. Golomb, and D. Safadi, "Investigation of the Relationship between Clinical and Tissue Response to Chemotherapeutic Agents on Human Cancer," *New England Journal of Medicine* 257 (1957): 1207–1211. The report also indicated that some drugs which were successful in research animals did not achieve the desired results with human beings. And in some cases, drugs that showed good results on the cultured cells extracted from a malignant growth did not always inhibit the cancer in the patient. Generally, however, the report indicated that this new method was a much better predictor of effect than had been available to date. (It should be noted that "effect" in oncology is often termed "response rate," which doesn't necessarily mean that the patient is permanently cured or that he or she will not have a recurrence.)

31. Cobb, written communication. Cazort and Jackson both indicate that Wright con-

tinued her directorship of the Cancer Research Foundation at Harlem Hospital while she was director of the NYU Medical Center's cancer chemotherapy research program (Cazort, p. 1285; Jackson, p. 184).

32. Cazort, "Jane C. Wright"; Jackson, *Black Women Makers*; Kessler, *Distinguished African American Scientists*. See also articles on Wright by Fern Marja Eckman, *New York Post*, 22 April 1957; Hope Johnson, *New York World Telegram and Sun*, 16 December 1959; May Okon, *New York Sunday News*, 16 November 1967; *New York Sunday News*, 19 November 1967.

33. Cazort, "Jane C. Wright," p. 1284.

34. F. M. Golomb, J. C. Wright, J. P. Cobb, S. L. Gumport, A. Postal, and D. Safadi, "The Chemotherapy of Human Solid Tumors by Perfusion Techniques," *Proceedings of the American Association for Cancer Research* 3 (1960).

35. Such tumors would be ones that were located too deep in the body for surgery to be an option or ones that were too spread out.

36. Golomb, "Chemotherapy of Human Solid Tumors." See also Kessler, *Distinguished African American Scientists*, p. 353; and Hayden, *Nine Black American Doctors*, pp. 116–117. According to Hayden, "The perfusion technique involved locating the major artery and vein serving the cancerous area. Using a special needle, the artery and vein were connected. Tourniquets were used to prevent the chemical from leaking out of the circulatory detour into the rest of the body. Then the chemical was pumped into the needle connecting the two blood vessels" (p. 117).

37. Kessler, *Distinguished African American Scientists*, p. 352; Jackson, *Black Women Makers*, p. 184.

38. Jackson, *Black Women Makers*, p. 184.

39. Jackson puts the number of Black women physicians at 100 (*Black Women Makers*, p. 185); see also Kessler, *Distinguished African American Scientists*; and Cazort, "Jane C. Wright." However, in a 1964 *Ebony* magazine article the number is listed as 300. "Despite the number of Negro women doctors in positions of great medical importance, they number no more than 300 of the 13,000 women doctors in the U.S." ("Of 13,000 Women Doctors in the U.S., 300 Are Negroes," *Ebony*, May 1964, pp. 72–73).

40. "Homecoming for Jane Wright."

41. Cazort, "Jane C. Wright," pp. 1284–1285; Kessler, *Distinguished African American Scientists*, pp. 353–354.

42. Her papers include a multi-part series of review articles beginning in 1984 in the *Journal of the National Medical Association* on new developments in chemotherapy for the treatment of gastrointestinal, breast, lung, and genitourinary cancers. See the appendix.

43. *Ebony Success Library*. See also *American Men and Women of Science: The Physical and Biological Sciences*, 14th ed. (New York: Bowker, 1971–), p. 5647.

44. See J. C. Wright, "Cancer Chemotherapy: Past, Present, and Future—Part I," *Journal of the National Medical Association* 76 (1984): 773–784.

JOSEPHINE A. SILONE YATES

A Path-Breaker Moving Upward and Onward

Careful study . . . forces one to the conclusion that, although the natural outcome of bondage is a cowardly, thieving, brutal, and otherwise totally degraded specimen of humanity, there were even in the darkest hours of the systems [of slavery], many high born souls, who, at the price of life itself, if necessary, maintained their integrity, rose superior to their surroundings, taught and transmitted to their posterity the same lofty principles that governed their own lives and that formed a goodly heritage for the generations yet unborn.[1]

— *Josephine A. Silone Yates*

Josephine A. Silone Yates became a science professor at Lincoln Institute (now Lincoln University) in Jefferson City, Missouri, in 1880, and retained the position until 1889. Born around 1859, in Mattituck, New York, Yates was the second daughter of a locally prominent Black family that had lived in the area for at least four generations.[2] A precocious child, she received her first lessons at home from her mother, Parthenia Reeve Silone, who was a well-educated woman for the time. Sent out to school at about age six, Yates's teachers quickly recognized her unusual preparedness in the standard subjects and immediately placed her in the upper grades. According to Monroe A. Majors, "Because of her eagerness and readiness to learn, she soon became a favorite with her teachers, although [she was] the only colored pupil in the school."[3] By age nine, Yates reportedly studied physiology and physics and possessed advanced mathematical ability.

As a result of her early academic success, when Yates was eleven years old, her maternal uncle, the Reverend John Bunyan Reeve of Philadelphia, suggested she come to that city to attend the Institute for Colored Youth, conducted by Fanny Jackson-Coppin.[4] Yates studied at the Institute for a year, until Reeve left Philadelphia to join the Theology Department at Howard University.[5] Jackson-Coppin, however, continued to "manifest a great deal of interest in her welfare, and often referred to her as a brilliant example of what a girl may do."[6]

A year later, around 1871, Yates went to live with her maternal aunt, Francis L. Girard, of Newport, Rhode Island, who had suggested she attend school there. At age fourteen, she entered "the highest grade of grammar school there," and was again "the only colored pupil of the school."[7] The following year she entered Rogers High School, again as the only Black student. Her science teacher considered her his brightest pupil, and she showed a strong interest in chemistry. She did additional laboratory work under her teacher, and as a result became an efficient chemist. In 1877, after only three years of high school, she graduated at the top of her class, delivered the valedictory address, and received a medal for scholarship.[8]

Yates's teachers had urged her to enter a university, but she decided instead to enroll at the Rhode Island State Normal School to prepare for a career as a teacher. She graduated two years later in 1879; she was the only Black student in a class of twenty, and the first Black person in the school's history to earn a degree there.[9] She then took and passed the Rhode Island state teacher certification exams, moved to Kansas, and began her teaching career.

Yates taught chemistry, and occasionally the other natural sciences, at Lincoln Institute for eight years. She was the first woman elected to a full professorship at Lincoln, and in 1888 she became head of the school's Department of Natural Sciences. During her tenure at Lincoln, she worked to keep abreast of developments in her field and improve her teaching skills. To that end, she spent her summers in the East, where she was active in teachers' associations and summer schools.[10] As word of her teaching skills spread, she became fairly well known in Black educational circles. In 1886 Booker T. Washington offered her the position of "lady principal" at Tuskegee, but she declined, choosing instead to remain at Lincoln.[11]

Yates read both French and German and had an interest in Russian literature. She began to write during the 1880s under the pen name R. K. Potter. In 1898, she married W. W. Yates, the principal of the Wendel Phillips School in Kansas City, Missouri, at which point she retired from teaching.[12] At the time of her resignation, she was earning $1,000 a year, and was likely the only Black woman in the country to have been a full professor and head of a college science department.[13]

After her marriage, Yates raised two children, Josephine and William. During those years she wrote for newspapers, lectured on educational topics, gave private lessons at home, taught intermittently at Lincoln Institute and at Lincoln High School in Kansas City, and wrote a series of poems. She also became very active in the nascent Black women's club movement, and organized the Kansas City Women's League in 1893.[14] When the National Association of Colored Women's Clubs was established in 1896, she became an active member, serving as the organization's second president, from 1901 to 1906.

Yates's husband died in 1910, and to support her children, Yates returned to teaching full time at Lincoln High School. She died two years later, in 1912, at age fifty-three. Yates was a forerunner in science. She married a successful, well-educated man who shared her strong interests in education, she became a science teacher at the college level, and she was active in the Black women's club movement and had a strong interest in writing.[15] Both of her children graduated from the University of Kansas—Josephine became a teacher and William became a physician.

NOTES

1. Josephine Silone Yates in 1904, quoted from an address before the National Association of Colored Women's Clubs Convention in St. Louis, Missouri; see Charles Harris Wesley, *The History of the National Association of Colored Women's Clubs* (Washington, D.C.: National Association of Colored Women's Clubs, 1984), pp. 61–62.

2. Robert L. Johns, "Josephine Silone Yates," in *Notable Black American Women*, ed. Jessie Carney Smith, vol. 2 (Detroit: Gale Research, 1992), pp. 1286–1287.

3. Monroe A. Majors, *Noted Negro Women: Their Triumphs and Activities* (Chicago: Donohue & Henneberry [1893]; reprint, The Black Heritage Library Collection, Freeport, N.Y.: Books For Libraries Press, 1971), p. 44.

4. Fanny Jackson Coppin was the second Black woman to earn an A.B. degree (Oberlin College, 1865); she was named principal of the Institute for Colored Youth in 1869—the first Black woman to head an institution of this type in the United States. An early opponent of gender discrimination, in 1897 Coppin spoke at the First Hampton Negro Conference in support of the idea of industrial education. See Jamie Hart and Elsa Barkley Brown, comps., "Black Women in the United States: A Chronology," in *Black Women*

in America: An Historical Encyclopedia, ed. Darlene Clark Hine, Elsa Barkley Brown, and Rosalyn Terborg-Penn, vol. 2 (New York: Carlson, 1993; Bloomington: Indiana University Press, 1994), pp. 1309–1332; also see pp. 898, 968.

5. Johns, "Josephine Silone Yates," p. 1286.

6. Majors, *Noted Negro Women,* p. 45. In Majors, Coppin's first name is spelled "Fannie" rather than "Fanny."

7. Ibid., p. 45

8. Ibid., p. 46. See also Johns, "Josephine Silone Yates," p. 1286.

9. Majors, *Noted Negro Women,* pp. 46–47.

10. Ibid., p. 48.

11. Ibid.

12. Ibid.

13. Ibid., p. 47

14. Hallie Q. Brown, comp., *Homespun Heroines and Other Women of Distinction* (Xenia, Ohio: Aldine Publishing, 1926).

15. Compare the similarities between Yates and Beebe Steven Lynk, who is also profiled in this volume. Yates appears to have had more opportunity to become known outside her local geographical area, but it is not clear whether she was a teacher or mentor of other Black women interested in the sciences. There may have been articles written about both women in the Black popular press of the time, and this might have given them a certain visibility as role models, particularly with regard to their activities in the women's club movement and their writing careers.

ROGER ARLINER YOUNG

A Cautionary Tale

Despite tremendous promise and a measure of early recognition, zoologist Roger Arliner Young's career path provides a chilling example of how much could go wrong when a woman entered the loftier reaches of the scientific professions.[1] Born in Clifton Forge, Virginia, in 1889, Young graduated from Burgetstown High School in Pennsylvania and entered Howard University in 1916 to major in music. According to historian of science Kenneth Manning, Young did not enroll in a science course (general biology) until the spring of 1921.[2] Ernest Everett Just, the eminent Black biologist, taught the course. Young's Howard University classmates would dedicate their yearbook to E. E. Just, "whose exemplary scholarship has been a source of inspiration to all of us."[3] Although Young only earned a grade of C in his class, it apparently changed both her career plans and her life.[4]

In 1923, seven years after entering Howard, Young earned her baccalaureate degree. Young's yearbook picture shows her to have been a young woman of great beauty—she was called "Sweetheart" by her friends, was "lively and vivacious," and participated actively in the college's Glee Club, the Young Women's Christian Association, and the Howard University Players. The editors commented that "Arliner is one of those girls whose company we always enjoy, and never tire of."[5] In the yearbook, Young said in her personal statement that "not failure, but low aim is a crime," and she indicated that she would probably do "social service work."[6] Instead, upon graduating, she became a scientist, although she had yet to earn a degree in the field.

After her graduation, Just hired Young as an assistant professor of zoology in his department. Just chose Young over other candidates for the position for reasons that, as Manning notes, were "not clear":

> [Young had] an undistinguished record overall and no strong preparation for work in the sciences. Aside from general zoology, she had taken only two other science courses: vertebrate and invertebrate embryology. Her grade was B in both.[7]

Marguerite Thomas Williams, a science major who was far more qualified for the position than Young, graduated in the same year, and obtained a position at Miner Teachers College, also in Washington, D.C. Whatever the reasons for Just's interest in Young,[8] he was clearly her mentor and she blossomed as a scientist under his tutelage. A year later, in 1924, she published her first article.[9] Her research was important, and received both national and international recognition:

> The publication of her paramecium work in 1924 had pre-dated by two months similar work by Dimitry Nasconov, the eminent Russian cell physiologist who had been chosen to visit the United States as a Rockefeller International Fellow. Young had, according to Just, come up with "some interesting conclusions . . . and . . . hitherto undescribed results." Her study received notice alongside Nasconov's in several European journals.[10]

Given that Young was not at that time a trained scientist or the holder of an advanced degree, her achievement was significant.

Apparently, Young and Just worked well together; he tried, unsuccessfully, to obtain funding for her to attend graduate school. Beginning in the summer of 1924, she enrolled on a part-time basis at the University of Chicago, and in 1926 she earned her master's degree in zoology. Although Young had been a mediocre student during her undergraduate years at Howard, she appeared to hit her stride at the University of Chicago. She was elected to the Sigma Xi honorary society, an unusual achievement for a non-doctoral candidate.[11] At this point, she had embarked upon a promising career in science.

Young had already made herself indispensable to Just at Howard University, where she helped ease his heavy teaching burden and worked closely with him in the lab. He had embarked on a new area of research—the effect of ultraviolet light on the development of marine eggs such as *Nereis*, *Platvnereis*, *Arbacia*, and *Chaetopterus*. Between 1927 and 1936, Young would spend every summer doing research at the Marine Biological Laboratory at Woods Hole, Massachusetts,[12] but during her first two summers, 1927 and 1928, she accompanied Just as his research assistant; she helped him in his work on fertilization, examining cortical changes in marine eggs to gain a better understanding of the kinetics of fertilization.[13]

Young was the first Black woman to join the Woods Hole scientific community.[14] Despite the interracially tense atmosphere of Woods Hole during this period,[15] Young was able to carve out a place for herself there. At the same time, she gained a strong measure of respect from Just, who increasingly mentioned her work and expertise in his funding proposals, describing her as having a "real genius in zoology" and as a "skillful worker" whose work surpassed his own in "technical excellence."[16] Her star was on the rise, and by the summer of 1929, Young had secured a niche at Woods Hole in which her scientific reputation no longer depended entirely on her work with Just.

Just, meanwhile, had plans of his own, which allowed him even less time to attend to his teaching duties at Howard. In January 1929, he left for a year of travel and research in Europe, made possible by a Julius Rosenwald Fund grant.[17] During the spring semester, Young stood in for Just as head of the zoology department and performed the lion's share of the teaching duties left vacant by his absence. In her spare time she wrote proposals to secure funding for her doctoral studies. With Just out of the country during the summer of 1929, Young went to Woods Hole alone, and worked with University of Chicago embryologist Frank Lillie, who earlier had been Just's mentor. Young was finally awarded a General Education Board fellowship; immediately upon Just's return from Europe at the end of the summer, she left for the University of Chicago to study with Lillie and prepare for her qualifying examinations.[18]

In January 1930, a week before Young was to take her qualifying examinations, Just left for Europe yet again—this time for a planned six-month stay in Germany. At this point things began to fall apart for Young. She failed the qualifying examination—unable even to answer questions on topics she had successfully addressed in her course work, her teaching at Howard, and perhaps more surprisingly, in her own research. She was denied acceptance into the doctoral program, and left Chicago immediately.

Not only was her failure a tremendous blow to Young herself and to Just, it was also a cause for consternation for Mordecai Johnson, the president of Howard University. The Julius Rosenwald Fund had already awarded Just's zoology department a grant of $80,000, to be dispensed over a period of five years toward developing Just's research program at Howard and training advanced students in zoology.[19] The grant meant a great deal of prestige for the university, and Johnson was a fierce protector of Howard's reputation. With Just out of the country once again, the zoology department was woefully short of qualified personnel. If Young had succeeded in obtaining the doctorate, it was assumed that she would return to Howard, assist in running the department, and step into the new professorship that had been proposed under the Rosenwald grant. Thus, Young's failure at the University of Chicago caused a major glitch in both Just's and Johnson's plans. The two men may have been disappointed, but Young was devastated.

In a farewell note to Frank Lillie, whom she was too upset to face in person, Young tried to explain what had gone wrong:

> The trouble is that for two years I've tried to keep going under responsibilities that were not wholly mine but were not shared and the weight of it has simply worn me out. I have forced myself on so long that I automatically accepted arrangements for examination which I knew the first of last August I would fail unless there was some relief. Instead of relief the situation has become worse

since I've come here. It is not exactly and [sic] outside thing—it does concern
my work at Howard—if I took an examination a month from now under the
same conditions I'd more than likely do worse. I could go on as long as there
was any hope of satisfying Dr. Just by getting the degree before my return. I
could keep on trying but now that I've made that impossible I think I would
gain more and create less embarrassment by giving up the whole thing for a
while at least.[20]

Not only did Young feel tremendously overworked, due largely to Just's extended
absences, but she also felt pressured to undertake the program at Chicago at a
time when she felt she was not ready to do so. And, as her note reveals quite poign-
antly, she felt a need to please Just, and was strongly dependent on him, both intellectually and emotionally. In fairness to Just, he was apparently very supportive
of all his female students, among them Lillian Burwell Lewis, Dorothy Young,
Leona Gray, and Caroline Silence.[21]

Young also experienced other problems—physical, emotional, and financial.
During her disastrous year at Chicago, she complained that her eyes bothered
her. As Manning indicates, she had done "permanent damage to her eyes while
performing experiments using ultraviolet rays . . . a sacrifice, in a way, for Just and
for science."[22] Young was also the sole support and succor for her invalid mother.
Their separation while Young was at the University of Chicago—to say nothing of
the demands of her teaching, research, and administrative duties while at Howard and at Woods Hole—had no doubt produced a great deal of additional stress.
Money was a continuing problem for her as well. Although she had the General Education Board fellowship and some Julius Rosenwald money through Just,
the combined sum, along with her small Howard University salary, was insuf-
ficient for her research, study, and travel expenses, or her familial obligations. The
last lines of her note to Frank Lillie provide an indication of her mental and emotional state: "I'm sorry I couldn't talk to you but I seemed to have lost my grip all
around."[23]

Young returned to Howard in disgrace in Just's eyes—and probably in her own
as well. Just needed her to perform her usual duties in the zoology department for
the present, but their friendship, as well as his mentorship of her, had ended.
Nevertheless, Young appeared to pull herself together. In the summer of 1930,
Young returned to Woods Hole without Just, who was again in Europe. She secured a place in the lab with V. L. Heilbrunn of the University of Pennsylvania.
She now began, both independently and with Heilbrunn and Donald P. Costello,
to follow up the marine egg studies she had pursued with Just in the 1920s. The
trio published a series of articles on their findings. At Woods Hole, at least, Young
was holding her career together.[24]

Despite Just's now poor relationship with Young, he was well aware that she
was needed at Howard. Thus, according to Manning, "he played a part in keeping
her at Howard and persuading her not to accept job offers at other institutions,
notably at Spelman College around 1931."[25] Nevertheless, Young's career at
Howard was on the decline: "Just ignored her as much as possible, giving his attention instead to the two other junior professors in the department, Louis A.
Hansborough and Hyman Y. Chase."[26] Matters reached a head in the spring of

1935. Howard president Mordecai Johnson had begun pressuring Just to tighten up the zoology department's budget by reducing the number of its junior professors from three to two. So when Young finally confronted Just with regard to his negative attitude toward her, it was no contest: in the spring of 1936 Young was fired.[27] Just described her work as

> far below standards . . . [she was] careless with scientific equipment and would not return it promptly for the use of others. . . . she often missed her classes . . . [and] students went to the administration on their own to complain.[28]

However, opinions on Young's performance and the quality of her work during that period vary, often inexplicably.[29] Perhaps one of the most crucial elements in Young's fate at Howard University was that the school's power elite was not particularly supportive of women during this period.[30] Certainly Young did not feel she had been supported, or even fairly treated: "The situation here is so cruel and cowardly that every spark of sentiment that I have held for Howard is cold. I wish I did not have to go back on the campus again this year or ever."[31]

Nor do Young's achievements during the next few years of her career appear to support Just's negative assessment of her performance. Between 1930 and 1936 she engaged in publishable research at the Marine Biological Laboratory with Heilbrunn and Costello. In the fall of 1937, Heilbrunn accepted her into the doctoral program at the University of Pennsylvania, and, despite her earlier failure at Chicago, the General Education Board awarded her a two-year grant to study at Pennsylvania.[32] Young earned her doctorate in 1940; her dissertation was "The Indirect Effects of Roentgen Rays on Certain Marine Eggs."[33] Young had had to borrow money to pay for her final year of study. The lack of money was to be a crucial factor during the remainder of her life. She sank deeper and deeper into debt, and the stress of her always uncertain financial situation, coupled with her tremendous responsibilities in caring for her mother, began to exacerbate the degeneration of her emotional health.

Immediately after earning her doctorate, Young secured an assistant professorship at North Carolina College for Negroes, in Raleigh. She then moved to Shaw University where she became head of the biology department. At this point, however, Young's career began to decline precipitously. Her financial problems worsened: pay was particularly poor at second-tier historically Black colleges, and she fell further and further into debt. Lack of money also hastened the decline of her research opportunities. Historically Black colleges such as North Carolina College and Shaw University had little equipment and lacked funds for even the most basic efforts. Since she was no longer able to afford to go to Woods Hole during the summers, her research ended — now her career in science was solely as a teacher. In 1947, she left Shaw and returned to North Carolina College as a professor of biology.

In 1953 Young's mother died, but instead of lessening her difficulties this hastened her emotional breakdown. In her final years, Young's mother had lost all ability to reason, and Young worried that a similar fate was in her own future. She experienced difficulty in holding a position for even a moderate period of time.[34] As she bounced from one position to another, Young became increasingly fearful and depressed.[35] In 1955, two years after her mother's death, she wrote to Peter Marshall Murray, a Black New York physician who had served on the Howard University Board of Trustees during her time there:

> I have lost three jobs in a row. In November after 2 months I had lost the job here at Paul Quinn [College]. The reasons are never quite clear to me and in each case I've had the legal technicality on my side, which means nothing. . . .
>
> I am terrifically afraid I'm going the same way [as my mother] with the difference that I have spent so much [time] in school and moving that I not only have no friends, no insurance, and a rapidly increasing inability to sew a seam, read a page, write, sleep, remember. I feel that my brain is like a (switch) sign board, with some wire reversed and now the bulbs are going out one by one.
>
> I have no money for medical care. No relatives and a deep fear of the institutions down here. I've read in the papers that they are inadequate for whites. I have a little 3 hour job which I know the priest gave me as a rescue because I had nothing. It ends in May—kindergarten. I'm not Catholic— can expect nothing more from this very poor mission. . . . I am so scared I'm numb. . . .
>
> I haven't been criticized on the teaching but I can only teach college biology. I seemed never to get in the related sciences. I've tried to get several people to help me with handling my salary, but they thought I was joking. What can I do? I've driven myself for 25 years. Until last year I was certain I could work ten years more and save enough to pay to be taken care of, but the only thing I seem to have is some cloth by the yard which is now wasted and about four rooms of furniture which is being used for its storage in Galveston. . . .
>
> I shouldn't even send this letter, but I don't want to go to anyone in Waco and confirm a rumor that President Frank Veal has apparently disseminated here to the effect that I'm "off." I have practically nothing here. The President asked me not to bring my things until the dormitory was finished then fired me two weeks before that.[36]

As her letter indicates, Young was reduced to an almost hand-to-mouth existence. In addition, gossip about her mental state was now circulating among college administrators. After securing a temporary position at Jackson State, her condition had deteriorated to such an extent that she had to quit and check herself into the Mississippi Mental Asylum in Whitfield. She was discharged in December 1962.[37] Her last position, in which she had been reduced to the status of temporary visiting lecturer in biology, was at Southern University in Louisiana.

Young and Her Mentors

Roger Arliner Young's life illustrates the tremendous effect, for both good and ill, that a mentor-mentoree relationship can have on a woman scientist's career. Without Ernest Everett Just's interest in her, particularly in light of her lack of a strong scientific background as an undergraduate, it is doubtful that Young would have become a scientist.[38] Just was her mentor, role model, and entree into science. He secured an assistantship for her in the zoology department at Howard University immediately following her graduation, and perhaps more importantly, introduced her to and included her in his own research projects, both at Howard and at the Marine Biological Laboratory at Woods Hole. At the Marine Biological Laboratory, where she came in contact with eminent biologists from around the country, Young had the opportunity to expand her horizons.

The mentor relationship with Just lasted for at least ten years—which seems an inordinately long period. Nevertheless, it was initially quite useful, particularly in terms of Just's expert tutelage of Young in laboratory work. Just was also instrumental in placing her under the wing of his own former mentor, Frank Lillie.

As both the pattern and the duration of her relationship with Just indicates, Young was never really able to outgrow her dependence on her mentor—as all young scholars and scientists must eventually do if they are to blossom in their own right. In some respects, Just's support of Young served his own purposes more than hers: she was an indefatigable, trustworthy, and quite able worker in the laboratory; she handled a variety of duties for Just when he was otherwise occupied; and, thanks to her own early work and publication, she provided him with a star pupil to display—at Howard, at Woods Hole, and in his grant proposals. Placing Young with Lillie at the University of Chicago was supposed to provide Just with another doctorate-level scientist for his department, which would have added to Howard University's and Just's prestige.

Although ostensibly it was Young's failure at Chicago that ended the relationship with Just, parallel events most certainly played a part. Just had made repeated attempts to disassociate himself from Howard—Europe was where he wanted to be. He also severed his ties, quite forcefully and quite publicly, with the scientists at Woods Hole. As Manning recounts, in June of 1930, the members of the Marine Biological Laboratory gathered to honor Frank Lillie; as Lillie's most illustrious protégé and close friend, Just was expected to deliver a glowing account of his old mentor's accomplishments. Instead, he used the occasion to castigate his former colleagues: "I have received more in the way of fraternity and assistance in my one year at the Kaiser-Wilhelm Institut [in Germany] than in all my other years at Woods Hole put together."[39] With that remark, Just publicly severed all ties with the Marine Biological Laboratory. Given these parallel events, one must question if there wasn't more to the demise of Just's relationship with Young than her failure to perform at Chicago as expected. One must also consider the effect that Just's rejection of the Woods Hole scientists must have had on Young.

Although she worked with Lillie at Woods Hole and he was her advisor at Chicago, Young does not appear to have ever developed the close mentor-mentoree relationship with him that she had experienced with Just (nor, in fact, did she appear to achieve the level of the relationship that Just had experienced with Lillie). For almost six years after her fall from grace, Young continued to work for Just, and one can only speculate whether she tried to win back her former place in his eyes. When she was finally fired, she expressed the depths of her reliance on Just: she hoped "the 'jolt' of being fired might 'break the spell' of hypnosis, might help her surface from the condition of abject submission and control that in her view Just had imposed on her."[40]

Her subsequent attempt to replace Just's mentorship with that of V. L. Heilbrunn appears to have been on a par with her relationship with Lillie, although with a more favorable end result—a doctorate. Nevertheless, after she left the University of Pennsylvania in 1940, her relationship with Heilbrunn, for all intents and purposes, ended—amicably, it appears. As a result, Young, who had never learned to truly function on her own, became, at age forty-one, a woman alone in a man's field.

NOTES

1. Roger Arliner Young was the first Black woman to earn a doctorate in zoology.

2. Kenneth R. Manning, "Roger Arliner Young, Scientist," *Sage* 6 (fall 1989): 3–7.

3. *The Bison*, 1923 edition, Howard University Yearbook. Moorland-Spingarn Research Center, Howard University, Washington, D.C. The pages are not numbered.

4. Young was one of three women in this volume who took courses at Howard University with Ernest Everett Just during the 1920s. The other two were Marguerite Thomas Williams and Lillian Burwell Lewis.

5. *The Bison.*

6. *The Bison.*

7. Manning, "Roger Arliner Young," p. 4.

8. In his biography of Just, Manning hints very strongly that Young and Just were having a love affair: "There were rumors that they were involved in a love affair, but there is not much evidence to support this beyond the fact that as time went on they could often be seen together working late at night in the laboratory." See Kenneth W. Manning, *Black Apollo of Science: The Life of Ernest Everett Just* (New York: Oxford University Press, 1983), pp. 147, 217–219, 220. In his later *Sage* article on Young, Manning appears to back off from that assessment, giving another possible interpretation of Just's interest in Young: "Just probably had another motive, a practical one, for bringing Young into the field. He often lamented the fact that men competent in zoology pursued medicine as a career rather than academic zoology. . . . On the other hand, women who aspired to careers were more likely to go into teaching than medicine. Still, it was unusual for a woman, especially a Black woman, to be involved in the teaching of science at the university level, instead of music or English or home economics" ("Roger Arliner Young," pp. 3–4). This assessment does not explain why Just hired Young rather than Williams.

9. Roger Arliner Young, "On the Excretory Apparatus in Paramecium," *Science*, 12 September 1924.

10. Manning, "Roger Arliner Young," pp. 3–4.

11. Ibid., p. 3; Manning, *Black Apollo*, p. 147.

12. Evelyn Hammonds, "Roger Arliner Young," in *Black Women in America: An Historical Encyclopedia*, ed. Darlene Clark Hine, Elsa Barkley Brown, and Rosalyn Terborg-Penn, vol. 2 (New York: Carlson, 1993; Bloomington: Indiana University Press, 1994), pp. 1298–1299.

13. Manning, "Roger Arliner Young," p. 3

14. Manning, *Black Apollo*, p. 148.

15. Ibid., pp. 108–110.

16. Ibid.

17. Ibid., p. 167.

18. Ibid., pp. 218–220.

19. Ibid., pp. 211–212, 373, f.n. 12.

20. Ibid., p. 218. Manning cites a letter from Young to Frank R. Lillie, January 1930 (Frank R. Lillie Correspondence, box 6, folder 27).

21. Ibid.

22. Ibid., p. 219; Manning is quoting his April 1977 interview with Donald P. Costello (fn. 37, p. 374), who was Young's co-worker at Woods Hole.

23. Ibid., p. 218.

24. Young returned to Woods Hole every summer through 1936.

25. Manning, "Roger Arliner Young," p. 5.

26. Manning, *Black Apollo*.

27. Manning, "Roger Arliner Young," p. 6.

28. Ibid.

29. For example, a former student and later colleague of Young's who wishes to remain anonymous said she was "an early teacher . . . [who] generated [my] desire for perfection. She was like a trigger point in bringing out [my] desire for rigor and perfection." Yet in the next breath, this same individual describes her as "not as accurate and meticulous as she should have been." This information was provided by Margaret Strickland Collins (telephone interview by author, 15 August 1995). Dr. Collins spoke to Young's former student and colleague, now retired and living in the Washington, D.C. area. I acknowledge that this is essentially hearsay information, but I consider Dr. Collins an unimpeachable third party. In addition, the anonymous individual is known to me through other sources, both primary and secondary, so I have no doubt as to the veracity of the quoted statements.

30. Wini Mary Edwina Warren, "Hearts and Minds: Black Women Scientists in the United States, 1900–1960," Ph.D. dissertation, Department of History and Philosophy of Science, Indiana University, 1997.

31. Manning, "Roger Arliner Young," p. 6, possibly citing a letter from Young to Heilbrunn dated April 21, 1936 (*Black Apollo of Science*, fn., p. 381).

32. Manning, "Roger Arliner Young," p. 6; see also the sections on Young in *Black Apollo*. The General Education Board was surely aware of her earlier failure at Chicago because it had also funded that attempt.

33. Archival Collection, Moorland-Spingarn Research Center, Howard University, Washington, D.C. See also Harry W. Greene, *Holders of Doctorates among American Negroes: An Educational and Social Study of Negroes Who Have Earned Doctoral Degrees in Course, 1876–1943* (Boston: Meador, 1946), p. 197; and James M. Jay, *Negroes in Science: Natural Science Doctorates, 1876–1969* (Detroit: Belamp, 1971), p. 60.

34. Greene, *Holders of Doctorates*.

35. Paul Quinn College in Texas, Jackson State College in Mississippi, and Southern University in Louisiana.

36. Manning, "Roger Arliner Young," pp. 6–7. Manning quotes Young's letter to Murray.

37. Ibid., p. 7.

38. I believe that speculating on whether or not Young and Just were romantically involved is pointless. If such an affair did occur there is no written record of it, and relying on snippets of gossip does not facilitate understanding.

39. Manning, *Black Apollo of Science*, p. 194.

40. Manning, "Roger Arliner Young," p. 6.

Appendix: Publications

This appendix contains representative publications of some, but not all, of the women in this volume. Only selected publications for each woman are listed, in chronological order. In some cases, I have listed titles but have been unable to track down publication information; these entries are starred (*).

Gloria Long Anderson

Anderson, Gloria L. "^{19}F Chemical Shifts for Aromatic Molecules." Ph.D. dissertation, University of Chicago, 1968.

Anderson, Gloria L., and L. M. Stock. "Chemical Shifts for Bicyclic Fluorides." *Journal of the American Chemical Society* 90 (January 3, 1968): 212–213.

Anderson, Gloria L., and L. M. Stock. "^{19}F Chemical Shifts for Bicyclic and Aromatic Molecules." *Journal of the American Chemical Society* 91 (1969): 68–64.

Anderson, Gloria L., Roger C. Parish, and Leon M. Stock. "Transmission of Substituent Effects in Anthracene: Acid Dissociation Constants of 10-Substituted-9-anthroic Acids and Substituent Chemical Shifts of 10-Substituted-9-fluoroanthracenes: Evidence for the π Inductive Effect." *Journal of the American Chemical Society* 93 (December 15, 1971): 6984–6988.

Anderson, Gloria L. *^{19}F Chemical Shifts and Infrared C-F Stretching Frequencies for Bridgehead Fluorides.* Technical Report No. 1, Office of Naval Research, Grant Number NONR (G)-00021-73, 5 August 1974.

Anderson, Gloria L. *Substituent Effect on the C-F Stretching Frequencies in Some Substituted Aryl Fluorides.* Technical Report No. 2, Office of Naval Research, Grant Number NONR (G)-00021-73, 2 April 1976.

Anderson, Gloria L. *New Synthetic Techniques for Advanced Propellant Ingredients: Selective Chemical Transformations and New Structures—Bis-Fluorodinitroethylamino Derivatives.* Final Technical Report, Southeastern Center for Electrical Engineering Education/Air Force Office of Scientific Research, Contract Number F49620-82-C-0035, 1984.

Anderson, Gloria L., and Issifu I. Harruna. "Synthesis of Triflate and Chloride Salts of Alkyl N,N-Bis (2,2,2-Trifluoroethyl) Amines." *Synthetic Communications* 17 (1987): 111–114.

Anderson, Gloria L., Winifred A. Burks, and Issifu I. Harruna. "Novel Synthesis of 3-Fluoro-1-Aminoadamantane and Some of Its Derivatives." *Synthetic Communications* 18 (1988): 1967–1974.

Anderson, Gloria L., Betty J. Randolph, and Issifu I. Harruna. "Novel Synthesis of Some 1-N-(3-Fluoroadamantyl) Ureas." *Synthetic Communications* 19 (1989): 1955–1963.

PRESENTATIONS

Anderson, Gloria L., and V. G. Appling. "Reactions of Alcohols with 1-(N,N-Diethylamino)-1, 1, 2-Trifluoro-2-Chloroethane." Paper presented at the National Convention of the National Institute of Science, Dillard University, New Orleans, 17 April 1974.

Anderson, Gloria L., and E. J. B. Tutwan. "Infrared Frequency Shifts in Alicyclic Bridgehead Fluorocarbons." Paper presented at the National Convention of the National Institute of Science, Dillard University, New Orleans, 18 April 1974.

Anderson, Gloria L., and W. A. Burks. "The Synthesis of Medicinal Compounds Containing Fluorine." Paper presented at the National Convention of the National Institute of Science, Dillard University, New Orleans, 18 April 1974.

Anderson, Gloria L., and V. R. Shah. "Infrared C-F Substituent Frequency Shifts (SPS) for Some Aryl Fluorides." Paper presented at the Annual Meeting of the Georgia Academy of Science, Savannah State College, Savannah, Ga., 26 April 1975.

Anderson, Gloria L., and E. J. B. Tutwan. "^{19}F NMR Studies on Some Alicyclic Bridgehead Fluorides." Paper presented at the Annual Meeting of the Georgia Academy of Science, Valdosta State College, 27 April 1974.

Anderson, Gloria L. "^{19}F Substituent Chemical Shifts (SCS)." Paper presented at the Department of Chemistry, Jackson State University, Jackson, Miss., 1975.

Anderson, Gloria L. "Structure Activity Studies on Some Substituted Amantadines." Paper presented at Benedict College, Columbia, S.C., 19 October 1976.

Anderson, Gloria L. "The Synthesis of Some 1,3-Disubstituted Adamantanes." Paper presented at Kellogg Foundation Lecture Series, University of Arkansas at Pine Bluff, 29 April 1977.

Anderson, Gloria L. "Novel Synthetic Methods for Preparing Some Potential Antiviral Drugs: 1,3-Disubstituted Adamantanes." Paper presented at the 1990 Minority Biomedical Research Symposium, National Institute of General Medical Sciences, Nashville, Tenn., 13 October 1990.

Anderson, Gloria L. "Novel Synthetic Methods for Preparing Some Potential Antiviral Drugs: 1,3-Disubstituted Adamantanes." Paper presented at the Department of Pharmacology, Meharry Medical College, Nashville, Tenn., 31 May 1991.

Patricia Erna Bath

Primm, B. J., and P. E. Bath. "Pseudoheroinism." *International Journal of Addiction* 8 (1973): 231–242.

Inwang, E. E., N. S. Patel, B. J. Primm, A. McBride, and P. E. Bath. "Evidence for the Stimulant and Depressant Central Effects of L-Alpha-Acetyl Methadol." *Experientia* 31 (1975): 203–205.

Bath, P. E. "Rationale for a Program in Community Ophthalmology." *Journal of the National Medical Association* 71 (1979): 145–148.

Bath, P. E. "Keratoprosthesis: An Alternative in Anterior Segment Reconstruction." *Journal of the American Intraocular Implant Society* 6 (1980): 126–128.

Stark, W. J., D. M. Worthen, J. T. Holladay, P. E. Bath, et al. "The FDA Report on Intraocular Lenses." *Ophthalmology* 90 (1983): 311–317.

Mondino, B. J., P. E. Bath, R. Y. Foos, L. Apt, and G. M. Rajacich. "Absent Meibomian Glands in the Ectodactyly, Ectodermal Dysplasia, Cleft Lip-Palate Syndrome." *American Journal of Ophthalmology* 97 (1984): 496–500.

Stark, W. J., D. M. Worthen, J. T. Holladay, P. E. Bath, et al. "The FDA Report on Intraocular Lenses." *Australian Journal of Ophthalmology* 12 (1984): 61–69.

Bath, P. E., D. L. Fridge, K. Robinson, and R. C. McCord. "Photometric Evaluation of YAG-Induced Polymethylmethacrylate Damage in a Keratoprosthesis." *Journal of the American Intraocular Implant Society* 11 (1985): 253–256.

Bath, P. E., Y. Dang, and W. H. Martin. "Comparison of Glare in YAG-Damaged Intraocular Lenses: Injection-Molded versus Lathe-Cut." *Journal of Cataract and Refractive Surgery* 12 (1986): 662–664.

Bath, P. E., and F. Fankhauser. "Long-Term Results of Nd:YAG Laser Posterior Capsulotomy with the Swiss Laser." *Journal of Cataract and Refractive Surgery* 12 (1986): 150–153.

Bath, P. E., A. B. Romberger, and P. Brown. "A Comparison of Nd:YAG Laser Damage Thresholds for PMMA and Silicone Intraocular Lenses." *Investigative Ophthalmology and Visual Science* 27 (1986): 795–798.

Bath, P. E., A. Romberger, P. Brown, and D. Quon. "Quantitative Concepts in Avoiding Intraocular Lens Damage from the Nd:YAG Laser in Posterior Capsulotomy." *Journal of Cataract and Refractive Surgery* 12 (1986): 262–266.

Ullman, H. E., A. A. Gonzalez, P. E. Bath, K. Prendiville, K. C. Cox, and A. Alston. "Posterior Chamber Intraocular Lens Implantation in Ethnic Minorities by Resident Housestaff." *Journal of Cataract and Refractive Surgery* 12 (1986): 40–43.

Bath, P. E., C. F. Boerner, and Y. Dang. "Pathology and Physics of YAG-Laser Intraocular Lens Damage." *Journal of Cataract and Refractive Surgery* 13 (1987): 47–49. Published erratum appears in *Journal of Cataract and Refractive Surgery* 13 (1987): 217.

Bath, P. E., K. J. Hoffer, D. Aron-Rosa, and Y. Dang. "Glare Disability Secondary to YAG Laser Intraocular Lens Damage." *Journal of Cataract and Refractive Surgery* 13 (1987): 309–313.

Bath, P. E., G. Mueller, D. J. Apple, and R. Brems. "Excimer Laser Lens Ablation." *Archives of Ophthalmology* 105 (1987): 1164–1165.

Tetz, M. R., D. J. Apple, F. W. Price Jr., K. L. Piest, M. C. Kincaid, and P. E. Bath. "A Newly Described Complication of Neodymium-YAG Laser Capsulotomy: Exacerbation of an Intraocular Infection. Case Report." *Archives of Ophthalmology* 105 (1987): 1324–1325.

Dugel, P. U., G. N. Holland, H. H. Brown, T. H. Pettit, J. D. Hofbauer, K. B. Simons, H. Ullman, P. E. Bath, and R. Y. Foos. "Mycobacterium Fortuitum Keratitis." *American Journal of Ophthalmology* 105 (1988): 661–669.

Prendiville, K. J., and P. E. Bath. "Lateral Cantholysis and Eyelid Necrosis Secondary to *Pseudomonas aeruginosa*." *Annals of Ophthalmology* 20 (1988): 193–195.

Rosa, D. S., C. R. Boerner, M. Gross, J. C. Timsit, M. Delacour, and P. E. Bath. "Wound Healing Following Excimer Laser Radial Keratotomy." *Journal of Cataract and Refractive Surgery* 14 (1988): 173–179.

Zakka, K. A., P. E. Bath, and K. Robinson. "Removal of Anterior Chamber Intraocular Lenses: A Surgical Technique." *Annals of Ophthalmology* 20 (1988): 476–477.

Bath, P. E. "Blacks at Greater Risk for Blindness." *Archives of Ophthalmology* 108 (1990): 1377–1378.

Bath, P. E. "Laserphaco: Fiberoptics Plus Irrigation/Aspiration." *Journal of Cataract and Refractive Surgery* 16 (1990): 525–527.

Bath, P. E. "Questionable Authenticity and Accuracy of Maguen Excimer Experiments." *Journal of Cataract and Refractive Surgery* 17 (1991): 519–521.

Bath, P. E. "Cataract Surgery Training of Residents in an Urban and Virtual Environment." *Journal of Cataract Refractive Surgery* 24 (1998): 727–729.

Sylvia Trimble Bozeman

Bozeman, Sylvia T. "Representations of Generalized Inverses of Fredholm Operators." Ph.D. dissertation, Emory University, Atlanta, Ga., 1980.

Bozeman, Sylvia T., and Louis Kramarz. "Approximating Eigenfunctions of Fredholm Operations in Banach Spaces." *Journal of Mathematics Analysis and Applications* 89 (1982): 612.

Bozeman, Sylvia T. "Black Women Mathematicians: In Short Supply." *Sage* 6 (fall 1989): 18–23.

Mamie Phipps Clark

Clark, Mamie Phipps. "The Development of Consciousness of Self in Negro Pre-school Children." *Archives of Psychology.* Washington, D.C.: Howard University, 1939.

Clark, Mamie Phipps, and Kenneth Clark. "The Development of Consciousness of Self and the Emergence of Racial Identification in Negro Preschool Children." *Journal of Social Psychology* 10 (1939): 591–599.

Clark, Mamie Phipps, and Kenneth B. Clark. "Segregation as a Factor in the Racial Identification of Negro Pre-school Children." *Journal of Experimental Education* 8 (1939): 161–165.

Clark, Mamie Phipps, and Kenneth B. Clark. "Skin Color as a Factor in Racial Identification of Negro Preschool Children." *Journal of Social Psychology* 11 (1940): 159–169.

Clark, Mamie Phipps. "Changes in Primary Mental Abilities with Age." Ph.D. dissertation, Columbia University, New York, 1944. Reprinted in *Archives of Psychology,* no. 291.

Clark, Mamie Phipps, and Kenneth B. Clark. "Racial Identification and Preference in Negro Children." In *Readings in Social Psychology,* edited by T. M. Newcomb and E. L. Hartley. New York: Holt, 1947.

Clark, Mamie Phipps, and Kenneth B. Clark. "Emotional Factors in Racial Identification and Preference in Negro Children." *Journal of Negro Education* 19 (1950): 341–350.

Clark, Mamie Phipps, and Jeanne Karp. "A Report on a Summer Remedial Program." *Elementary School Journal* 61 (1960): 137–142.

Clark, Mamie Phipps. "Changing Concepts in Mental Health, a Thirty-Year View." *Conference Proceedings, Thirtieth Anniversary Conference.* New York: Northside Center for Child Development, 1970.

Clark, Mamie Phipps, and Kenneth B. Clark. "What Do Blacks Think of Themselves?" *Ebony,* November 1980, p. 170.

Jewel Plummer Cobb

Antikajian, G., L. T. Wright, J. Plummer, and S. Weintraub. "The Effect of Triethylene Melamine, Aureomycin, and Some 4-Amino Derivatives of Folic Acid on Tissues in Vitro." *Journal of the National Cancer Institute* 12 (1951): 269–274.

Plummer, J. "The in Vitro Effects of A-Methopterin." Proceedings of Second Conference on Folic Acid Antagonists in Leukemia Treatment. *Blood* 7 (1952 suppl.): 152–190.

Plummer, J., L. T. Wright, G. Antkajian, and S. Weintraub. "Triethylene Melamine in Vitro Studies. I. Mitotic Alterations Produced in Chick Fibroblast Tissue Cultures." *Cancer Research* 12 (1952): 796–800.

Plummer, J., and M. J. Kopac. "The in Vitro Production of Pigment Granules: Pigment Cell Growth." In *Pigment Cell Growth: Proceedings of the Third Conference on the Biology of Normal and Atypical Pigment Cell Growth,* pp. 305–306. New York: Academic Press, 1953.

Wright, J. C., J. I. Plummer, R. S. Coidan, and L. T. Wright. "The in Vivo and in Vitro Effects of Chemotherapeutic Agents on Human Neoplastic Diseases." *The Harlem Hospital Bulletin* 6 (1953): 58–63.

Cobb, J. P. "Evaluation of Variation in Transplantability and Growth of Pigmented and

Pale Fragments of the Cloudman S91 Mouse Melanoma Following X-Irradiation in Vitro." *Proceedings of the American Association of Cancer Research* 2 (1955): 10.

Cobb, J. P. "Tissue Culture Observations of the Effects of Chemotherapeutic Agents on Human Tumors." *Transactions of the New York Academy of Sciences* 17 (1955): 237–249.

Cobb, J. P., J. H. Keifer, and H. Woods. "Human Bladder Neoplastic Cells in Tissue Culture." *Journal of Urology* 73 (1955): 1039–1044.

Cobb, J. P. "Effects of in Vitro X-Irradiation on Pigmented and Pale Slices of Cloudman S91 Mouse Melanoma as Measured by Subsequent Proliferation in Vivo." *Journal of the National Cancer Institute* 17 (1956): 657–666.

Cobb, J. P., and D. G. Walker. "Biological Activity in Tissue Culture of Actinomycin D on Normal and Neoplastic Cells." *Proceedings of the American Association for Cancer Research* 2 (1957).

Wright, J., J. P. Cobb, S. L. Gumport, F. M. Golomb, and D. Safadi. "Investigation of the Relationship between Clinical and Tissue Response to Chemotherapeutic Agents on Human Cancer." *New England Journal of Medicine* 257 (1957): 1207–1211.

Wright, J. C., P. Foster, B. Billow, S. S. Gumport, and J. P. Cobb. "The Effect of Triethylene Thiophosphoramide on Fifty Patients with Incurable Neoplastic Diseases." *Cancer* 10 (1957): 239–245.

Cobb, J. P., and D. G. Walker. "Effect of Actinomycin D on Tissue Cultures of Normal and Neoplastic Cells." *Journal of the National Cancer Institute* 21 (1958): 263–277.

Cobb, J. P., and J. C. Wright. "Studies on a Craniopharyngioma in Tissue Culture." *Journal of Neuropathology and Experimental Neurology* 18 (1959): 563–568.

Wright, J. C., J. P. Cobb, F. M Golomb, S. L. Gumport, D. Lyall, and D. Safadi. "Chemotherapy of Disseminated Carcinoma of the Breast." *Annals of Surgery* 150 (1959): 221–240.

Cobb, J. P. "The Comparative Cytological Effects of Several Alkylating Agents on Human Normal and Neoplastic Cells in Tissue Culture." *Annals of New York Academy of Sciences* 84 (1960): 513–542.

Cobb, J. P., and D. G. Walker. "Studies on Human Melanoma Cells in Tissue Cultures. I. Growth Characteristics and Cytology." *Cancer Research* 20 (1960): 858–867.

Cobb, J. P., D. G. Walker, and J. C. Wright. "Observations on the Action of Triethylene Thiophosphoramide within Individual Cells." In *Acta of VII International Cancer Congress,* 1960, pp. 567–583.

Golomb, F. M., J. C. Wright, J. P. Cobb, S. L. Gumport, A. Postal, and D. Safadi. "The Chemotherapy of Human Solid Tumors by Perfusion Techniques." *Proceedings of the American Association for Cancer Research* 3 (1960).

Cobb, J. P. "Cells Tell a Story." *Sarah Lawrence Alumnae Magazine,* winter 1961, pp. 6–10.

Cobb, J. P., and D. G. Walker. "Comparative Chemotherapy Studies on Primary Short-Term Cultures of Human Normal, Benign, and Malignant Tumor Tissues—A Five Year Study." *Cancer Research* 21 (1961): 583–590.

Cobb, J. P., and D. G. Walker. "Effect of Heterologous, Homologous, and Autologous Serums on Human Normal and Malignant Cells in Vitro." *Journal of the National Cancer Institute* 27 (1961): 1–15.

Golomb, F. M., J. P. Cobb, D. G. Walker, and J. C. Wright. "In Vitro Selection of Chemotherapeutic Agents for Perfusion Therapy of Human Cancer." *Surgery* 51 (1962): 639–644.

Wright, J. C., J. P. Cobb, S. L. Gumport, D. Safadi, D. G. Walker, and F. M. Golomb.

"Further Investigation of the Relation between the Clinical and Tissue Culture Response to Chemotherapy Agents on Human Cancer." *Cancer* 15 (1962): 284–293.

Cobb, J. P., and D. G. Walker. "Studies on Human Melanoma Cells in Tissue Culture. II. Effects of Several Cancer Chemotherapeutic Agents on Cytology and Growth." *Acta Union International Center de Cancer* 20 (1964): 206–208.

Nadolney, C. H., and J. P. Cobb. "Melanin Mobilization in Cultured Cloudman S91 Mouse Melanocytes." *Excerpta Medica*, Tissue Culture Meetings, 1964.

Cobb, J. P., and E. S. Rose. "Significance of Large Molecular Protein Molecules in the Nutrition of Mouse Melanoma Organ Cultures." *Excerpta Medica* 9 (1965): 33.

Cobb, J. P., and D. G. Walker. "Time Lapse Cinematography in Medical Research." *Bolex Magazine*, 1966.

Cobb, J. P., and D. G. Walker. "Cytologic Studies on Human Melanoma Cells in Tissue Culture after Exposure to Five Chemotherapeutic Agents." *Cancer Chemotherapy Reports* 52 (1968): 543–552.

Cobb, J. P. "The Impact of the Black Experience on Higher Education in New England." In *An Occasional Paper of the School of Education, University of Connecticut*, edited by G. C. Atkyns, pp. 100–115. Storrs, Conn.: School of Education, University of Connecticut, 1969–1970.

Cobb, J. P., and A. McGrath. "S91 Mouse Melanoma Sublimes Following Total in Vitro versus Alternate in Vivo Passages." *Journal of the National Cancer Institute* 48 (1972): 885–891.

Cobb, J. P. "Cancer—A Solution?" *Connecticut College Alumni Magazine*, winter 1972, pp. 10–11.

Cobb, J. P. "I Am Woman, Black, Educated." *Hartford Courant Sunday Supplement*, 4 February 1973.

Cobb, J. P., and A. McGrath. "In Vitro Effects of Melanocyte-Stimulating Hormone, Adrenocorticotropic Hormone, 17B-Estradiol, or Testosterone Propionate on Cloudman S91 Mouse Melanoma Cells." *Journal of the National Cancer Institute* 52 (1974): 567–570.

Cobb, Jewel Plummer, and Carolyn McDew, editors. *The Morning After—A Retrospective View of a Selected Number of Colleges and Universities with Increased Black Student Enrollment in the Past Five Years. The Report of a Conference at the University of Connecticut.* Storrs, Conn., 30 April 1973. [Racine Printing of Connecticut], 1974.

Cobb, J. P., A. McGrath, and N. Willetts. "Brief Communication: Response of Cloudman S91 Melanoma Cells to Melanocyte Stimulating Hormone: Enhancement by Cytochalasin B." *Journal of the National Cancer Institute* 56 (1976): 1079–1081.

Cobb, J. P. "Toward Licking Cancer." *Douglass College [Rutgers University] Alumnae Bulletin*, spring 1977.

Cobb, J. P. "Postbaccalaureate Premedical Programs for Minority Students." In *Minorities in Science: The Challenge for Change in Biomedicine*, edited by Vijaya L. Melnick and Franklin D. Hamilton, pp. 236–248. New York: Plenum Press, 1977.

Cobb, J. P. "Black Women and Higher Education: A Brief History." *The Black Woman, Myths and Realities: Selected Papers*, edited by Doris J. Mitchell and Jewell H. Bell, pp. 114–119. Cambridge, Mass.: Radcliffe Symposium, 1978.

Cobb, J. P. "Breaking Down Barriers to Women Entering Science." *Physics Today* 32 (1979): 78.

Cobb, J. P. "Filters for Women in Science." *Annals of the New York Academy of Sciences* 323 (1979): 236–248.

Benezet, Louis, Joel Conarroe, Jewel P. Cobb, Phyllis Keller, Robert E. Marshak, John Ratte, Henry Rosovsky, Frederick Rudolph, Edward G. Sparrow, Ilja Wachs, and Henry

R. Winkler. "Issues and Problems: A Debate." In *The Great Core Curriculum Debate: Education as a Mirror of Culture*, pp. 51–78. New Rochelle, N.Y.: Change Magazine Press, 1979.

Cobb, J. P. "Planning the Academic Future for Women and Minorities." *American Association of State Colleges and Universities, Memo to the President* 30, no. 20 (18 May 1990).

Cobb, J. P. "The Role of Women Presidents/Chancellors in Intercollegiate Athletics." In *Women at the Helm: Pathfinding Presidents at State Colleges and Universities*, edited by Judith A. Sturnick, Jane E. Milley, and Catherine A. Tisinger, pp. 42–51. Washington, D.C.: AASCU Press, 1991.

Cobb, J. P. "Societal Barriers and Strategies for Succeeding in the Technical Sciences." In *Women in Engineering Conference, A National Initiative, Conference Proceedings*, edited by Jane Zimmer Daniels, pp. 3–7. [West Lafayette, Ind.]: Women in Engineering Program Advocates Network, 1991.

PRESENTATIONS

Cobb, J. P., and D. G. Walker. "Observations on the Action of Triethylene Thiophosphoramide (TSPA) within Individual Cells." Paper presented at the Seventh International Cancer Congress, 1958.

Cobb, J. P., F. M. Golomb, S. L. Gumport, D. Safadi, and J. C. Wright. "Chemotherapy of Human Breast Cancer." Paper presented at the Seventh International Cancer Congress, 1958.

Cobb, J. P. "Environmental Influences on the Growth of Cloudman S91 Mouse Melanoma in Organ Culture." Proceedings of an Oral Research Seminar, 1967–1968.

Cobb, J. P. "Melanoma Cancer Research." Proceedings of the 145th National Meeting, American Association for the Advancement of Science, Houston, Texas, January 1979.

Johnnetta Betsch Cole

Cole, Johnnetta B., editor. *Anthropology for the Eighties: Introductory Readings*. New York: The Free Press; London: Collier Macmillan, 1982.

Cole, Johnnetta B., editor. *All American Women: Lines That Divide, Ties That Bind*. New York: Free Press; London: Collier Macmillan, 1986.

Cole, Johnnetta B. *Race toward Equality*. Havana, Cuba: J. Marti Publishing House, 1986.

Cole, Johnnetta B., editor. *Anthropology for the Nineties: Introductory Readings*. New York: Free Press; London: Collier Macmillan, 1988. Revised edition of *Anthropology for the Eighties*.

Cole, Johnnetta B. *Conversations: Straight Talk with America's Sister President*. New York: Doubleday, 1993.

Cole, Johnnetta B. *Dream the Boldest Dreams: And Other Lessons of Life*. Atlanta, Ga.: Longstreet Press, 1997.

Margaret James Strickland Collins

Strickland, M. J. "Differences in Toleration of Drying between Species of Termites (*Reticulitermes*)." Ph.D. dissertation, 1950. Published in *Ecology* 31 (1950): 373–385.

Dunmore, L., and M. S. Collins. "Caste Differences in Toleration of Drying in *Reticulitermes flavipes* (Kollar)." *Anatomical Record* 111 (1951). Abstract.

Collins, M. S. "Differences in Toleration of Drying and Rate of Water Loss between Species of Termites (*Reticulitermes, Kalotermes, Neotermes, Cryptotermes*)." *Anatomical Record* 132 (1958): 423. Abstract.

Collins, M. S. "Studies on Water Relations in Florida Termites. I. Survival Time and Rate of Water Loss During Drying." *Quarterly Journal Florida Academic Science* 21 (1959): 341–352.

Collins, M. S., and A. Glenn Richards. "Studies on Water Relations in North American Termites. I. Eastern Species of the Genus *Reticulitermes* (Isoptera, Rhinotermitidae)." *Ecology* 44 (1963): 600–604.

Collins, M. S. "Water Loss and Cuticular Structure in North American Termites — A Preliminary Report." In *Proceedings 2nd Workshop Termite Research, Biloxi, Mississippi*, pp. 25–33. Washington, D.C.: National Academy of Natural Resources Council, 1965.

Collins, M. S., and A. Glenn Richards. "Studies on Water Relations in North American Termites. II. Water Loss and Cuticular Structure in Eastern Species of the Kalotermitidae (Isoptera)." *Ecology* 47 (1966): 328–331.

Collins, M. S. "Water Relations in Termites." Chapter 14 in *Biology of Termites*, edited by K. Krishna and F. M. Weesner. Vol. 1, pp. 433–458. New York and London: Academic Press, 1969.

Collins, M. S. "Isoptera." In *The Encyclopedia of the Biological Sciences*, edited by Peter Gray. 2nd ed. New York: Van Nostrand Reinhold, 1970.

Collins, M., M. I. Haverty, J. P. LaFage, and W. L. Nutting. "High Temperature Tolerance in Two Species of Subterranean Termites from the Sonoran Desert of Arizona." *Environmental Entomology* 2 (1973): 1122–1123.

Collins, M. "The Insect in Art." *Black Art, An International Quarterly* 3 (1979): 13–28.

Collins, M. "Kartabo Revisited — Termite Studies in Guyana." *Sociobiology* (summer 1979).

Prestwich, G., J. Luher, and M. Collins. "Two New Tetracyclic Diterpenes from the Defense Secretion of the Neotropical Termite *Nasutitermes octopilis. Tetrahedron Letters* 40 (1979): 3827–3830.

Preer, J., H. Sekhon, and M. Collins. "Factors Affecting Heavy Metal Content of Garden Vegetables." *Environmental Pollution* 1 (1980): 95–104.

Prestwich, G., and M. Collins. "A Novel Enolic Beta-Ketoaldehyde in the Defense Secretion of the Termite *Rhinotermes hispidus." Tetrahedron Letters* 21 (1980): 5001–5002.

Collins, M., I. Wainer, and T. Bremner, editors. *Science and the Question of Human Equality*. Boulder, Colo.: Westview Press, 1981. (This book was chosen as one of the 100 outstanding titles in science and technology for 1981 by *Library Journal*.)

Prestwich, G., and M. Collins. "Chemotaxonomy of *Subulitermes* and *Nasutitermes* Termite Soldier Defense Secretions: Evidence against the Diphyletic Hypothesis for Nasutitermitinae." *Biochemical Systematics and Ecology* 9 (1981): 83–88.

Prestwich, G., and M. Collins. "Macrocyclic Lactones as the Defense Substances of the Termite Genus *Armitermes. Tetrahedron Letters* 22 (1981): 4587–4590.

Prestwich, G., R. Jones, and M. Collins. "Terpene Biosynthesis by Nasute Termite Soldiers (Isoptera: Nasutitermitinae). *Insect Biochemistry* vol. 11 (1981): 331–336.

Prestwich, G., and M. Colllins. "3-Oxo-(Z)-9-Hexadecenal: An Unusual Enolic Beta-Ketoaldehyde from a Termite Soldier Defense Secretion." *Journal of Organic Chemistry* 46 (1981): 2383.

Collins, M. "Chemical Warfare of Guyana Termites." *Earthwatch*, spring 1982, pp. 19–20.

Prestwich, G., and M. Collins. "Chemical Defense Secretions of the Termite Soldiers of *Acorhinotermes* and *Rhinotermes* (Isoptera, Rhinotermitidae)." *Journal of Chemical Ecology* 8 (1982): 147–161.

Collins, M., and Prestwich, G. "Defense in *Nasutitermes octopilis* Banks (Isoptera, Termitidae, Nasutitermitinae): Comparative Effectiveness of the Soldier Secretion." *Insectes Sociaux* 30 (1983): 70–81.

Collins, M. "Problems in Termite Taxonomy." *Sociobiology* 14 (1988): 207–210.

Collins, M., and D. Nickle. "Keys to Kalotermitidae of Eastern United States, with the Description of a New Species of Kalotermitid (*Neotermes luykxi*) from Florida." 1988.

Nickle, D., and M. Collins. "The Termite Fauna (Isoptera) in the Vicinity of Chamela, state of Jalisco, Mexico." *Folia Entomológica Mexicana*, No. 77 (1988): 85–122.

Collins, M., and D. Nickle. "Keys to Kalotermitidae of Eastern United States, with the Description of a New Neotermes from Florida (Isoptera). *Proceedings Entomological Society of Washington* 91 (1989): 269–285.

Collins, M., and D. Nickle. "A Key to the Termites of the Chamela Area, Jalisco, Mexico." Folio Mexicana, Estración de Chamela, in press.

Collins, M., and D. Nickel. "The Termites of North America, Canada through Panama." (Summary treatment of the biology and taxonomy of the termites of the area, with keys, detailed descriptions, and illustrations, in preparation at the time of Collins's death in 1996.)

PRESENTATIONS

"The Evolution of Desiccation Resistance in Termites." Paper presented at Department of Entomology, Florida A&M University, Tallahassee, Fla., February 1974.

Clark Lectureship at the Scripps Colleges, Claremont, Calif., May 1974. Several addresses on research and experiences.

"Pollution, Public Policy and the City Garden." Paper presented at the Urban Food Conference, Institute for Policy Studies, Washington, D.C., March 1974.

Addresses on termite research and pollution of the urban vegetation during the festivities marking the dedication of the new science buildings. Address on wild plants as alternative food sources. Florida A&M University, Tallahassee, Fla., April 1975.

"The Termites of Guyana." Address to the Conservation Society, Georgetown, Guyana. December 1977.

United States government sponsored interview indicating potential of Guyana as a host nation for scientific study groups. Georgetown, Guyana, January 1978.

Studies on Guyana termites. Entomological Society of Washington, 1979.

"Chemical Defenses in Termites." Paper presented at Woods Hole, Mass., Seminar for Mariculture Participants, May 1981.

"Chemical Defenses in Guyana Termites." Presentation at Earthwatch Seminar for Principal Investigators, May 1982.

"Toll of Termites in the Ecosystems." Presentation at Stockton State College, N.J., Fall College Biology Seminar Series: Research on Termites. November 1985.

"Current Problems in Termite Taxonomy." Paper presented at National Conference on Urban Entomology, University of Maryland, February 1986.

"Termite Biology and the Role of Termites in the Ecosystem." Workshop for Pest Control Operators and Company Representatives, Florida A&M University, Tallahassee, Fla., March 1987.

"Taxonomic Problems with Termites, Canada through Panama." Presentation at National Conference on Urban Entomology, College Park, Md., March 1988.

"Recently Described Taxa of United States Termites and Techniques Leading to Their Recognition." Paper presented at National Conference on Urban Entomology, College Park, Md., March 1990.

Patricia Suzanne Cowings

Blizzard, D., P. Cowings, and N. E. Miller. "Visceral Responses to Opposite Types of Autogenic Training Imagery." *Biological Psychology* 3 (1975): 49–55. Reprinted in *Biofeedback and Self-Control, 1975/76*, edited by T. X. Barber et al., pp. 164–171. Chicago: Aldine Publishing, 1976.

Cowings, P. S. "Combined Use of Autogenic Therapy and Biofeedback in Training Effective Control of Heart Rate in Humans." *Therapy in Psychosomatic Medicine* 4 (1977): 167–173. Reprinted in *Autogenic Methods: Application and Perspectives*, edited by W. Luthe and F. Antonelli, pp. 167–173. Rome: Edizioni Luigi Pozzi, 1977.

Cowings, P. S. "Observed Differences in Learning Ability of Heart Rate Self-Regulation as a Function of Hypnotic Susceptibility." *Therapy in Psychosomatic Medicine* 4 (1977): 221–226. Reprinted in *Autogenic Methods: Application and Perspectives*, edited by W. Luthe and F. Antonelli, pp. 221–226. Rome: Edizioni Luigi Pozzi, 1977.

Cowings, P. S., J. Billingham, and W. B. Toscano. "Learned Control of Multiple Autonomic Responses to Compensate for the Debilitating Effects of Motion Sickness." *Therapy in Psychosomatic Medicine* 4 (1977): 318–323. Reprinted in *Autogenic Methods: Application and Perspectives*, edited by W. Luthe and F. Antonelli, pp. 318–323. Rome: Edizioni Luigi Pozzi, 1977. (This paper was awarded "Best Paper by a Young Investigator" by the Space Medicine Branch of the Aerospace Medical Association, 47th Annual Meeting, Bal Harbor.)

Cowings, P. S., and W. B. Toscano. "Psychosomatic Health: Simultaneous Control of Multiple Autonomic Responses by Humans—A Training Method." *Therapy in Psychosomatic Medicine* 4 (1977): 184–190. Reprinted in *Autogenic Methods: Application and Perspectives*, edited by W. Luthe and F. Antonelli, pp. 184–189. Rome: Edizioni Luigi Pozzi, 1977.

Cowings, P. S., and W. B. Toscano. "A Theory on the Evolutionary Significance of Psychosomatic Disease." *Therapy in Psychosomatic Medicine* 4 (1977): 184–190. Reprinted in *Autogenic Methods: Application and Perspectives*, edited by W. Luthe and F. Antonelli, pp. 404–409. Rome: Edizioni Luigi Pozzi, 1977.

Stewart, J., B. Clark, P. S. Cowings, and W. B. Toscano. "Learned Regulation of Autonomic Responses to Control Coriolis Motion Sickness: Its Effects on Other Vestibular Functions." *Proceedings of the 49th Annual Scientific Meeting of the Aerospace Medical Association*, pp. 133–134. New Orleans, 1978.

Toscano, W. B., and P. S. Cowings. "Transfer of Learned Autonomic Control for Symptom Suppression across Opposite Directions of Coriolis Acceleration." *Proceedings of the 49th Annual Scientific Meeting of the Aerospace Medical Association*, pp. 132–133. New Orleans, 1978.

Cowings, P. S., and W. B. Toscano. "The Relationship of Motion Sickness Susceptibility to Learned Autonomic Control for Symptom Suppression." *Aviation, Space and Environmental Medicine* 53 (1982): 570–575.

Toscano, W. B., and P. S. Cowings. "Reducing Motion Sickness: Autogenic-Feedback Training Compared to an Alternative Cognitive Task." *Aviation, Space and Environmental Medicine* 53 (1982): 449–453.

Finger, H. J., L. Edsinger, W. A. Weeks, T. M. Hedges, P. S. Cowings, and E. Luizzi. "An Ambulatory Feedback System for Space Adaptation Syndrome." In *Proceedings of the 37th Annual Conference on Engineering in Medicine and Biology (ACEMB)*, p. 173. Los Angeles, 17–19 September 1984.

Cowings, P. S., W. B. Toscano, J. Kamiya, N. E. Miller, and J. C. Sharp. "Autogenic-Feedback Training as a Preventive Method for Space Adaptation Syndrome. NASA

Flight Experiment No. AFT23. Spacelab-3. Progress Report I." In *NASA Conference Publication #2429: Spacelab 3 Mission Science Review*, pp. 84–89. National Aeronautics and Space Administration, 1985.

Cowings, P. S., S. Suter, W. B. Toscano, J. Kamiya, and K. Naifeh. "General Autonomic Components of Motion Sickness." *Psychophysiology* 23 (1986): 542–551.

Cowings, P. S., K. Nafieh, and C. Thrasher. *A Computer Program for Processing Impedance Cardiographic Data: Improving Accuracy through User-Interactive Software*. Technical Memorandum 10120. Moffett Field, Calif.: National Aeronautics and Space Administration, Ames Research Center, 1988.

Cowings, Patricia S., W. Toscano, J. Kamiya, N. Miller, and J. Sharp. *Summary of Payload Integration Plan (PIP) for Starlab-1 Flight Experiment, Enclosure 3*. Technical Memorandum 89713. Moffett Field, Calif.: National Aeronautics and Space Administration, Ames Research Center, January 1988.

Cowings, P. S., W. B. Toscano, J. Kamiya, N. E. Miller, and J. C. Sharp. *Final Report. Spacelab-3 Flight Experiment No. 3AFT23: Autogenic-Feedback Training as a Preventive Method for Space Adaptation Syndrome*. Technical Memorandum 89412. Moffett Field, Calif.: National Aeronautics and Space Administration, Ames Research Center, 1988.

Cowings, P. S. "Autogenic-Feedback Training: A Preventive Method for Motion and Space Sickness." Chapter 17 in *Motion and Space Sickness*, edited by G. Crampton, pp. 354–372. Boca Raton, Fla.: CRC Press, 1990.

Cowings, P. S., K. H. Naifeh, and W. B. Toscano. "The Stability of Individual Patterns of Autonomic Responses to Motion Sickness Stimulation." *Aviation Space and Environmental Medicine* 61 (1990): 399–405.

Cowings, P. S., W. B. Toscano, J. Kamiya, N. E. Miller, T. Pickering, and D. Shapiro. "Autogenic-Feedback Training: Countermeasure for Orthostatic Intolerance." In *Proceedings of the First Joint NASA Cardiopulmonary Workshop*, pp. 145–153. Houston, Texas: Krug Life Sciences, 1990.

Cowings, Patricia S. "Autogenic Feedback Training Experiment: A Preventative Method for Space Motion Sickness." In *Marshall Space Flight Center, Spacelab J Experiment Descriptions*, pp. 227–248. Houston, Texas: National Aeronautical and Aerospace Administration, Lyndon B. Johnson Space Center, August 1993.

Cowings, P. S., and W. B. Toscano. *Autogenic-Feedback Training (AFT) as a Preventive Method for Space Motion Sickness: Background and Experimental Design*. Technical Memorandum 108780. Moffett Field, Calif.: National Aeronautics and Space Administration, Ames Research Center, 1993.

Kellar, M. A., R. A. Folen, P. S. Cowings, W. B. Toscano, and G. L. Hisert. *Autogenic-Feedback Training Improves Pilot Performance during Emergency Flying Conditions*. Technical Memorandum 104005. Moffett Field, Calif.: National Aeronautics and Space Administration, Ames Research Center, 1993. Reprinted in *Flight Safety Digest*, July 1993.

Cowings, P. S., W. B. Toscano, N. E. Miller, T. Pickering, and D. Shapiro. *Autogenic-Feedback Training: A Potential Treatment for Post-Flight Orthostatic Intolerance in Aerospace Crews*. Technical Memorandum 108785. Moffett Field, Calif.: National Aeronautics and Space Administration, Ames Research Center, 1993. Reprinted in *Journal of Clinical Pharmacology* 34 (1994): 599–608.

Stout, C. S., W. B. Toscano, and P. S. Cowings. *Reliability of Autonomic Responses and Malaise across Multiple Motion Sickness Stimulation Tests*. Technical Memorandum 108787. Moffett Field, Calif.: National Aeronautics and Space Administration, Ames Research Center, December 1993.

Stout, C. S., and P. S. Cowings. *Increasing Accuracy in the Assessment of Motion Sickness: A Construct Methodology.* Technical Memorandum 108797. Moffett Field, Calif.: National Aeronautics and Space Administration, Ames Research Center, 1993.

Cowings, P. S., W. B. Toscano, N. E. Miller, and S. M. Reynoso. *Autogenic-Feedback Training as a Treatment for Air-Sickness in High Performance Military Aircraft: Two Case Studies.* Technical Memorandum 108810. Moffett Field, Calif.: National Aeronautics and Space Administration, Ames Research Center, March 1994.

Toscano, W. B., and P. Cowings. *The Effects of Autogenic-Feedback Training on Motion Sickness Severity and Heart Rate Variability in Astronauts.* Technical Memorandum 108840. Moffett Field, Calif.: National Aeronautics and Space Administration, Ames Research Center, October 1994.

Stout, C. S., W. B. Toscano, and P. S. Cowings. "Reliability of Psychophysiological Responses across Multiple Motion Sickness Stimulation Tests." *Journal of Vestibular Research* 5 (1995): 25–33.

Cowings, P. S., C. Stout, W. B. Toscano, S. Reynoso, and C. DeRoshia. *The Effects of Promethazine on Human Performance, Mood States, and Motion Sickness Tolerance.* Technical Memorandum 110420. Moffett Field, Calif.: National Aeronautics and Space Administration, Ames Research Center, November 1996.

Cowings, P. S., C. Stout, W. B. Toscano, and C. DeRoshia. *An Evaluation of the Frequency and Severity of Motion Sickness Incidences in Personnel within the Command and Control Vehicle (C2V).* Technical Memorandum 112221. Moffett Field, Calif.: National Aeronautics and Space Administration, Ames Research Center, January 1998.

PRESENTATIONS

Cowings, P. S. "Volitional Control of Autonomic Response Patterns to Suppress Motion Sickness." Paper presented at the 16th Annual Meeting of the Society for Psychophysiological Research, San Diego, 20–23 October 1976.

Cowings, P. S. "The Development of Autogenic-Feedback Training as a Potential Countermeasure for Space Adaptation Syndrome." Paper presented at the 11th Joint US/USSR Symposium of the Biomedical Problems of Manned Spaceflight, Moscow, September 1980.

Cowings, P. S. "Autogenic-Feedback Training as a Treatment for Motion Sickness." Paper presented at the 15th Annual Meeting of the Biofeedback Society of America, Albuquerque, 23–28 March 1984.

Cowings, P. S., S. Suter, and K. Naifeh. "Autonomic Changes during Motion Sickness." Paper presented at the 15th Annual Meeting of the Biofeedback Society of America, Albuquerque, 23–28 March 1984.

Cowings, P. S., and W. B. Toscano. "Shuttle Flight Experiment #3AFT23: A Test of Autogenic-Feedback Training in Space." Paper presented at the 15th Annual Meeting of the Biofeedback Society of America, Albuquerque, 23–28 March 1984.

Cowings, P. S., W. B. Toscano, S. Suter, J. Kamiya, and K. Naifeh. "Visceral Learning and the Space Adaptation Syndrome." Paper presented at the 9th Annual Symposium on the Role of Psychology in the Department of Defense, USAF Academy, Colorado Springs, 18–20 April 1984.

Kamiya, J., P. S. Cowings, and J. L. Chen. "The Electrogastrogram as an Index of Motion Sickness." Paper presented at the Society for Psychophysiological Research, Monterey, Calif., 16–18 November 1984.

Morgan, G. M., and P. S. Cowings. "Stimulus Specificity and Idiosyncratic Autonomic Responses to Motion Stressors." Paper presented at the 25th Annual Meeting of the Society for Psychophysiological Research, Houston, 17–20 October 1985.

Cowings, P. S., W. B. Toscano, J. Kamiya, N. E. Miller, and J. C. Sharp. "Autogenic-Feedback Training as a Preventive Method for Space Adaptation Syndrome on Space-Lab 3." Paper presented at the Space Life Sciences Symposium: Three Decades of Life Science Research in Space, Washington, D.C., 21–26 June 1987.

Cowings, P. S., W. B. Toscano, J. Kamiya, and N. E. Miller. "A Behavioral Medicine Alternative: Autogenic-Feedback Training as a Treatment for Space Motion Sickness." Paper presented at the Annual Meeting of the Biofeedback Society of California, Los Angeles, 11 November 1989.

Toscano, W. B., and P. S. Cowings. "The Effect of Training Schedule on Learned Suppression of Motion Sickness Symptoms Using Autogenic-Feedback Training." Paper presented at the Annual Meeting of the Aerospace Medical Association, New Orleans, 13–17 May 1990.

Toscano, W. B., and P. S. Cowings. "Heart Rate Variability of Humans in Space." Paper presented at the 64th Annual Meeting of the Aerospace Medical Association, Toronto, 23–27 May 1993. See NASA Technical Memorandum 108840.

Cowings, P. S., W. B. Toscano, C. Sekiguchi, and M. Ishii. "Preflight AFT for Control of Space Motion Sickness: SPACELAB-J." Paper presented at the 64th Annual Meeting of the Aerospace Medical Association, Toronto, 23–27 May 1993.

Cowings, P. S., W. B. Toscano, and N. E. Miller. "A Behavioral Medicine Approach to Facilitating Adaptation to Space: Autogenic-Feedback Training." Poster presentation at the 33rd Annual Meeting of the Society for Psychophysiological Research, Germany, 27–31 October 1993.

Cowings, P. S., W. B. Toscano, N. E. Miller, C. Stout, and S. Reynoso. "The Effects of Promethazine on Psychological Responses, Performance and Susceptibility to Motion Sickness." Poster presentation at the 33rd Annual Meeting of the Society for Psychophysiological Research, Germany, 27–31 October 1993.

Stout, C. S., P. S. Cowings, and W. B. Toscano. "The Assessment of Motion Sickness Symptoms and Autonomic Responses: A Single Subject Design." Poster presentation at the 33rd Annual Meeting of the Society for Psychophysiological Research, Germany, 27–31 October 1993.

Toscano, W. B., P. S. Cowings, and N. E. Miller. "Monitoring Astronauts' Functional State: Autonomic Responses to Microgravity." Poster presentation at the 33rd Annual Meeting of the Society for Psychophysiological Research, Germany, 27–31 October 1993.

Cowings, P. S. "Psychophysiology of Humans in Space." Paper presented at the Goddard Space Flight Center's Engineering Colloquium, Greenbelt, Md., 24 January 1994.

Cowings, P. S. "Neurophysiologic Changes Occurring in Astronauts and How They Are Measured." Paper presented at the 41st Annual Meeting of Society of Nuclear Medicine, Orlando, Fla., 5–8 June 1994.

Cowings, P. S. "Autogenic-Feedback Training Applications for Man in Space." Paper presented at the Society of Women Engineers 1994 National Convention and Student Conference, Pittsburgh, 22–26 June 1994.

Cowings, P. S., W. B. Toscano, N. E. Miller, T. G. Pickering, and D. Shapiro. "A Potential Treatment for Post-Flight Orthostatic Intolerance in Aerospace Crews: Autogenic-Feedback Training." Paper presented at the Fifth International Symposium on the Autonomic Nervous System, Mayo Clinic, Rochester, Minn., 20–23 October 1994.

Cowings, P. S., and W. B. Toscano. "The Effects of Autonomic Conditioning on Motion Sickness Tolerance." Paper presented at the Fifth International Symposium on the Autonomic Nervous System, Mayo Clinic, Rochester, Minn., 20–23 October 1994.

Toscano, W. B., and P. S. Cowings. "Autonomic Responses to Microgravity." Paper presented at the Fifth International Symposium on the Autonomic Nervous System, Mayo Clinic, Rochester, Minn., 20–23 October 1994.

Cowings, P. S., and W. B. Toscano. "Effects of Autonomic Conditioning on Motion Sickness Tolerance." Paper presented at the Fifth International Symposium on the Autonomic Nervous System, American Autonomic Society, Mayo Clinic, Rochester, Minn., 20–23 October 1994. Abstract in *Clinical Autonomic Research* 4, no. 4 (1994): 198.

Toscano, W. B., and P. S. Cowings. "Heart Rate Variability during Early Adaptation to Space." Paper presented at the Fifth International Symposium on the Autonomic Nervous System, American Autonomic Society, Mayo Clinic, Rochester, Minn., 20–23 October 1994. Abstract in *Clinical Autonomic Research* 4 (1994): 216.

Cowings, P. S., C. Stout, W. B. Toscano, S. Reynoso, C. DeRoshia, and N. E. Miller. "The Effects of Promethazine on Human Performance, Autonomic Responses, and Motion Sickness Tolerance." Paper presented at the 66th Annual Aerospace Medical Association Scientific Meeting, Anaheim, Calif., 7–11 May 1995. Abstract in *Aviation Space and Environmental Medicine* 66 (1995): 466.

Cowings, P. S., and W. B. Toscano. "Monitoring and Correcting Autonomic Responses during Long-Duration Spaceflight with Autogenic-Feedback Training Exercise (AFTE): A NASA Technology Transfer Opportunity." Presented at the Seventh International Symposium on the Autonomic Nervous System, American Autonomic Association, Montreal, Canada, 1996. Abstract in *Clinical Autonomic Research* 6 (1996): 305.

Rashed, H., P. S. Cowings, W. B. Toscano, E. Rebello, T. Abell, and S. Cardoso. "A Case Report: Autogenic-Feedback Training (AFTE) as a Potential Treatment for Dysautonomia." Paper presented at the Eighth International Symposium on the Autonomic Nervous System, American Autonomic Association, Hawaii, November 1997.

Christine Voncile Mann Darden

Darden, C. V. M. "Mathematical Calculations of Light Scattering by Non-Spherical Objects." M.S. thesis, Virginia State College, 1967.

Beckwith, I. E., W. B. Harvey, and C. V. M. Darden. "Effects of Heat Losses by Conduction and Radiation on Bare Wire Thermocouple Probes." Appendix A of NASA TN D-6192, 1970.

Johnson, C. B., and C. V. M. Darden. *Flight Transition Data at Angles of Attack at Mach 22, Including Correlations of Data.* Technical Memorandum X3235. Washington, D.C.: National Aeronautics and Space Administration, Langley Research Center, August 1975.

Darden, Christine M. *Minimization of Sonic Boom Parameters in Real and Isothermal Atmospheres.* TN D-7842. Washington, D.C.: National Aeronautics and Space Administration, Langley Research Center, March 1975.

Darden, Christine M. "Comparison of Sonic Boom Minimization Results in Real and Isothermal Atmospheres." *Journal of Aircraft* 12 (1975): 496–497.

Darden, Christine M. *Sonic-Boom Theory: Its Status in Prediction and Minimization.* New York: American Institute of Aeronautics and Astronautics, 1976.

Darden, Christine M., and Robert J. Mack. "Current Research in Sonic Boom Minimization." *Proceedings of the SCAR Conference.* National Aeronautics and Space Administration, CP-001, Part 2. Langley Research Center, Hampton, Va., 9–12 November 1976.

Darden, Christine M. "Sonic Boom Studies." In *Proceedings, National Technical Association's 49th Annual Conference,* Hampton, Va., 2–5 August 1977.

Darden, Christine M. *Sonic Boom Minimization with Nose Bluntness Relaxation.* Technical Paper 1348. Washington, D.C.: National Aeronautics and Space Administration, Langley Research Center, January 1979.

Mack, Robert J., and C. M. Darden. *A Wind Tunnel Investigation of a Sonic Boom Minimization Concept.* Technical Paper 1421. Washington, D.C.: National Aeronautics and Space Administration, Langley Research Center, 1979.

Mack, Robert J., and Christine M. Darden. "Some Effects of Applying Sonic Boom Minimization to Supersonic Cruise Aircraft Design." *Journal of Aircraft* 17 (1980): 182.

Darden, Christine M. *Charts for Determining Potential Minimum Sonic-Boom Overpressures for Supersonic Cruise Aircraft.* Technical Paper 1820. Washington, D.C.: National Aeronautics and Space Administration, Langley Research Center, March 1981.

Darden, Christine M. "An Analysis of Shock Coalescence including Three-Dimensional Effects with Applications to Sonic Boom Prediction." D.Sc. dissertation, George Washington University, 1983.

Darden, Christine M. *An Analysis of Shock Coalescence Including Three-Dimensional Effects with Applications to Sonic-Boom Prediction.* Technical Paper 2214. Washington, D.C.: National Aeronautics and Space Administration, Langley Research Center, January 1984.

Darden, Christine M. *Spatial Derivatives of Flow Quantities behind Curved Shocks of All Strengths.* Technical Memorandum 85782. Washington, D.C.: National Aeronautics and Space Administration, Langley Research Center, July 1984.

Darden, Christine M. *First Derivatives of Flow Quantities behind Two-Dimensional Nonuniform Supersonic Flow over a Convex Corner.* Technical Report 86272. Washington, D.C.: National Aeronautics and Space Administration, Langley Research Center, February 1985.

Darden, Christine M. "The Influence of Leading Edge Load Alleviation on Supersonic Wing Design." *Journal of Aircraft* 22 (1985): 71–77.

Carlson, Harry W., Barrett L. Shrout, and Christine M. Darden. "Wing Design with Attainable Leading-Edge Thrust Considerations." *Journal of Aircraft* 22 (1985): 244–248.

Small, W. J., B. L. Shrout, G. D. Riebe, K. W. Hom, B. A. Campbell, H. W. Carlson, C. M. Darden, J. S. Jackson, and M. E. Johnson. "Studies of Several Advanced Supersonic Wing Designs for Subsonic and Supersonic Maneuver." NASA CP-2398, Vol. II: 179. *Proceedings, Langley Aerodynamics Symposium,* Hampton, Va., 23–25 April 1985.

Darden, Christine M. *The Effect of Leading Edge Load Constraints on the Design and Performance of Supersonic Wings.* Technical Paper 2446. Washington, D.C.: National Aeronautics and Space Administration, Langley Research Center, July 1985.

Carlson, Harry W., and Christine M. Darden. "Attached Flow Numerical Methods for Aerodynamic Design and Analysis of 'Vortex Flaps'." NASA CP 2416, Vol. II: 111. *Proceedings, Vortex Flow Aerodynamics Meeting,* Langley Research Center, Hampton, Va., 8–10 October 1985.

Carlson, Harry W., and Christine M. Darden. *Applicability of Linearized Theory Attached Flow Methods to Design and Analysis of Flap Systems at Low Speeds for Thin Swept Wings with Sharp Leading Edges.* Technical Paper 2653. Washington, D.C.: National Aeronautics and Space Administration, Langley Research Center, January 1987.

Carlson, Harry W., and Christine M. Darden. *Validation of a Pair of Computer Codes for Estimation and Optimization of Subsonic Aerodynamic Performance of Simple Hinged-Flap Systems for Thin Swept Wings.* Technical Paper 2828. Washington, D.C.: National Aeronautics and Space Administration, Langley Research Center, November 1988.

Darden, Christine M. "The Effect of Model Roughness Caused by Numerical-Control Milling Machines on Wind-Tunnel Force Data." *Journal of the National Technical Association* 61 (1988): 49–55.

Darden, Christine M. *Effect of Milling Machine Roughness and Wing Dihedral on the Supersonic Aerodynamic Characteristics of a Highly Swept Planform.* Technical Paper 2918. Washington, D.C.: National Aeronautics and Space Administration, Langley Research Center, August 1989.

Darden, Christine M., Clemans A. Powell, Wallace D. Hayes, Albert R. George, and Allan D. Pierce. *Status of Sonic Boom Methodology and Understanding.* NASA-CP-3027. Washington, D.C.: National Aeronautics and Space Administration, Langley Research Center, 1989. (Originally presented at the Langley Research Center Sonic Boom Workshop, Hampton, Va., 19–20 January 1988.)

Carlson, Harry W., Christine M. Darden, and Michael J. Mann. *Validation of a Computer Code for Analysis of Subsonic Aerodynamic Performance of Wings with Flaps in Combination with a Canard or Horizontal Tail and an Application to Optimization.* Technical Paper 2961. Washington, D.C.: National Aeronautics and Space Administration, Langley Research Center, January 1990.

Darden, Christine M. "The Importance of Sonic Boom Research in the Development of Future High Speed Aircraft." *Journal of the National Technical Association* 65 (1992): 54–62.

Darden, Christine M., and Kevin P. Shepherd. "Assessment and Design of Low Boom Configurations for Supersonic Transport Aircraft." In *Proceedings of the 14th DGLR/AIAA Aeroacoustics Conference,* Aachen, Germany, 11–14 May 1992. Vol. 1, pp. 334–341. Deutsche Gesellschaft fuer Luft und Raumfahrt, January 1992.

Darden, Christine M., editor. *High-Speed Research: Sonic Boom, Volume 1.* NASA-CP-3172. Washington, D.C.: National Aeronautics and Space Administration, Langley Research Center, October 1992. Papers presented at the Langley Research Center High-Speed Sonic Boom Workshop, Hampton, Va., 25–27 February 1992.

Darden, Christine M., Robert J. Mack, Kathy E. Needleman, Daniel G. Baize, Peter G. Coen, Raymond L. Barger, N. Duane Melson, Mary S. Adams, Elwood W. Shields, and Marvin E. Mcgraw. "Design and Analysis of Low Boom Concepts at Langley Research Center." In *First Annual High-Speed Research Workshop.* N94-33462 10-02. Part 2, pp. 675–699. Washington, D.C.: National Aeronautics and Space Administration, Langley Research Center, April 1992.

Darden, Christine M. "Limitations of Linear Theory Methods as Applied to Sonic Boom Calculations." *Journal of Aircraft* No. 3 (1993): 309–314.

Sinclair, M. J., and Christine M. Darden. "Euler Code Prediction of Near-Field to Midfield Sonic Boom Pressure Signatures." *Journal of Aircraft* 30 (1993): 911–917.

PRESENTATIONS

Darden, Christine M. "A Study of the Limitations of Linear Theory Methods as Applied to Sonic Boom Calculations." AIAA PAPER 90-0368. Paper presented at the AIAA, 28th Aerospace Sciences Meeting, Reno, Nev., 8–11 January 1990.

Needleman, Kathy E., Christine M. Darden, and Robert J. Mack. "A Study of Loudness as a Metric for Sonic Boom Acceptability." AIAA PAPER 91-0496. Paper presented at the AIAA, 29th Aerospace Sciences Meeting, Reno, Nev., 7–10 January 1991.

Darden, Christine M., Erik D. Olson, and Elwood W. Shields. "Elements of NASA's High-Speed Research Program." AIAA PAPER 93-2942. Paper presented at AIAA, 24th Fluid Dynamics Conference, Orlando, Fla., 6–9 July 1993.

Darden, Christine M. "Marketing Considerations for a High-Speed Civil Transport."

NASA Senior Executive Service Candidate Development Program (SESCDP), Lockheed-Martin Aeronautical Systems Company, Marietta, Ga., 15 December 1995.

Darden, Christine M. "Globalization of the Aerospace Industry: Offsets Speed Transfer of Technology." NASA Senior Executive Service Candidate Development Program (SESCDP), Lockheed-Martin Aeronautical Systems Company, 4 December 1995.

Georgia M. Dunston

Dunston, G. M., and H. Gershowitz. Further Studies of Xh, a Serum Protein Antigen in Man. *Vox Sanguinis* 24 (April 1973): 343–353.

Dunston, G. M., and H. Gershowitz. A Hormonally Influenced Human Serum Globulin: Elevation of Xh by Estrogen. *Journal of Laboratory and Clinical Medicine* 84 (August 1974): 187–190.

Ofosu, M.D., D. A. Saunders, G. M. Dunston, O. Castro, and L. Alarif. Association of HLA and Autoantibody in Transfused Sickle Cell Disease Patients. *American Journal of Hematology* 22 (May 1986): 27–33.

Dunston, G. M., C. K. Hurley, R. J. Hartzman, and A. H. Johnson. Unique HLA-D Region Heterogeneity in American Blacks. *Transplantation Proceedings* 19 (February 1987): 870–871.

Bonney, G. E., G. M. Dunston, and J. Wilson. Regressive Logistic Models for Ordered and Unordered Polychotomous Traits: Application to Affective Disorders. *Genetic Epidemiology* 6 (1989): 211–215.

Dunston, G. M., L. W. Henry, J. Christian, M. D. Ofosu, and C. O. Callender. HLA-DR3, DQ Heterogeneity in American Blacks Is Associated with Susceptibility and Resistance to Insulin Dependent Diabetes Mellitus. *Transplantation Proceedings* 21 (February 1989): 653–655.

Frederick, W. R., C. O. Callender, C. W. Saxinger, J. C. Flores, S. S. Alexander, S. E. Barnes, R. Flagg, C. S. Walters, G. M. Dunston, and W. L. Greaves. Serologic and Immunologic Correlates of Retroviral Infection in Transplant Recipients. *Transplantation Proceedings* 21 (February 1989): 2093–2096.

Johnson, A., T. F. Tang, G. M. Dunston, N. Steiner, C. K. Hurley. Relationship of DR3 T Cell Recognition Determinants to DNA Sequence. *Transplantation Proceedings* 21 (February 1989): 624–625.

Toussaint, R. M., C. O. Callender, G. M. Dunston, J. Flores, C. S. Walters, D. John, C. Yeager, O. Bond, C. Thompson, and J. C. Gear. Prednisone Used as Treatment for Rejection Correlates with Poor Outcome. *Transplantation Proceedings* 21 (February 1989): 1712–1715.

Toussaint, R. M., C. O. Callender, G. M. Dunston, J. Flores, C. S. Walters, D. John, C. Yeager, O. Bond, C. Thompson, and J. C. Gear. Prednisone When Used as Treatment for Rejection Correlates with Poor Outcome. *Journal of the National Medical Association* 81 (May 1989): 499–503.

Dunston, G. M., and R. M. Halder. Vitiligo Is Associated with HLA-DR4 in Black Patients. A Preliminary Report. *Archives of Dermatology* 126 (January 1990): 56–60.

Callendar, C. O., L. E. Hall, C. L. Yeager, J. B. Barber Jr., G. M. Dunston, and V. W. Pinn-Wiggins. Organ Donation and Blacks. A Critical Frontier. *New England Journal of Medicine* 8, no. 325 (August 1991): 442–444.

Hall, L.E., C. O. Callender, C. L. Yeager, J. B. Barber Jr., G. M. Dunston, and V. W. Pinn-Wiggins. Organ Donation in Blacks: The Next Frontier. *Transplantation Proceedings* 23 (October 1991): 2500–2504.

Freedman, B. I., B. J. Spray, G. M. Dunston, and E. R. Heise. HLA Associations in End-

Stage Renal Disease Due to Membranous Glomerulonephritis: HLA-DR3 Associations with Progressive Renal Injury. Southeastern Organ Procurement Foundation. *American Journal of Kidney Diseases* 23 (June 1994): 797–802.

Dunston, G. M., O. Akinsete, and F. S. Collins. Diabetes Project. *Science* 16, no. 276 (May 1997): 1013.

Agurs-Collins, T., K. S. Kim, G. M. Dunston, and L. L. Adams-Campbell. Plasma Lipid Alterations in African-American Women with Breast Cancer. *Journal of Cancer Research and Clinical Oncology* 124 (1998): 186–190.

Hizawa, N., L. R. Freidhoff, Y. F. Chiu, E. Ehrlich, C. A. Luehr, J. L. Anderson, D. L. Duffy, G. M. Dunston, J. L. Weber, S. K. Huang, K. C. Barnes, D. G. Marsh, and T. H. Beaty. Genetic Regulation of Dermatophagoides pteronyssinus–Specific IgE Responsiveness: A Genome-wide Multipoint Linkage Analysis in Families Recruited through 2 Asthmatic Sibs. Collaborative Study on the Genetics of Asthma (CSGA). *Journal of Allergy and Clinical Immunology* 102 (September 1998): 436–442.

Hizawa, N., L. R. Freidhoff, E. Ehrlich, Y. F. Chiu, D. L. Duffy, C. Schou, G. M. Dunston, T. H. Beaty, D. G. Marsh, K. C. Barnes, and S. K. Huang. Genetic Influences of Chromosomes 5q31-q33 and 11q13 on Specific IgE Responsiveness to Common Inhaled Allergens among African American Families. Collaborative Study on the Genetics of Asthma (CSGA). *Journal of Allergy and Clinical Immunology* 102 (September 1998): 449–453.

Barbara Jeanne Dyce

Haverback, B. J., M. I. Stubrin, and B. J. Dyce. "Relationship of Histamine to Gastrin and Other Secretagogues." *Federation Proceedings* 24 (1965): 1326–1330.

Adham, N. F., B. J. Dyce, M. C. Geokas, and B. J. Haverback. "Stool Chymotrypsin and Trypsin Determinations." *American Journal of Digestive Diseases* 12 (1967): 1272–1276.

Saunders, R., B. J. Dyce, W. E. Vannier, and B. J. Haverback. "The Separation of Alpha-2 Macroglobulin into Five Components with Differing Electrophoretic and Enzyme-Binding Properties." *Journal of Clinical Investigation* 50 (1971): 2376–2383.

Dyce, B. J., and S. P. Bessman. "A Rapid Nonenzymatic Assay for 2,3-DPG in Multiple Specimens of Blood." *Archives of Environmental Health* 27 (1973): 112–115.

Mersch, J., B. J. Dyce, B. J. Haverback, and R. P. Sherwin. "Diphosphoglycerate Content of Red Blood Cells." *Archives of Environmental Health* 27 (1973): 94–95.

DiSaia, P., B. J. Haverback, B. J. Dyce, and M. Morrow. "Carcinoembryonic Antigen in Patients with Squamous Cell Carcinoma of the Cervix Uteri and Vulva." *Surgery, Gynecology, and Obstetrics* 138 (1974): 542–544.

Dyce, B. J., and B. J. Haverback. "Free and Bound Carcinoembryonic Antigen in Neoplasms and in Normal Adult and Fetal Tissue." *Immunochemistry* 11 (1974): 423–430.

Haverback, B. J., and B. J. Dyce. "Gastrointestinal Cancer Syndromes: Gastrins, Multiple Endocrine Adenomatosis, and the Zollinger-Ellison Syndrome." *Annals of the New York Academy of Science* 230 (1974): 297–305.

DiSaia, P. J., B. J. Haverback, B. J. Dyce, and C. P. Morrow. "Carcinoembryonic Antigen in Patients with Gynecologic Malignancies." *American Journal of Obstetrics and Gynecology* 15 (1975): 159–163.

DiSaia, P. J., C. P. Morrow, B. J. Haverback, and B. J. Dyce. "Carcinoembryonic Antigen in Cervical and Vulvar Cancer Patients: Serum Levels and Disease Progress." *Obstetrics and Gynecology* 47 (1976): 95–99.

Kido, D. K., B. J. Dyce, B. J. Haverback, and C. L. Rumbaugh. "Carcinoembryonic

Antigen in Patients with Untreated Central Nervous System Tumors." *Bulletin of the Los Angeles Neurological Society* 41 (1976): 47–54.

DiSaia, P. J., C. P. Morrow, B. J. Haverback, and B. J. Dyce. "Carcinoembryonic Antigen in Cancer of the Female Reproductive System: Serial Plasma Values Correlated with Disease State." *Cancer* 39 (1977): 2365–2370.

Annie Easley

Easley, A. J., and A. F. Kascak. *Effect of Turbulent Mixing on Average Fuel Temperatures in a Gas-Core Nuclear Rocket Engine.* Technical Note D-4882. Cleveland, Ohio: National Aeronautics and Space Administration, Lewis Research Center, November 1968.

Kascak, A. F., and A. J. Easley. *Bleed Cycle Propellant Pumping in a Gas-Core Nuclear Rocket Engine System.* Technical Memo X-2517. Cleveland, Ohio: National Aeronautics and Space Administration, Lewis Research Center, March 1972.

Nainiger, Joseph J., Raymond K. Burns, and Annie J. Easley. *Performance and Operational Economics Estimates for a Coal Gasification Combined-Cycle Cogeneration Powerplant.* Technical Memo 82729. Cleveland, Ohio: National Aeronautics and Space Administration, Lewis Research Center, March 1982.

Cecile Hoover Edwards

Belton, W. E., and C. A. Hoover. "Investigations on the Mung Bean (*Phaseolus areus* Roxburgh). I. The Determination of Eighteen Amino Acids in the Mung Bean Pydrolyzate by Chemical and Microbiological Methods." *Journal of Biological Chemistry* 175 (1948): 377.

Hoover, C. A., and M. C. Coggs. "Food Intakes of Fifty College Women." *Journal of Home Economics* 40 (1948): 193.

Edwards, C. H., J. R. Mitchell, F. McEnge, L. Trigg, and L. Jones. "Influence of Dietary Supplements on Hemoglobin, Erythrocytes." *Journal of the National Medical Association* 45 (1953): 180.

Richardson, F. R., and C. H. Edwards. "Relation of Dietary Habits to Hemoglobin, Erythrocytes, and Minor Physical Ailments of College Students." *Journal of the National Medical Association* 45 (1953): 201.

Edwards, C. H. "Book Review: Food Selection and Preparation." *Journal of Home Economics* 46 (1954): 608.

Edwards, C. H., H. McSwain, and B. Haire. "Odd Dietary Practices of Women." *Journal of the American Dietetic Association* 30 (1954): 976.

Edwards, C. H. "Women in Science." *Alpha Kappa Mu Journal*, 11, 7. No. 1 (1955).

Edwards, C. H., L. P. Carter, and C. E. Outland, "Cystine, Tyrosine and Essential Amino Acid Contents of Selected Foods." *Journal of Agricultural and Food Chemistry* 3 (1955): 952.

Edwards, C. H., J. A. Lomax, and G. Grimmett. "Effect of a Dietary Supplement on the Height and Weight of Children." *Journal of Home Economics* 48 (1956): 363.

Edwards, C. H., G. Grimmett, and J. A. Lomax. "Influence of a Low Cost Dietary Supplement on Scholastic Achievement and Personality Factors of Elementary School Children." *Journal of the National Medical Association* 48 (1956): 244.

Edwards, C. H., S. C. McDonald, J. R. Mitchell, A. M. Kemp, K. M. Laing, L. Jones, L. L. Mason, and L. Trigg. "Relation of Certain Odd Dietary Practices to Medical and Biochemical Findings and the Outcome of Pregnancy in Women." *Ivy Leaf* 36 (1957): 6.

Edwards, C. H., and C. H. Allen, "Cystine, Tryosine and Essential Amino Acid Content of Selected Foods of Plant and Animal Origin." *Journal of Agricultural and Food Chemistry* 6 (1958): 219.

Edwards, Gerald A., Cecile H. Edwards, and Evelyn L. Gadsden. "An Inexpensive Metabolism Apparatus for the Simultaneous and Intermittent Collection of Expired Carbon Dioxide, Excreta, and Blood from Small Experimental Animals." *International Journal of Applied Radiation and Isotypes* 4 (1959): 264.

Edwards, C. H., Evelyn L. Gadsden, Lolla P. Carter, and Gerald A. Edwards. "Paper Chromatography of Amino Acids and Other Organic Compounds in Selected Solvents." *Journal of Chromatography* 2 (1959): 188.

Edwards, C. H., S. McDonald, J. R. Mitchell, L. Jones, L. Mason, A. M. Kemp, D. Laing, and L. Trigg. "Clay- and Cornstarch-eating Women." *Journal of the American Dietetic Association* 35 (1959): 810.

Edwards, C. H. "The Need for Scholars in This Age of Science." *Alpha Kappa Mu Journal* 16, no. 1 (1960).

Edwards, C. H., E. L. Gadsden, and G. A. Edwards. "Utilization of Methionine by the Adult Rat. I. Distribution of the *alpha*-Carbon of DL-Methionine-2C14 in Tissues, Tissue Fractions, Expired Carbon Dioxide, Blood and Excreta." *Journal of Nutrition* 72 (1960): 185.

Gadsden, E. L., C. H. Edwards, and G. A. Edwards. "Paper Chromatography of Vitamins in Phenol and Butanol; Propionic Acid: Water Solvents." *Analytical Chemistry* 32 (1960): 2415.

Edwards, C. H., E. L. Gadsden, and G. A. Edwards. "Utilization of DL-Methionine-2-C14 and L-Methionine-Methyl-C14 by the Adult Rat." In *Proceedings, Fifth International Congress on Nutrition*, Washington, D.C., 1960.

Edwards, C. H., E. L. Gadsden, and G. A. Edwards. "Effect of Irradiation on the Uptake of Methionine by Rat Tissues." In *Proceedings, Sixth International Congress of Nutrition, Edinburgh, Scotland*, 1963.

Edwards, C. H., E. L. Gadsden, and G. A. Edwards. "Utilization of Methionine by the Adult Rat. II. Absorption and Tissue Uptake of L- and DL-Methionine." *Journal of Nutrition* 80 (1963): 69.

Edwards, C. H., E. L. Gadsden, and G. A. Edwards. "Utilization of Methionine by the Adult Rat. III. Early Incorporation of Methionine-Methyl-C14 and Methionine-2-C14 into Rat Tissues." *Journal of Nutrition* 80 (1963): 211.

Edwards, C. H., E. L. Gadsden, and G. A. Edwards. "Utilization of Methionine by the Adult Rat. IV. Distribution of the Methyl Carbon of Methionine in Tissues of Expired Carbon Dioxide and Excreta Metabolism." *Metabolism* 12 (1963): 951.

Edwards, C. H., E. L. Gadsden, and G. A. Edwards. "Utilization of Methionine by the Adult Rat. VII. The Methyl Carbon of Methionine as a Source of Carbon in Cholesterol." *Journal of the Elisha Mitchell Scientific Society* 79 (1963): 108.

Edwards, C. H., E. L. Gadsden, and G. A. Edwards. "Utilization of Methionine by the Adult Rat. VIII. Uptake of Methionine-2-C14 and Methionine-Methyl-C14 by Tissues of the Gastrointestinal Tract." *American Journal of Gastroenterology* 40 (1963): 471.

Edwards, C. H., E. L. Gadsden, and G. A. Edwards. "Chromatography of Compounds of Biological Interest on Glass Fiber, Paraffin-Coated, and Untreated Cellulose Paper." *Journal of Chromatography* 2 (1963): 343.

Edwards, C. H., E. L. Gadsden, and G. A. Edwards. "Tomatine and Digitonin as Precipitating Agents in the Estimation of Cholesterol." *Analytical Chemistry* 36 (1964): 420.

Edwards, C. H., E. L. Gadsden, and G. A. Edwards. "Utilization of Methionine by the Adult Rat. VI. Influence of Anesthesia and Surgery on the Uptake of Methionine-2-C14 and Methionine-Methyl-C14." *American Journal of Surgery* 4 (1964): 118.

Edwards, C. H., M. R. Ruffin, E. L. Gadsden, and G. A. Edwards. "Moisture Contents of Rat Tissues." *Journal of the Elisha Mitchell Scientific Society* (spring 1964).

Edwards, C. H., M. R. Ruffin, E. L. Gadsden, and G. A. Edwards. "Effect of Clay and Cornstarch Intake on Women and Their Infants." *Journal of the American Dietetic Association* 44 (1964): 109.

Edwards, C. H., M. R. Ruffin, I. M. Woolcock, and A. W. Rica. "Efficiency of Vegetarian Diets as a Source of Protein for Growth and Maintenance." *Journal of Home Economics* 56 (1964): 164.

Edwards, C. H., E. L. Gadsden, and G. A. Edwards. "Methionine and Homocysteine as Protective Agents against Irradiation Damage." *Metabolism* 13 (1964): 373.

Edwards, C. H., E. L. Gadsden, and G. A. Edwards. "Effect of Irradiation on the Tissue Uptake of Methionine-2-C14 and Methionine-Methyl-C14." *Radiation Research* 22 (1964): 116.

Edwards, C. H. "Behind the Scenes of Nutrition Surveys." *Nutrition News*, October 1964.

Edwards, C. H., G. Hogan, and S. Spahr. "Nutrition Survey of 6,200 Teen Age Youth in Greensboro Public Schools." *Journal of the American Dietetic Association* 45 (1965): 543.

Gadsden, E. L., C. H. Edwards, A. Webb, and G. A. Edwards. "Autoradiographic Patterns of Methionine-2-C14 and Methionine-Methyl-C14 in Tissues." *Journal of Nutrition* 87 (1965): 139–147.

Edwards, C. H., E. L. Gadsden, and G. A. Edwards. "Utilization of Methionine by the Adult Rat. V. Incorporation of Methionine into Tissue Proteins." *Journal of the Elisha Mitchell Science Society* 82 (1965): 12.

Edwards, C. H., L. K. Booker, C. H. Rumph, and S. N. Gnapathy. "Nitrogen Metabolism of Young Men Receiving Diets Containing Wheat and Wheat Supplemented with Pinto Beans, Rice and Peanut Butter." *Cereal Science Today* 11 (1966): 154.

Edwards, C. H., E. S. Thompson, and M. H. Tyson. "Nitrogen Balances and Growth of Rats Fed Vegetable-Protein Diets." *Journal of the American Dietetic Association* 48 (1966): 38–44.

Gnapathy, S. N., L. Booker, and C. H. Edwards. "Plasma Proteins of Adult Men Fed Diets Containing Wheat and Wheat Supplemented with Pinto Beans, Rice, or Peanut Butter." In *Proceedings, Seventh International Congress on Nutrition*, Hamburg, Germany, 1967.

Edwards, G. A., S. S. Rawalay, and C. H. Edwards. "Metabolites of Methionine-S35." In *Proceedings, Seventh International Congress on Nutrition*, Hamburg, Germany, 1967.

Edwards, C. H. "Contribution and Obligations of Preprimary Education in a Democratic Society." In *College Annual*. Mysore, India: Institute of Education, 1968.

Edwards, C. H. "Protein Supplementation and Its Application to Indian Diets." *Journal of Nutrition and Dietetics* 5 (1968): 257.

Edwards, C. H. "The Importance of Teaching Nutrition in Secondary Schools in India." *Indian Journal of Home Science* 2 (1968): 74.

Gnapathy, S. N., L. Booker, R. Craven, and C. H. Edwards. "Retention of Copper, Iron, Molybdenum, Selenium and Zinc in Men Receiving Diets Containing Wheat and Wheat Supplemented with Pinto Beans, Rice or Peanut Butter." In *Proceedings, Eighth International Congress on Nutrition*, Prague, Czechoslovakia, 1969.

Edwards, C. H. "Progress Report on the White House Conference." *Journal of the American Dietetic Association* 55 (1970).

Edwards, C. H. "Taking Up the Gauntlet." In *Proceedings, Ninth Annual Meeting, Association of Administrators of Home Economics,* Kansas City, Mo., 1971.

Edwards, C. H., L. K. Booker, G. H. Rumph, W. G. Wright, and S. N. Gnapathy. "Utilization of Wheat by Adult Man: Nitrogen Metabolism, Plasma Amino Acids and Lipids." *American Journal of Clinical Nutrition* 24 (1971): 181–193.

Edwards, C. H., L. K. Booker, G. H. Rumph, W. G. Wright, and S. N. Gnapathy. "Utilization of Wheat by Adult Men: Vitamins and Minerals." *American Journal of Clinical Nutrition* 24 (1971): 547–555.

Edwards, C. H., G. A. Edwards, and E. G. Jones. "A Method of Correcting for the Absorption of 14C in Animal Tissues." *International Journal of Applied Radiation and Isotopes* 22 (1971): 309–311.

Edwards, C. H., W. D. Wade, M. M. Freeburne, et al. "Formation of Methionine from Alpha-Amino-N-Butyric Acid and 5'-Methylthioadenosine in the Rat." *Journal of Nutrition* 107 (1977): 1927–1936.

Gnapathy, S. N., L. K. Booker, R. Craven, and C. H. Edwards. "Trace Minerals, Amino Acids, and Plasma Proteins in Adult Men Fed Wheat Diets." *Journal of the American Dietetic Association* 78 (1981): 490–497.

Edwards, C. H. "African American Women and Their Pregnancies. Introduction." *Journal of Nutrition* 124 (1994 suppl.): v–viii.

Edwards, C. H., O. J. Cole, U. J. Oyemade, et al. "Maternal Stress and Pregnancy Outcomes in a Prenatal Clinic Population." *Journal of Nutrition* 124 (1994 suppl.): 1006S–1021S.

Edwards, C. H., E. M. Knight, A. A. Johnson, et al. "Demographic Profile, Methodology, and Biochemical Correlates during the Course of Pregnancy." *Journal of Nutrition* 124 (1994 suppl.): 917S–926S.

Edwards, C. H., E. M. Knight, A. A. Johnson, et al. "Multiple Factors as Mediators of the Reduced Incidence of Low Birth Weight in an Urban Clinic Population." *Journal of Nutrition* 124 (1994 suppl.): 927S–935S.

Edwards, C. H., A. A. Johnson, E. M. Knight, et al. "Pica in an Urban Environment." Journal of Nutrition 124 (1994 suppl.): 954S–962S.

Johnson, A. A., E. M. Knight, C. H. Edwards, et al. "Dietary Intakes, Anthropometric Measurements and Pregnancy Outcomes." *Journal of Nutrition* 124 (1994 suppl.): 936S–942S.

Johnson, A. A., E. M. Knight, C. H. Edwards, et al. "Selected Lifestyle Practices in Urban African American Women—Relationships to Pregnancy Outcome, Dietary Intakes and Anthropometric Measurements." *Journal of Nutrition* 124 (1994 suppl.): 963S–972S.

Knight, E. M., B. G. Spurlock, C. H. Edwards, et al. "Biochemical Profile of African American Women during Three Trimesters of Pregnancy and at Delivery." *Journal of Nutrition* 124 (1994 suppl.): 943S–953S.

Knight, E. M., H. James, C. H. Edwards, et al. "Relationships of Serum Illicit Drug Concentrations during Pregnancy to Maternal Nutritional Status." *Journal of Nutrition* 124 (1994 suppl.): 973S–980S.

Nolan, G. H., M. Nahavandi, C. H. Edwards, et al. "Deoxyribonucleic Acid, Ribonucleic Acid, and Protein in the Placentas of Normal and Selected Complicated Pregnancies." *Journal of Nutrition* 124 (1994 suppl.): 1022S–1027S.

West, W. L., E. M. Knight, C. H. Edwards, et al. "Maternal Low Level Lead and Pregnancy Outcomes." *Journal of Nutrition* 124 (1994 suppl.): 981S–986S.

Edwards, C. H. "Emerging Issues in Lifestyle, Social, and Environmental Interventions to Promote Behavioral Change Related to Prevention and Control of Hypertension in the African-American Population." *Journal of the National Medical Association* 87 (1995 suppl.): 642–646.

Anna Cherrie Epps

Cherrie, A. "The Sickling Phenomenon in a College Population." *Journal of the National Medical Association* 55 (1963): 142.

Taylor, C. R., A. Cherrie, and B. Kopp. "Experimental Schistosomiasis. 1. Electrophoretic Studies." *Journal of the National Medical Association* 55 (1963): 208.

Cherrie, A., and A. D. Ferguson. "The Incidence of Hemoglobin S in Human Blood Groups." *Medical Annals of the District of Columbia* 35 (1966): 13–14.

Shelton, T. G., A. L. Cherrie, and K. A. Harden. "Blood and Sarcoidosis." *Journal of the National Medical Association* 58 (1966): 99.

Akdamar, K., A. C. Epps, L. Maumus, and R. Sparks. "Immunoglobulin Changes in Liver Disease (IN)." *Proceedings of the National Council on Alcoholism,* 1971.

Akdamar, K., L. Maumus, A. Cherrie Epps, R. Leach, and S. Warren. "S.H. Antigen in Bile." *Lancet* 1, no. 7705 (1971): 909.

Epps, A. Cherrie. "Immune Response in the Chick Embryo to Grafts." *Proceedings of Oral Research Seminar* 3 (1971).

Akdamar, K., A. C. Epps, L. T. Maumus, and R. D. Sparks. "Immunoglobulin Changes in Liver Disease." *Annals of the New York Academy of Science* 197 (1972): 101–107.

Akdamar, K., L. Maumus, A. Cherrie Epps, R. Font, and R. Sparks. "Serum Alpha-Phetoprotein, Immunoglobulin (IgG: IgM: IgA) and Hepatitis Induced Antigen in Patients with Liver Disease in the New Orleans Area." In *72nd Annual Meeting, American Gastroenterological Association,* 1972. Abstract.

Epps, A. Cherrie, E. J. Blasini, K. Akdamar, and R. D. Sparks. "The Detection of Alpha-1 Fetoprotein." *American Journal of Medical Technology* 38 (1972): 302–305.

Akdamar, K., A. C. Epps, E. Mufdi, and R. Agrawal. "Alpha 1-Fetoprotein in Human Hepatic Bile—A Normal Constituent." In *L'alpha-foeto-proteine,* edited by R. Masseyeff, pp. 173–180. Paris, INSERM, 1974.

Hocking, W., A. C. Epps, and K. Akdamar. "The Detection of an Antigen Present in Gastric Carcinoma." *The American Journal of Digestive Diseases* 19 (1974): 537–546.

Epps, A. C. "The Howard-Tulane Challenge: A Medical Education Reinforcement and Enrichment Program." *Journal of the National Medical Association* 67 (1975): 55–60.

Epps, A. C. "Improving Methods of Early Minority Student Identification and Education Preparation for Dental School." In *Workshop on Minority Dental Student Recruitment, Retention, and Education, 1975,* edited by R. W. Sumnicht and C. F. Anderson, pp. 37–40. 27 May 1977. Monograph.

Cadbury, W., C. Cadbury, A. C. Epps, and J. Pisano. *Medical Education: Responses to a Challenge.* Mount Kisco, N.Y.: Futura Publishing, 1979.

Epps, Anna Cherrie, Joseph C. Pisano, and J. Allender. *Strategies to Increase Participation of Minorities in Medicine.* AAAS Selected Symposium 66, 1982.

Pisano, J. C., and A. C. Epps. "The Impact of MCAT Intervention Efforts on Medical Student Acceptance Rates." *Journal of the National Medical Association* 75 (1983): 773–777.

Pisano, J.C., and A. C. Epps. "The Impact of a Medical-School-Based Summer Program on the Acceptance of Minority Undergraduate Students into Health Professional Schools." *Journal of the National Medical Association* 75 (1983): 17–23.

Epps, A. C., and J. C. Pisano. "Medical Schools and Medical Education Cost Reduction." *Journal of Medical Education* 59 (1984): 439–440.

Pisano, J.C., and A. C. Epps. *MEdREP at Tulane: Effectiveness of a Medical Education Reinforcement and Enrichment Program for Minorities in the Health Professions*. Mount Kisco, N.Y.: Futura Publishing, 1985.

Boucree, M. C., A. C. Epps, and J. C. Pisano. "Parental Educational Background and Residency Training Selection of Minority and Nonminority Medical Students." *Journal of the National Medical Association* 80 (1988): 23–33.

Epps, A. Cherrie. "Recruitment and Retention of Minority Students in American Medical Schools." *Journal of the Student National Medical Association* 1, no. 1 (1989).

Epps, A. C. "National Boards and the Minority Medical Student." *Journal of the National Medical Association* 82 (1990): 161–163.

Epps, Anna Cherrie. *Increasing Minority Participation in the Health Professions—The Problem of Underrepresentation and an Inventory and Analysis of Effective Strategies and Problems*. Washington, D.C.: Institute of Medicine, National Academy of Sciences, March 1993.

Epps, A. C., M. T. Cureton-Russell, and H. Kitzman. "Effective Strategies and Programs to Increase Minority Participation in the Health Professions for the Twenty-First Century." *Journal of the Association for Academic Minority Physicians* 4 (1993): 116–126.

Bettye Washington Greene

Greene, Bettye Washington. "Determination of Size Distributions in Emulsions Using Scattering Spectra." *NASA Accession*, 1965.

Heller, W., M. H. Cheng, and B. W. Greene. "Surface Tension Measurements by Means of a 'Microcone Tensiometer.'" *Journal of Colloid and Interface Science* 22 (1966): 179–194.

Greene, B. W., D. P. Sheetz, and T. D. Filer. "In Situ Polymerization of Surface-Active Agents on Latex Particles. 1. Preparation and Characterization of Styrene-Butadiene Latexes." *Journal of Colloid and Interface Science* 32, no. 1 (1970): 90.

Greene, B. W., and D. P. Sheetz. "In Situ Polymerization of Surface-Active Agents on Latex Particles. 2. Mechanical Stability of Styrene Butadiene Latexes." *Journal of Colloid and Interface Science* 32, no. 1 (1970): 96.

Greene, B. W., and F. L. Saunders. "In Situ Polymerization of Surface-Active Agents on Latex Particles. 3. Electrolyte Stability of Styrene-Butadiene Latexes." *Journal of Colloid and Interface Science* 32, no. 3 (1970): 393.

Greene, B.W., and F. L. Saunders. "Effect of Added Salt on Adsorbability of a Synthetic Polyelectric." *Journal of Colloid and Interface Science* 37, no. 1 (1971): 144.

Greene, B. W. "Quantitative Determination of Surface Carboxyl Groups in Vinyl Acid Modified Styrene-Butadiene Copolymer Latexes." In *Abstracts of Papers of the American Chemical Society*, p. 86. Washington, D.C.: American Chemical Society, 1972.

Greene, B.W., and F. L. Saunders. "Quantitative Determination of Surface Carboxyl Groups in Vinyl Acid Modified Styrene-Butadiene Copolymer Latexes. 1. Latexes Prepared with Acrylic Acid." *Journal of Colloid and Interface Science* 43, no. 2 (1973): 449–461.

Greene, B.W., and F. L. Saunders. "Quantitative Determination of Surface Carboxyl Groups in Vinyl Acid Modified Styrene-Butadiene Copolymer Latexes. 2. Latexes Prepared with Methacrylic Acid." *Journal of Colloid and Interface Science* 43, no. 2 (1973): 462–472.

Greene, B. W., and A. S. Reder. "Electrokinetic and Rheological Properties of Calcium Carbonate Dispersions Used in Paper Coatings." *TAPPI Journal* 57, no. 5 (1974): 101–106.

Greene, B. W., and A. S. Reder. "Electrokinetic and Rheologic Properties of Calcium-Carbonate Dispersion Paper Coatings." *Papier* 29, no. 4 (1975): 156.

Gloria Conyers Hewitt

Hewitt, Gloria C. "The Existence of Free Unions in Classes of Abstract Algebras." *Proceedings American Mathematics Society* 14 (1963): 417–422.

Hewitt, Gloria C. "Limits in Certain Classes of Abstract Algebras." *Pacific Journal of Mathematics* 22 (1967): 109–115.

Hewitt, Gloria C. "Women in Mathematics." Monthly, *MAA*, November 1971. Abstract.

Hewitt, Gloria C. "A One Model Approach to Group Theory." Report, University of Montana, 1978.

Hewitt, Gloria C. "Emmy Noether's Notions of Finiteness Conditions—Revisited." Report, University of Montana, 1979.

Hewitt, Gloria C. "On $\aleph$-Noetherian Conditions." *Notices of the American Mathematical Society* 26 (January 1979): A-55.

Hewitt, Gloria C. "The Status of Women in Mathematics." *Annals of the New York Academy of Science* 323 (1979).

Hewitt, Gloria C. "Characterizations of Generalized Noetherian Rings." *Acta Mathematica Hungarica* 53 (1989).

Ruth Winifred Howard

Howard, Ruth W. "Fantasy and Play Interview." *Character and Personality* 13 (1944): 151–165.

Howard, Ruth W. "Intellectual and Personality Traits of a Group of Triplets." *Journal of Psychology* 21 (1946): 25–36.

Howard, Ruth W. "The Developmental History of a Group of Triplets." *Journal of Genetic Psychology* 70 (1947): 191–204.

Howard, Ruth W. *Faith of the Young Child.* New York: Hearthstone, 1955.

Howard, Ruth W. "Predicting Success in Nurse Training at Provident Hospital." Unpublished manuscript, 1963.

Howard, Ruth W. "Two Early Black Psychologists: Beckham and Howard." A speech reprinted 12 August 1976.

Deborah J. Jackson

Jackson, D. J. "The Superconducting Energy Gap in Single Crystal Tantalum." Thesis, Massachusetts Institute of Technology, 1974.

Bianconi, A., D. J. Jackson, and K. Monahan. "Intrinsic Luminescence Excitation Spectrum and Extended X-Ray Absorption Fine Structure above the K-Edge in CaF_2." *Physical Review* B 17 (1978): 2021.

Jackson, D. J., J. E. Lawler, and T. W. Hansch. "Broadly Tunable Pulsed Laser for the Infrared Using Color Centers." *Optics Communications* 29 (1979): 352.

Lawler, J. E., A. L. Ferguson, J. E. M. Goldsmith, D. J. Jackson, and A. L. Schawlow. "Doppler-Free Intermodulated Optogalvanic Spectroscopy." *Physical Review Letters* 42 (1979): 1046.

Lawler, J. E., A. L. Ferguson, J. E. M. Goldsmith, D. J. Jackson, and A. L. Schawlow. "Doppler-Free Optogalvanic Spectroscopy." In *Proceedings of the International Conference on Laser Spectroscopy IV*, edited by H. Walther and K. W. Rothe, p. 188. Rottach-Egren, Germany: Springer-Verlag, 1979.

Carlsen, N., D. J. Jackson, A. L. Schawlow, M. Gross, and S. Haroche. "Superradiance Triggering Spectroscopy." *Optics Communications* 32 (1980): 350.

Jackson, D. J. "Excited State Spectroscopy on Helium Using a Color Center Laser." Thesis, GL Report No. 3207, Stanford University, 1980.

Jackson, D. J., H. Gerhardt, and T. W. Hansch. "Doppler-Free Optogalvanic Spectroscopy in the Helium Positive Column Using an F-Center Laser." *Optics Communications* 33 (1980): 51.

Jackson, D. J. "Atomic Spectroscopy: Teaching Old Atoms New Tricks." In *Proceedings of the Annual Day of Lectures of the National Society of Black Physicists*, edited by H. B. White. Batavia, Ill.: Fermi National Laboratory, 1981.

Jackson, D. J., and J. J. Wynne. "The Case of the Disappearing 6s Multiphoton Ionization Signal in Xenon." *Bulletin of the American Physical Society*, D.E.A.P. (1982): 1316.

Jackson, D. J., and J. J. Wynne. "Interference Effects between Different Optical Harmonics." *Physical Review Letters* 49 (1982): 543.

Jackson, D. J., and J. J. Wynne. "What Is the Signature of a Two-Photon Laser?" In *Proceedings of the 12th International Quantum Electronics Conference*, p. 238. Munich, Germany, 1982.

Jackson, D. J., J. J. Wynne, and P. H. Kes. "The Case of the Disappearing 6s Multiphoton Ionization Signal in Xe I." In *Proceedings of the 12th International Quantum Electronics Conference*, Munich, Germany, 1982.

Jackson, D. J., and J. J. Wynne. "Applications of New Interference Effects in Optical Harmonic Generation." In *Proceedings of the 1983 Conference on Lasers and Electro-Optics*, p. 218. Baltimore, May 1983.

Jackson, D. J., J. J. Wynne, and P. H. Kes. "Resonance-Enhanced Multiphoton Ionization: Interference Effects Due to Harmonic Generation." *Physical Review* A 28 (1983): 781.

Wynne, J. J., and D. J. Jackson. "The Marriage of Multiphoton Excitation Spectroscopy and Optical Harmonic Generation." In *Proceedings of the Sixth International Conference on Laser Spectroscopy*, Interlaken, Switzerland, June 1983.

Jaung, F. Y., W. Li, P. K. Bhattacharya, U. Das, A. Chin, D. J. Jackson, and D. L. Persechini. "III-V Superlattice Photodiodes." In *Proceedings of the 13th International Symposium on Gallium Arsenide and Related Compounds*, Las Vegas, September 1986.

Jackson, D. J., and D. L. Persechini. "Large Area High Speed PIN Detectors in GaAs." *Electronics Letters* 21 (1986): 202.

Jackson, D. J., and D. L. Persechini. "Monolithically Integrable High Speed Photodetectors." In *Proceedings of the SPIE Conference on High Frequency Optical Communication*, vol. 716, p. 104. Cambridge, Mass., September 1986.

Jackson, D. J., and D. L. Persechini. "Monolithically Integrable Lateral PIN (MILPIN) Detectors." In *OFC/IGWO Conference Joint Session on Detectors*, Atlanta, Ga., February 1986.

Jackson, D. J., and D. L. Persechini. "Waveguide Formation in Bulk GaAs and InP Materials." *Electronics Letters* 22 (1986): 44.

Jackson, D. J., J. Y. Josefowicz, D. B. Rensch, and D. L. Persechini. In *Proceedings of SPIE*, vol. 839 OE Fibers 87. San Diego, Calif., August 1987.

Jackson, D. J., and D. L. Persechini. "Applications of Monolithic Detectors." In *Proceedings of the SPIE Conference on Optical Technologies for Space Communications Systems*, vol. 756, p. 150. Los Angeles, January 1987.

Jackson, D. J., J. Y. Josefowicz, D. B. Rensch, and D. L. Persechini. "Detectors for Monolithic Optoelectronics." *Fiber and Integrated Optics* 7 (1988): 229.

Huth, G., E. Bedrosian, A. Brewer, D. Jackson, and R. Frick. *SIE Project Outreach Evaluations: Communications (U)*. WD-5197-AF/NASA. December 1990.

Jackson, D. J. *Two-Way Laser Communications from Military Satellites (U)*. IN-25915-DARPA. September 1991.

Jackson, D. J. *A Structural Approach to the Photonic Processor (U)*. N-3399-RC. January 1992.

Clauss, R. C., M. I. Herman, D. L. Rascoe, H. Hemmati, S. S. Shiak, P. Estabrook, M. K. Sue, D. J. Jackson, and J. R. Lesh. *Final Report: A Point Design Comparison of Miniature Ka-band and Optical Communications Systems for Planetary Missions*. JPLD-11583. Jet Propulsion Laboratory, October 1993.

Jackson, D. J. *High Data Rate Photonic Switching for Satellite Crosslinks*. JPL D-11983. Jet Propulsion Laboratory, August 1994.

Jackson, D. J. "A Systems Approach to Photonic Processors." *Applied Optics* 33 (1994): 5451–5466.

PRESENTATIONS

Tangonan, G. L., V. Jones, J. Pikulski, D. J. Jackson, D. L. Persechini, G. Thornebooth, and S. R. Forrest. "An 8×8 Optoelectronic Cross-Bar Switch." Paper presented at the Conference on Optical Fiber Communications, New Orleans, 25–8 January 1988.

PATENTS

Jackson, D. J., and M. D. Clark. "Waveguide Fabrication in Bulk Materials." U.S. Patent 4,733,927. 23 September 1987.

CLASSIFIED DOCUMENTS

Jackson, D.J. IN-25842, upgraded to WD-5369-DARPA. February 1990.

Jackson, D. J. N-3341. December 1991.

Lempert, R., P. Jamison, R. Greene, M. Gallagher, G. Daniels, T. Blaschke, G. Parnell, D. Jackson, and R. Smith. R-4122-AF. February 1992.

Schaffer, M. B., K. Amer, E. Bedrosian, A. Brewer, M. Gallagher, J. Grossman, D. Jackson, W. Sollfrey, and G. Taylor. R-4109-DARPA. February 1992.

Jacquelyne Johnson Jackson

Clarke, Jacquelyne (Johnson). *These Rights They Seek: A Comparison of Goals and Techniques of Local Civil Rights Organizations*. Washington, D.C.: Public Affairs Press [1962].

Jackson, J. J. "Public Relations That Would Be Helpful to Local Public Relations Committees." *Bulletin of the Academy of General Dentistry* 9 (September 1966): 9.

Jackson, J. J. "Social Gerontology and the Negro: A Review." *Gerontologist* 7 (September 1967): 168–178.

Jackson, J. J. "Aged Negroes: Their Cultural Departures from Statistical Stereotypes and Rural-Urban Differences." *Gerontologist* 10 (summer 1970): 140–145.

Jackson, J. J. "Negro Aged: Toward Needed Research in Social Gerontology." *Gerontologist* 11 (spring 1971): S52–57.

Jackson, Jacquelyne Johnson, editor. *Proceedings*. Research Conference on Minority Group Aged in the South at Durham, N.C., 1971. Durham, N.C.: Center for the Study of Aging and Human Development, Duke University Medical Center, 1972.

Jackson, J. J. "Social Impacts of Housing Relocation upon Urban, Low-Income Black Aged." *Gerontologist* 12 (spring 1972): 32–37.

Jackson, J. J. "Face to Face, Mind to Mind, It Sho' Nuff Ain't No Zombie Jamboree." *Journal of the National Medical Association* 64 (March 1972): 145–150.

Jackson, J. J. "The National Center on Black Aged: A Challenge to Gerontologists." *Gerontologist* 14 (June 1974): 194, 196.

Jackson, Jacquelyne Johnson, consultant. *Old, Black, and Alive!—Some Contrasts in Aging.* 16 mm, 27 min. National Center on Black Aged/New Film Co., 1974.

Jackson, Jacquelyne Johnson, editor. *Aging Black Women: Selected Readings for NCBA.* [Washington: College and University Press], 1975.

Jackson, J. J. "Some Special Concerns about Race and Health: An Editorial Finale." *Journal of Health and Social Behavior* 16 (December 1975): 342, 429.

Jackson, J. J. "New Roles for Women in Health Care Delivery: A U.S. Response to the Cameroonian Experience." In *Proceedings of the International Conference on Women in Health,* pp. 133–135. Washington, D.C.: Department of Health, Education, and Welfare, 1976.

Jackson, Jacquelyne Johnson. "But Where Are the Men?" *Black Scholar* 2 (March 1977): 30–41.

Jackson, Jacquelyne Johnson. *Minorities and Aging.* Belmont, Calif.: Wadsworth Publishing, 1980.

Cooper, Catherine, Jacquelyne Jackson, Margarita Azmitia, Edward Lopez, and Nora Dunbar. "Bridging Students' Multiple Worlds: African American and Latino Youth in Academic Outreach Programs." In *LMRI Anthology,* edited by Reynaldo Macías and Reyna García Ramos. Santa Barbara, Calif.: University of California Linguistic Minority Research Institute, 1985.

Patrick, S. L., and J. J. Jackson. "Further Examination of the Equity Sensitivity Construct." *Perceptual and Motor Skills* 73 (December 1991): 1091–1106.

Jackson, Jacquelyne J., and James S. Jackson, compilers. *Ethnogerontology and American Blacks.* Washington, D.C.: Association for Gerontology in Higher Education, 1992.

Jackson, Jacquelyne Johnson. *Social Characteristics of Black Elders in Pitt County, North Carolina.* Raleigh, N.C.: Shaw Divinity School, [1993].

Jackson, Jacquelyne Johnson. *The MTEARBE Black Church Manual on Eldercare for At-Risk Black Elders in Pitt County, North Carolina.* Raleigh, N.C.: Shaw Divinity School, [1993].

Jackson, Jacquelyne Johnson. *Final Project Report Ministerial Training in Eldercare to At-Risk Black Elders, Administration on Aging Grant Number 90AT0516.* Raleigh, N.C.: Shaw Divinity School, [1994].

Charlene Drew Jarvis

Blake, L., C. D. Jarvis, and M. Mishkin. "Pattern Discrimination Thresholds after Partial Inferior Temporal and Lateral Striate Lesions in Monkeys." *Society for Neuroscience Abstracts* (1974): 174.

Jarvis, Charlene D. "Visual Discrimination and Spacial Localization Deficits after Lesions of the Tectofugal Pathway in Pigeons." *Brain, Behavior and Evolution* 9 (1974): 195–228.

Robinson, D. C., and C. D. Jarvis. "Superior Colliculus Neurons Studied during Head and Eye Movements of the Behaving Monkey." *Journal of Neurophysiology* 37 (1974): 533–540.

Jarvis, C. D., and M. Mishkin. "Responses of Inferior Temporal Neurons to Visual Discriminanda." *Society for Neuroscience Abstracts* 1 (1975): 61.

Jarvis, C. D., and M. Mishkin. "Responses of Cells in the Inferior Temporal Cortex of Monkeys during Discrimination Reversal." *Society for Neuroscience Abstracts* 3 (1977): 564.

Jarvis, C. D., and M. Mishkin. "Responses of Inferior Temporal Neurons to Visual Cortical Lesions in Monkeys." *Brain Research* 120 (1977): 209–220.

Jarvis, C. D., M. Mishkin, M. Shinohara, O. Sakurada, M. Miyaoka, and C. Kennedy. "Mapping the Primate Visual System with the 2-(3143C)deoxyglucose Technique." *Neuroscience Abstracts* 4 (1978): 632.

Kennedy, C., C. D. Jarvis, O. Sakurada, and M. Mishkin. "A Delineation of the Visually Responsive Loci of the Temporal Lobe by Means of 2-(3143C)deoxyglucose." *Neurology* 28 (1978): 366.

Shinohara, M., M. Miyaoka, O. Sakurada, C. D. Jarvis, M. Mishkin, C. Kennedy, and L. Sokoloff. "Mapping the Primate Visual System with 2-(3143C)deoxyglucose." *Acta Neurology Scandinavia* 60 (1979): 14–15.

Kennedy, C., M. Miyaoka, S. Suda, K. Macko, C. D. Jarvis, M. Mishkin, and L. Sokoloff. "Local Metabolic Responses in Brain Accompanying Motor Activity." *Annals Neurology* 8 (1980): 90.

Macko, K., C. D. Jarvis, C. Kennedy, M. Miyaoka, M. Shinohara, L. Sokoloff, and M. Mishkin. "Mapping the Primate Visual with (2-¹⁴C) Deoxyglucose." *Science* 218 (1982): 392–396.

Katherine Coleman Goble Johnson

Skopinski, T. H., and Katherine G. Johnson. *Determination of Azimuth Angle of Burnout for Placing a Satellite over a Selected Earth Position.* Technical Note D-233. Springfield, Va.: National Aeronautics and Space Administration, Scientific and Technical Information Branch, 1959.

Westrick, Gertrude C., and Katherine G. Johnson. *The Orbital Behavior of the Echo I Satellite and Its Rocket Casing during the First 500 Days.* Technical Note D-1366. Springfield, Va.: National Aeronautics and Space Administration, Scientific and Technical Information Branch, 1962.

White, Jack A., and Katherine G. Johnson. *Approximate Solutions for Flight-Path Angle of a Reentry Vehicle in the Upper Atmosphere.* Technical Note D-2379. Springfield, Va.: National Aeronautics and Space Administration, Scientific and Technical Information Branch, 1964.

Blackshear, Thomas W., and Katherine G. Johnson. A *Study of Solar-System Geometric Parameters for Use as Interplanetary Navigation Aids.* Technical Note D-2890. Springfield, Va.: National Aeronautics and Space Administration, Scientific and Technical Information Branch, July 1965.

Hamer, Harold A., Katherine G. Johnson, and Thomas W. Blackshear. *Midcourse-Guidance Procedure with Single Position Fix Obtained from Onboard Optical Measurements.* Technical Note D-4246. Springfield, Va.: National Aeronautics and Space Administration, Scientific and Technical Information Branch, December 1967.

Hamer, Harold A., and Katherine G. Johnson. *Effect of Gravitational Model Selection on Accuracy of Lunar Orbit Determination from Short Data Arcs.* Technical Note D-5105. Springfield, Va.: National Aeronautics and Space Administration, Scientific and Technical Information Branch, March 1969.

Hamer, Harold A., and Katherine G. Johnson. *An Approach Guidance Method Using a Single Onboard Optical Measurement.* Technical Note D-5963. Springfield, Va.: National Aeronautics and Space Administration, Scientific and Technical Information Branch, October 1970.

Hamer, Harold A., and Katherine G. Johnson. *Midcourse and Approach Guidance Requirements for Simplified Control of Moon-to-Earth Trajectories.* Technical Note D-6343. Springfield, Va.: National Aeronautics and Space Administration, Scientific and Technical Information Branch, July 1971.

Hamer, Harold A., and Katherine G. Johnson. *Fixed-Angle Translunar Guidance Procedures Using Onboard Optical Measurements*. Technical Note D-6461. Springfield, Va.: National Aeronautics and Space Administration, Scientific and Technical Information Branch, September 1971.

Hamer, Harold A., and Katherine G. Johnson. *Simplified Interplanetary Guidance Procedures Using Onboard Optical Measurements*. Technical Note D-6752. Springfield, Va.: National Aeronautics and Space Administration, Scientific and Technical Information Branch, May 1972.

Hamer, Harold A., and Katherine G. Johnson. *Effects of Errors on Decoupled Control Systems*. Technical Paper 1184. Springfield, Va.: National Aeronautics and Space Administration, Scientific and Technical Information Branch, July 1978.

Young, J. W., A. A. Schy, and Katherine G. Johnson. "Prediction of Jump Phenomena in Rotationally-Coupled Maneuvers of Aircraft, Including Nonlinear Aerodynamic Effects." *Journal of Guidance and Control* 1 (1978): 26–31.

Tabak, D., A. A. Schy, D. P. Giesy, and K. G. Johnson. "A Multiple Objective Optimization Approach to Aircraft Control Systems Design." *Automatica* 15 (September 1979): 595–600.

Hamer, Harold A., and Katherine G. Johnson. *Decoupled Control of a Long Flexible Beam in Orbit*. Technical Paper 1740. Springfield, Va.: National Aeronautics and Space Administration, Scientific and Technical Information Branch, December 1980.

Young, John W., Harold A. Hamer, and Katherine G. Johnson. *Pseudosteady-State Analysis of Nonlinear Aircraft Maneuvers*. Technical Paper 1758. Springfield, Va.: National Aeronautics and Space Administration, Scientific and Technical Information Branch, December 1980.

Young, John W., Harold A. Hamer, and Katherine G. Johnson. *Decoupled Control Analysis of a Large Flexible Space Antenna with Linear Regulator Comparisons*. Technical Paper 2293. Springfield, Va.: National Aeronautics and Space Administration, Scientific and Technical Information Branch, May 1984.

Hamer, Harold A., Katherine G. Johnson, and John W. Young. *Decoupled and Linear Quadratic Regulator Control of a Large Flexible Space Antenna with an Observer in the Control Loop*. Technical Paper 2484. Springfield, Va.: National Aeronautics and Space Administration, Scientific and Technical Information Branch, 1985.

Hamer, Harold A., and Katherine G. Johnson. *Effects of Model Error on Control of Large Flexible Space Antenna with Comparisons of Decoupled and Linear Quadratic Regulator Control Procedures*. Technical Paper 2604. Springfield, Va.: National Aeronautics and Space Administration, Scientific and Technical Information Branch, 1986.

PRESENTATIONS

Schy, A. A., W. M. Adams Jr., and Katherine G. Johnson. "Computer-Aided Design of Control Systems to Meet Any Requirements." Paper presented at the 17th Meeting of the AGARD Guidance and Control Panel, Geilo, Norway, September 1973.

Young, J. W., A. A. Schy, and Katherine G. Johnson. "Prediction of Jump Phenomena in Rotationally-Coupled Maneuvers of Aircraft, Including Nonlinear Aerodynamic Effects." Paper presented at the AIAA 4th Atmospheric Flight Mechanics Conference, Hollywood, Fla., August 1977.

Tabak, D., A. A. Schy, D. P. Giesy, and K. G. Johnson. "A Multiple Objective Optimization Approach to Aircraft Control Systems Design." Presented at the 17th Annual IEEE Conference on Decision and Control, San Diego, Calif., January 1979.

Schy, A. A., D. P. Giesy, and K. G. Johnson. "Pareto-Optimal Multi-Criteria Design of

Airplane Control Systems." Presented at the 1980 Joint Automatic Control Conference, San Francisco, 13–15 August 1980.

Young, J. W., A. A. Schy, and Katherine G. Johnson. "Pseudosteady State Analysis of Nonlinear Aircraft Maneuvers." Presented at the AIAA 7th Atmospheric Flight Mechanics Conference, Danvers, Mass., 11–13 August 1980.

Reatha Clark King

King, Reatha Clark, O. J. Kleppa, and L. S. Hersh. "Studies of Fused Salts. III. Heats of Mixing Silver Nitrate Mixtures." *Journal of Chemistry and Physics* 35 (1961): 1975.

King, Reatha Clark, and O. J. Kleppa. "Heats of Formation of the Solid Solutions of Zinc, Gallium, and Germanium in Copper." *Acta Metallurgica et Materialia* 10 (1962): 1183.

King, Reatha Clark, and G. T. Armstrong. "Heat of Combustion and Heat of Formation of Aluminum Carbide." *National Bureau of Standards Journal of Research* 68A (1964): 661.

King, R. C., and O. J. Kleppa. "A Thermochemical Study of Some Selected Laves Phases." *Acta Metallurgica et Materialia* 12 (1964): 87–97.

King, Reatha Clark. "The Heat of Formation of Aluminum Carbide." *National Bureau of Standards Technical News Bulletin*, February 1965.

King, Reatha Clark, and G. T. Armstrong. "Constant Pressure Flame Calorimetry with Fluorine." II. The Heat of Formation of Oxygen Difluoride." *National Bureau of Standards Journal of Research* 72A (1968): 113.

King, Reatha Clark, and G. T. Armstrong. "Fluorine Flame Calorimetry. III. The Heat of Formation of Chlorine Trifluoride." *National Bureau of Standards Journal of Research* 74A (1970): 769.

King, Reatha Clark, and G. T. Armstrong. "Fluorine Flame Calorimetry." Chapter 15 in *Experimental Thermochemistry*, edited by S. Sunner. International Union of Pure and Applied Chemistry, Division of Physical Chemistry, Commission on Thermodynamics and Thermochemistry, 1976.

King, Reatha Clark. *Equity Issues for Women in Higher Education: A Progress Report.* St. Paul, Minn.: Metropolitan State University, President's Office, 1984.

King, Reatha Clark. *Two Risks, Perfectly Suited for Each Other.* St. Paul, Minn.: Metropolitan State University, President's Office, 1984.

Flemmie Pansey Kittrell

Kittrell, Flemmie P. "A Study of the Situations Little Children Meet, the Responses They Make, and the Guidance They Receive." Ph.D. dissertation, Cornell University, 1935.

Kittrell, Flemmie P. "Homemaking Education in the College Program." *The Quarterly Review of Higher Education among Negroes* (January 1937).

Kittrell, F. P., and Rosetta L. Quach. "A Comparative Study of the Proteins in Soybean Meal and Cows' Milk," 1946.*

Kittrell, Flemmie P. *Food and Nutrition Survey of Liberia, West Africa.* ER&T 241. Washington, D.C.: Federal Extension Service, U.S. Department of Agriculture, 1963 (reprinted in 1967 and 1968 for Extension Services in West Africa). Originally written for U.S. Department of State, African Division in 1948.

Kittrell, Flemmie P. "The Contribution of Research to the Field of Child Development and Family Relations." *International Journal of Home Economics*, 1949.*

Kittrell, Flemmie P. "The Negro Family as a Health Agency." *Journal of Negro Education* 18, no. 3 (summer 1949): 422–428.

Kittrell, Flemmie P. "University of Baroda Establishes Home Economics in Higher Education." *Journal of Home Economics* 44 (February 1952): 97–100.

Kittrell, Flemmie P. "Science in Home Economics." In *Proceedings of All-India Home Science Association,* Madras, India, 1954.*

Kittrell, Flemmie P. "Frontiers in Uganda." *Young Women's Christian Association Magazine,* December 1958.

Kittrell, F. P., and Majumder. "Changes in the Eating Habits of Foreign Born Children as Observed in the Nursery School." In *Research Related to Children.* Washington, D.C.: Howard University/U.S. Department of Health, Education and Welfare, 1958.

Kittrell, Flemmie P. *Survey on Nutrition Training in India.* New York: United Nations (FAO), Food and Agriculture, 1960.

Kittrell, Flemmie P. "International Education with Special Reference to Home Economics." Home Economics Advisory Council Meeting, Cornell University, 4 May 1962.*

Kittrell, F. P., and M. Leolia Spaugh. "The School Lunch Program in the Elementary Schools of the District of Columbia." 1962.*

Kittrell, F. P., and Hazel S. Holton. "A Study of the Food Habits and Food Preferences of a Group of Obese and Non-Obese Women." 1963.*

Kittrell, Flemmie P. "Home Economics Philosophy Underlying the New Home Economics Building." *Howard University Magazine,* March 1963.

Kittrell, Flemmie P. "Family Needs I Have Seen around the World." *Methodist Woman,* June 1963.

Kittrell, F. P. "Nutritional Status of Pre-School Children in Washington, D.C. and Parental Involvement." 1965.*

Kittrell, Flemmie P. "The Family Is Central, Cornerstone for Human Development." *Journal of the American Association of University Women,* March 1966.

Kittrell, Flemmie P. *A Nursery School Program within Day Care Hours for Culturally Deprived Children and Parents.* HE-119. Washington, D.C.: Division of Home Economics, Federal Extension Service, U.S. Department of Agriculture, October 1967.

Kittrell, Flemmie P. "Supplementation of Protein Diets with Selected Essential Amino Acids." 1967.*

Kittrell, Flemmie P. *Interrelated Services for the Planned Community.* Montpelier, Vt.: Johnson State College, August 1968.

Kittrell, Flemmie P. "Enriching the Preschool Experience of Children from Age 3." Department of Health, Education and Welfare, Social Rehabilitation Service, Children's Bureau, *Children,* July-August 1968.

Kittrell, Flemmie P., Ivor Kraft, and Jean Fuschillo. "Prelude to School: An Evaluation of an Inner-City Preschool Program." *Children's Bureau Research Reports,* No. 3 (1968).

Kittrell, Flemmie P. *Nutritional Status of Pre-School Children and Parental Involvement.* Research Project with Low Income Families. Children's Bureau Report, 1968.

Kittrell, Flemmie P. "Growing Up." *National 4-H Club Foundation Journal,* 1968.

Kittrell, Flemmie P., and Lola M. Lewis. "Various Federal Feeding Programs for the Poor in the United States, 1933–1969." *Journal of Home Economics,* 1969.

Kittrell, F. P., and Patricia A. Gentry. "The Contributions of the Children's Bureau to Child Nutrition from 1912 Onward." 1969.*

Kittrell, Flemmie P. "Program for the Implementation of a School of Human Ecology." In-house report, Howard University, 8 June 1970.*

Kittrell, Flemmie P. "A Preliminary Report, The Parent and Child Center." In-house report, Howard University, Kittrell, Chairman, 4 January 1971.*

Kittrell, Flemmie P. "Proposal for a Program Leading to the Doctor of Philosophy Degree in Nutrition." In-house proposal, Howard University, 6 May 1971.*

PRESENTATIONS

Kittrell, Flemmie P. "The Supplement of Protein Poor Diets with Selected Essential Amino Acids." Paper presented at the Second World Food Congress, FAO, The Hague, Netherlands, June 1970.

Kittrell, Flemmie P. "Report on Invitational Visit to Zaire." USAID Mission, 14 May–20 June 1972.

Kittrell, Flemmie P. "Regional Seminar—Child of the North." Presentation at University of Alaska at Fairbanks, 11–18 July 1973.

Margaret Morgan Lawrence

Lawrence, Margaret Morgan. *The Mental Health Team in the Schools.* New York: Behavioral Publications, 1971.

Lawrence, Margaret Morgan. *Young Inner City Families: The Development of Ego Strength under Stress.* New York: Behavioral Publications, 1975.

Lawrence, Margaret Morgan. *Out of Chaos: Living Stones.* Unpublished manuscript, 1990.*

Katheryn Emanuel Lawson

Lawson, Katheryn Emanuel. "Optical Absorption Spectroscopy of Transition Metal Complexes."*

Lawson, Katheryn E. *Infrared Absorption of Inorganic Substances.* New York: Reinhold, 1961.

Carolyn R. Mahoney

Mahoney, Carolyn R. "On the Unimodality of Independent Set Numbers of a Class of Matroids." *Journal of Combinatorial Theory* Series B, 39 (1985): 77–85.

Mahoney, Carolyn R., and Franklin Demana. "Filling the Math and Science Pipeline with Young Scholars." Special Issue on Women in Mathematics. *Notices of the American Mathematical Society* 38, no. 7 (February 1991).

Vivienne Lucille Malone Mayes

Mayes, Vivienne Malone. *Some Steady State Properties of $(\int_0^x f(t)dt)/f(x)$.* Department of Mathematics Preprint Series 80. Norman: University of Oklahoma, 1968.

Mayes, Vivienne Malone. "Some Steady State Properties of $(\int_0^x f(t)dt)/f(x)$." Proceedings of the American Mathematical Society 22 (1969): 672–677.

Mayes, Vivienne Malone, and Howard Rolf. *Pre-Calculus; Algebraic and Trigonometric Functions.* Individual Learning Systems Publication, 1971.

Mayes, Vivienne Malone. "Some Properties of the Leininger Generalized Hausdorff Matrix." *Houston Journal of Mathematics* 6 (1989): 287–299.

Lenora Moragne

Moragne, Lenora. *Your Diet Manual.* Hines, Ill.: Veterans Administration Hospital Press, 1955.

Moragne, Lenora. *Manual of Diets.* Evanston, Ill.: Garnett Printing Service, 1957.

Moragne, Lenora, Karla Longree, and James C. White. "Heat Transfer in Refrigerated Foods." 33rd Annual Report, *New York State Milk Sanitarians,* December 1959.

Moragne, Lenora, Karla Longree, and James C. White. "Heat Transfer in White Sauce

Cooled in Flowing Water." *Journal of the American Dietetic Association* 35 (1959): 1275–1282.

Moragne, Lenora, and Karla Longree. "Defrosting Times of Frozen Turkeys." *School Lunch Journal* 15 (1960): 23–26.

Moragne, Lenora, Karla Longree, and James C. White. "Cooling Starch-Thickened Food Items with Cold Tube Agitation." *Journal of Milk and Food Technology* 23 (1960): 330–336.

Moragne, Lenora, Karla Longree, and James C. White. "The Effect of Some Selected Factors on the Cooling of Food under Refrigeration." *Journal of Milk and Food Technology* 23 (1960): 141–150.

Moragne, Lenora, Karla Longree, and James C. White. "Cooling Custards and Puddings with Cold-Tube Agitation." *Journal of Milk and Food Technology* 24 (1961): 207–210.

Moragne, Lenora, and Karla Longree. "Defrosting Times of Frozen Turkeys." *Hospital Management* 92 (1961): 64–67.

Moragne, Lenora, Karla Longree, Nancy Lawrence Fuller, and James C. White. "Time-Temperature Relationships of Beef Patties Made with Whole Egg Solids." *Journal of Milk and Food Technology* 25 (1963): 274–276.

Moragne, Lenora, Karla Longree, Betty A. Bell, and James C. White. "Time-Temperature Studies of Baked Loaves." *Journal of the American Dietetic Association* 42 (1963): 500–504.

Moragne, Lenora, Karla Longree, and James C. White. "Effect of a Scraper-Lifter Agitator on Cooling Time of Food." *Journal of Milk and Food Technology* 26 (1963): 182–184.

Moragne, Lenora, Karla Longree, and James C. White. "Cooling Menu Items by Agitation under Refrigeration." *Journal of Milk and Food Technology* 26 (1963): 317–322.

Moragne, Lenora. "Influence of Household Differentiation on Food Habits among Low-Income Urban Negro Families." Thesis, Cornell University, 1969.

Moragne, Lenora. *Our Baby's Early Years: A Guide to Infant Care for New Parents, with Emphasis on Feeding, Health Visits, and Financial Planning for the Newborn*. Washington, D.C.: Len Champs Publishers, 1974.

Moragne, Lenora. *Focus on Food*. New York: McGraw-Hill, 1974.

Joan Murrell Owens

Murrell, Joan, L. Austin, A. Grant, C. Morrison, and E. Taylor. *College Reading Skills*. New York: Alfred A. Knopf, 1966.

Murrell, Joan. *To Gladly Learn*. Boston: Education Services Incorporated, 1966.

Murrell, Joan, and Carolyn Fitchett-Binns. *Ideas and Their Expressions: Teacher's Manuals and Accompanying Instructional Materials*. Boston: Institute for Services to Education, 1966–1972.

Owens, Joan Murrell. "Microstructural Changes in the Scleractinian Families Micrabaciidae and Fungiidae and Their Taxonomic and Ecologic Implications." Ph.D. thesis, George Washington University, Washington, D.C., 1984.

Owens, Joan Murrell. "Microstructural Changes in the Micrabaciidae and their Ecologic and Taxonomic Implications." *Palaeontographica Americana*, No. 54 (1984), pp. 519–522.

Owens, Joan Murrell. "Evolutionary Trends in the Micrabaciidae: An Argument in Favor of Preadaptation." *Geologos* 2 (1984): 87–93.

Owens, Joan Murrell. "*Rhombopsammia*, A New Genus of the Family Micrabaciidae (Coelenterata: Scleractinia)." *Proceedings of the Biological Society of Washington* 99 (1986): 248–256.

Owens, Joan Murrell. "On the Elevation of the *Stephanophyllia* subgenus *Letepsammia* to Generic Rank (Coelenterata: Scleractinia: Micrabaciidae)." *Proceedings of the Biological Society of Washington* 99 (1986): 486–488.

Owens, Joan Murrell. "*Letepsammia franki,* A New Species of Deep-Sea Coral (Coelenterata: Scleractinia: Micrabaciidae)." *Proceedings of the Biological Society of Washington* 107 (1994): 586–590.

Gwendolyn Washington Pla

Pla, Gwendolyn Ann Washington. "Rat Kidney Acid Phosphatase Isozymes." M.S. thesis, George Washington University, 1968.

Pla, G. W., N. M. Papadoupoulos, and S. Rosen. "Renal Acid Phosphatase Isoenzymes." *Enzymologia* 34 (1968): 40–44.

Fritz, J. C., G. W. Pla, T. Roberts, J. W. Boehne, and E. L. Hove. "Biological Availability in Animals of Iron from Common Dietary Sources." *Journal of Agricultural and Food Chemistry* 18 (1970): 647–651.

O'Neal, R.M., G. W. Pla, M. R. S. Fox, F. Gibson, and B. Fry. "Effect of Zinc Deficiency and Restricted Feeding on Protein and Ribonucleic Acid Metabolism of the Brain." *Journal of Nutrition* 100 (1970): 491–497.

Pla, G. W., and J. C. Fritz. "Availability of Iron." *Journal—Association of Official Analytical Chemists* 53 (1970): 791–800.

Fritz, J. C., G. W. Pla, and J. W. Boehne. "Influence of Chelating Agents on Utilization of Calcium, Iron and Manganese by the Chick." *Journal of Poultry Science* 50 (1971): 1444–1450.

Pla, G. W., and J. C. Fritz. "Collaborative Study of the Hemoglobin Repletion Test in Chicks and Rats for Measuring Availability of Iron." *Journal—Association of Official Analytical Chemists* 54 (1971): 13–17.

Fritz, J. C., and G. W. Pla. "Vitamins and Other Nutrients: Application of the Animal Hemoglobin Repletion Test to Measurement of Iron Availability in Foods." *Journal—Association of Official Analytical Chemists* 55 (1972): 1128–1132.

Pla, G. W., and J. C. Fritz. "Human Plasma Iron Responses Following Test Doses of Iron from Unknown Sources." *Journal—Association of Official Analytical Chemists* 55 (1972): 197–199.

Fritz, J. C., P. B. Mislivec, G. W. Pla, B. N. Harrison, C. F. Weeks, and J. G. Dantzman. "Toxogenicity of Moldy Feed for Young Chicks." *Poultry Science* 52, 4 (1973): 1523–1530.

Pla, G. W., B. N. Harrison, and J. C. Fritz. "Comparison of Chicks and Rats as Test Animals for Studying Bioavailability of Iron with Special Reference to Use of Reduced Iron in Enriched Bread." *Journal—Association of Official Analytical Chemists* 56 (1973): 1369–1373.

Fritz, J. C., G. W. Pla, B. N. Harrison, and G. A. Clark. "Collaborative Study of the Rat Hemoglobin Repletion Test for Bioavailability of Iron." *Journal—Association of Official Analytical Chemists* 57 (1974): 513–517.

Fritz, J. C., G. W. Pla, B. N. Harrison, and G. A. Clark. "Estimation of the Bioavailability of Iron." *Journal—Association of Official Analytical Chemists* 58 (1975): 902–905.

Fritz, J. C., G. W. Pla, and C. L. Rollinson. "Iron for Enrichment." *Bakers Digest* 49 (1975): 46–49.

Harrison, B. N., G. W. Pla, G. A. Clark, and J. C. Fritz. "Selection of Iron Sources of Cereal Enrichment." *Cereal Chemistry* 53 (1976): 78–84.

Pla, G. W., J. C. Fritz, and C. L. Rollinson. "Relationship between the Biological Availability and Solubility Rate of Reduced Iron." *Journal—Association of Official Analytical Chemists* 59 (1976): 582–583.

Fritz, J. C., G. W. Pla, B. N. Harrison, G. A. Clark, and E. A. Smith. "Measurement of the Bioavailability of Iron, Using the Rat Hemoglobin Repletion Test." *Journal—Association of Official Analytical Chemists* 61 (1978): 709–714.

Pla, Gwendolyn Ann Washington. "Factors Affecting the Biological Availability of Elemental Iron Powders." Thesis, University of Maryland at College Park, 1978.

Pla, G. W., et al. *Nutrition Interventions Manual for Professionals Caring for Older Americans: Nutrition Screening Initiative.* Washington, D.C.: Greer, Margolis, Mitchell, Grunwald & Associates, 1992.

Pla, G. W. *Dental Care in Nursing Care Ready Reference: Resident Assessment Protocols,* pp. 40–51. Columbus, Ohio: Ross Products Division/Abbott Laboratories, 1993.

White, J., J. Dwyer, N. Wellman, G. Blackburn, A. Barrocas, R. Chernoff, D. Cohen, L. Lysen, B. More, G. Pla, and D. Roe. "Beyond Nutrition Screening: A Systems Approach to Nutrition Intervention." *Journal American Dietetic Association* 93 (1993): 405.

Pla, G. W. "Oral Health and Nutrition." *Primary Care: Nutrition in Old Age* 21 (1994): 121–133.

Pla, Gwendolyn W. "Nutrition in Dental Education." *Perspectives in Applied Nutrition* 2 (1995): 27–28.

Muriel E. Poston

Poston, M. E., and H. J. Thompson. "Cytotaxonomic Observations in Loasaceae subfamily Loasoideae." *Systematic Botany* 2 (1977): 28–35.

Poston, Muriel Eloise. "A Systematic Examination of the Loasoideae (Loasaceae): A Re-Alignment of the Genera." Ph.D. dissertation, University of California at Los Angeles, 1979.

Poston, M. E. "Flora of Panama, Part V. Leguminosae Subfamily Papilionoidea. Canjanus." *Annals of the Missouri Botanical Garden* 67 (1980): 555–557.

Poston, M. E. "Flora of Panama, Part V. Leguminosae Subfamily Papilionoideae. Eriosema." *Annals of the Missouri Botanical Garden* 67 (1980): 681–686.

Poston, M. E. "Flora of Panama, Part V. Leguminosae Subfamily Papilionoideae. Flemingia." *Annals of the Missouri Botanical Garden* 67 (1980): 697–699.

Poston, M. E. "Flora of Panama, Part V. Leguminosae Subfamily Papilionoideae. Rhynochosia." *Annals of the Missouri Botanical Garden* 67 (1980): 453–460.

D'Arcy, W. G., and M. E. Poston. "Flora of Panama, Part V. Leguminosae Subfamily Papilionideae. Lablab." *Annals of the Missouri Botanical Garden* 67 (1980): 714–715.

Poston, Muriel E., and S. G. Shetler. "Systematic Importance of Seed Surface Features in Caiophora (Loasaceae)." *Association of Southeastern Biologists Bulletin* 31 (1984): 77.

Goldblatt, P., and M. E. Poston. "Observations on the Chromosome Cytology of Velloziaceae." *Annals of the Missouri Botanical Garden* 75 (1988): 192–195.

Lawry, J., M. Poston, and D. Young. "Research Support Liaison Committee in Ecology, Evolution and Systematics." *Plant Science Bulletin* 34 (1988): 2–3.

Poston, M. E., and G. A. Middendorf. "Maturation Characteristics of the Fruit of *Rubus pennsylvanicus* or Are Red Fruit and 'Black' Fruit the Same?" *Oecologia* 77 (1988): 69–72.

Tucker, A. O., M. E. Poston, and H. Iltis. "History of LCU: The Herbarium of the Catholic University of America." *Taxon* 38 (1989): 196–203.

Poston, M. E., and J. W. Nowicke. "A Re-evaluation of *Klaprothia* and *Sclerothrix* (Loasaceae: Klaprothieae)." *Systematic Botany* 15 (1990): 671–678.

Poston, M. E., and J. W. Nowicke. "Pollen Ultrastructure of Loasoideae (Loasaceae)." *American Journal of Botany* 77 (1990): 151. Abstract.

Middendorf, G. Z., and M. E. Poston. "The Natural History Sciences: Addressing the Issue of Minority Representation." *Plant Science Bulletin* 38 (1992): 6–7.

Poston, Muriel E., and Joan W. Nowicke. "Pollen Morphology, Trichome Types, and Relationships of the Gronovioideae (Loasaceae)." *American Journal of Botany* 80, 6 (1993): 689–704.

Poston, M. E. "Checklist of Loasaceae. Flora of Ecuador." *Annals of the Missouri Botanical Garden*, in review.

Poston, M. E. "Loasaceae. Flora of Nicaragua." *Annals of the Missouri Botanical Garden*, in review.

Poston, M. E., and S. Shelter. "Systematics of Seed Surface Microsculpture in Caiphora (Loasaceae)." *Systematic Botany*, in review.

Jessie Isabelle Price

Price, J. I. "Morphological and Cultural Studies of Pleuropneumonia-like Organisms and Their Variants Isolated from Chickens." Master's thesis, Cornell University, 1958.

Price, J. I. "Studies on *Pasteurella anatipestifer* Infection in White Pekin Ducklings." Ph.D. thesis, Cornell University, 1959.

Price, J. I. "Studies on *Pasteurella anatipestifer* Infection in White Pekin Ducks." *Avian Diseases* 3 (1959): 486–487.

Dougherty, E., III, and J. I. Price. "Eastern Encephalitis in White Pekin Ducklings on Long Island." *Avian Diseases* 4 (1960): 247–258.

Price, J. I., and E. Dougherty III. "Studies with Eastern Encephalitis in White Pekin Ducklings." *Avian Diseases* 4 (1960): 444–449.

Price, J. I., E. Dougherty III, and D. W. Bruner. "*Salmonella* Infections in White Pekin Ducks: A Short Summary of the Years 1950–1960." *Avian Diseases* 6 (1962): 145–147.

Brunner, D. W., C. I. Angstrom, and J. I. Price. "*Pasteurella anatipestifer* Infection in Pheasants: A Case Report." *Cornell Veterinarian* 60 (1970): 491–494.

Dean, W. F., J. I. Price, and L. Leibovitz. "Effect of Feed Medicaments on Bacterial Infections in Ducklings." *Poultry Science* 52 (1973): 549–558.

Price, J. I. "Immunization Trials Using a Low-Virulent *Pasteurella anatipestifer* Culture in White Pekin Ducklings." In *1974 Duck Research Progress Report*, pp. 77–105. Cornell University Duck Research Laboratory, 1974.

Price, J. I. "Immunization of Ducklings with Low-Virulent *Pasteurella anatipestifer.*" In *1975 Duck Research Progress Report*, pp. 95–123. Cornell University Duck Research Laboratory, 1975.

Price, J. I. "*Pasteurella multocida* Studies." In *1975 Duck Research Progress Report*, pp. 124–132. Cornell University Duck Research Laboratory, 1975.

Price, J. I., and C. J. Brand. "Persistence of *Pasteurella multocida* in Nebraska Wetlands under Epizootic Conditions." *Journal of Wildlife Diseases* 20 (1984): 90–94.

Price, J. I. "Immunizing Canada Geese against Avian Cholera." *Wildlife Society Bulletin* 13 (1985): 508–515.

Jensen, W. I., and J. I. Price. "The Global Importance of Type C Botulism in Wild Birds." In *Avian Botulism: An International Perspective*, edited by M. W. Ekland and V. R. Dowell, pp. 33–54. Springfield, Ill.: Charles C. Thomas, 1987.

Siegel, L. S., and J. I. Price. "Ineffectiveness of 3,4-Diaminopyridine as Therapy for Type C Botulism." *Toxicon* 25 (1987): 1015–1018.

Price, J. I. "Vaccination Prevents Avian Cholera in Waterfowl." In *U.S. Federal Wildlife Service Research Information Bulletins*, No. 87–113, 1987.

Price, J. I. "Recovery of Avian Tuberculosis Bacteria from Sandhill Cranes and Snow

Geese." In *U.S. Federal Wildlife Service Research Information Bulletins*, No. 90–109, 1990.

Price, J. I., B. S. Yandell, and W. P. Porter. "Chemical Ions Affect Survival of Avian Cholera Organisms in Pond Water." *Journal of Wildlife Management* 56 (1992): 274–278.

Samuel, M.D., D. R. Goldberg, D. J. Shadduck, J. I. Price, and E. G. Cooch. "*Pasteurella multocida* Serotype 1 Isolated from a Lesser Snow Goose: Evidence of a Carrier State." *Journal of Wildlife Diseases* 33 (1997): 332–335.

PRESENTATIONS

Price J. I. "Studies on *Pasteurella anatipestifer* Infection in White Pekin Ducks." Paper presented at 31st Northeastern Conference in Avian Diseases, Storrs, Conn., 1959.

Price, J. I., and E. Dougherty III. "Some New Aspects of Eastern Encephalitis." Presentation at 32nd Northeastern Conference on Avian Diseases, Orono, Maine, 1960.

Price, J. I., and E. Dougherty III. "*Salmonella* Isolations from White Pekin Ducklings." Presentation at 33rd Northeastern Conference on Avian Diseases, Morgantown, W.Va., 1961.

Price, J. I., E. Dougherty III, and M. Pokorny. "Developments in the Production of a Bacterin for *anatipestifer* Infection in White Pekin Ducklings." Presentation at 35th Northeastern Conference on Avian Diseases, Amherst, Mass., 1963.

Price, J. I. "*Pasteurella* Infections in Ducks." Paper presented at Symposium for Wildlife Disease Specialists Interested in Avian Cholera and Related Diseases, Patuxent, Md., 1964.

Price, J. I. "Growth and Serological Characteristics of *Pasteurella anatipestifer* Isolated from Ducks." Paper presented at the American Society for Microbiology, Los Angeles, 1966.

Price, J. I. "Studies on a Bacterial Respiratory Infection in the White Pekin Duckling." Paper presented at the Ninth International Conference for Microbiology, Moscow, USSR, 1966.

Price, J. I., and M. Lummis. "Serological Studies of *Pasteurella anatipestifer*, *Escherichia* spp., and *Mycoplasma* spp. in the White Pekin Duckling." Presentation at the American Society of Microbiology, Detroit, 1968.

Price, J. I., and M. Lummis. "Immunization Studies with Broth Bacterins to Control *P. anatipestifer–E. coli* Infections in White Pekin Ducklings." Presentation at the 42nd Northeastern Conference on Avian Diseases, Blacksburg, Va., 1970.

Price, J. I., and M. Lummis. "Studies of the Serological Characteristics of *Pasteurella anatipestifer* and *Pasteurella multocida* Organisms Isolated from White Pekin Ducks." Presentation at the 45th Northeastern Conference on Avian Diseases, Morgantown, W.Va., 1973.

Price, J. I., and M. Lummis. "Immunization Trials Using a Low-Virulence *Pasteurella anatipestifer* in White Pekin Ducklings—A Preliminary Report." Paper presented at the 46th Northeastern Conference on Avian Diseases, University Park, Penn., 1974.

Price, J. I. "*Pasteurella anatipestifer*: Culture and Disease Characteristics in White Pekin Ducklings." Presentation to the Wildlife Disease Association, Asilomar, Pacific Grove, Calif., 1974.

Price, J. I. "Oral Vaccination in Domestic Ducks and Its Application to Waterfowl." Paper presented at the Avian Cholera Symposium of the Wildlife Disease Association, Colorado State University, Fort Collins, Colo., 1978.

Price, J. I. "How Birds Establish Immunity." Paper presented at Wisconsin Game Breeders Conference, Madison, Wis., 1979.

Price, J. I., C. Brand, J. J. Hurt, S. Hurley, L. Locke, and R. Windingstad. "A Second

Report of an Avian Cholera Epornitic in South Central Nebraska in 1979." Presentation to the Wildlife Disease Association, University of Oklahoma, Norman, Okla., 1979.

Price, J. I., and C. Brand. "Avian Cholera Die-off in the Nebraska Rainwater Basin. P. multocida Isolations from Water." Presentation to the Wildlife Disease Association, Louisiana State University, Baton Rouge, 1980.

Price, J. I., S. K. Schmeling, and L. Hanson. "Avian Cholera in Ring-Necked Pheasants." Presentation to the Wildlife Disease Association, Louisiana State University, Baton Rouge, 1980.

Siegfried, L. M., J. Price, R. Duncan, J. Hurt, and C. Brand. "Choleric Crows—What Goes?" Presentation to the Wildlife Disease Association, Louisiana State University, Baton Rouge, 1980.

Anderson, T. E., L. Hanson, and J. Price. "Evaluation of Avian Cholera Bacterins (Pasteurella multocida) in Ring-Necked Pheasants." Presentation to the Wildlife Disease Association, Louisiana State University, Baton Rouge, 1980.

Price, J. I. "Preciptin Development as an Index of Exposure to P. multocida in Waterfowl, Crows, Pheasants, and Raptors." Presentation to the Wildlife Disease Association, University of Wyoming, Laramie, 1981.

Price, J. I. "Immunization of Canada Geese against Avian Cholera." Presentation to the Wildlife Disease Association, Madison, Wis., 1982.

Price, J. I., and M. T. Collins. "Comparative Recoveries of Mycobacterium spp. from Wild Birds Using Conventional and Radiometric Culture Methods." Paper presented at the North Central Branch—American Society for Microbiology, Madison, Wis., 1988.

Price, J. I. "Recovery of a Variety of Mycobacteria spp. from Hunter-Killed Sandhill Cranes and Snow Geese." Wildlife Disease Association, Corvallis, Oregon, 1989.

Yvonne A. Reid

Reid, Y. A., and A. W. Hamburger. "Biosynthetic Radiolabeling and Immunoprecipitation of Monoclonal Antibodies." Journal of Tissue Culture Methods 8 (1983): 137–139.

Reid, Y. A., L. A. Breth, and A. W. Hamburger. "Analytical Agarose Isoelectric Focusing of Immunoglobulins for Characterization of Monoclonal Antibodies." Journal of Tissue Culture Methods 8 (1983): 131–135.

Hamburger, A. W., Y. A. Reid, et al. "Isolation and Characterization of a Monoclonal Antibody Specific for Epithelial Cells." Cancer Research 45 (1985): 783–790.

Hamburger, A. W., Y. A. Reid, et al. "Isolation and Characterization of Monoclonal Antibodies Reactive with Endothelial Cells." Tissue and Cells 17 (1985): 451–459.

Reid, Y. A., and I. M. Cour. "Cryopreservation of Hybridomas." Journal of Tissue Culture Methods 9 (1985): 163–165.

Reid, Yvonne A. "Monoclonal Antibody Directed against Human Endothelial Cells: Isolation, Characterization and Functional Studies." Ph.D. dissertation, Howard University, 1986.

Mann, D. L., D. A. Gilbert, Y. A. Reid, et al. "On the Origin of the HIV Susceptible Human CD4+ Cell Line H9." AIDS Research and Human Retroviruses 5 (1989): 253–255.

Gilbert, D. A., Y. A. Reid, et al. "Identification of Human Cell Lines through Use of DNA Hypervariable Probes." American Journal of Human Genetics 47 (1990): 499–514.

Reid, Y. A., D. A. Gilbert, and S. J. O'Brien, "The Use of DNA Hypervariable Probes for Human Cell Line Identification." ATCC Quarterly Newsletter 10 (1990): 103.

Caputo, J., A. Thompson, Y. Reid, et al. "An Effective Method for Establishing Human B

Lymphoblastic Cell Lines Using Epstein-Barr Virus." *Journal of Tissue Culture Methods* 13 (1991): 39–44.

Hamburger, A. W., D. Mehta, G. Pinnamaneni, L. C. Chen, and Y. A. Reid. "Density Dependent Regulation of Epidermal Growth Factor Receptor Expression." *Pathobiology* 59 (1991): 329–334.

C. H. Lee, Y. A. Reid, et al. "Lipopolysaccharide Differential Cell Surface Expression of Intercellular Adhesion Molecule-1 in Cultured Human Umbilical Chord Vein Endothelial Cells." *Shock* 3 (1995): 96–101.

Reid Y. A., L. McGuire, K. O'Neill, et al. "Cell Line Cross-Contamination of U-937 [Correction of U-397]." *Journal of Leukocyte Biology* 57 (1995): 804.

Hay, R. J., Y. A. Reid, P. R. McClintock, T. R. Chen, and M. L. Macy. "Cell Line Banks and Their Role in Cancer Research." *Journal of Cell Biochemistry Supplement* 24 (1996): 107–130.

MacLeod, R. A., W. G. Dirks, Y. A. Reid, R. J. Hay, and H. G. Drexler. "Identity of Original and Late Passage DAMI Megakaryocytes with HEL Erythroleukemia Cells Shown by Combined Cytogenetics and DNA Fingerprinting." *Leukemia* 11 (1997): 2032–2038.

Marianna Beck Sewell

Hayes, R. L., W. Booker, and M. B. Sewell. "Ascorbic Acid Metabolism in Patients with Certain Gingival Disturbances." *Federation Proceedings VIII* (March 1949): 70.

Booker, W. M., R. L. Hayes, M. B. Sewell, and F. M. Dent. "Experimental Studies on Ascorbic Acid Metabolism." *American Journal of Physiology* 166 (August 1951): 374.

Sewell, Marianna Beck. "An Evaluation of Food Habits of Nigerian Students in American Colleges and Universities." Ed.D. thesis, New York University, 1965.

Dolores Cooper Shockley

Ansah, T. A., L. H. Wade, and D. C. Shockley. "Effects of Calcium Channel Entry Blockers on Cocaine- and Amphetamine-Induced Motor Activities and Toxicities." *Life Science* 53 (1993): 1947–1956.

Shockley, D. C., L. H. Wade, and M. M. Williams-Johnson. "Effects of Alpha-2-Adrenoceptor Agonists on Induced Diuresis in Rats." *Life Science* 53 (1993): 251–259.

Ansah, T. A., L. H. Wade, and D. C. Shockley. "Changes in Locomotor Activity, Core Temperature, and Heart Rate in Response to Repeated Cocaine Administration." *Physiology and Behavior* 60 (1996): 1261–1267.

Mills, K., T. A. Arsah, S. F. Ali, and D. C. Shockley. "Calcium Channel Antagonist Isradipine Attenuates Cocaine-Induced Motor Activity in Rats: Correlation with Brain Monoamine Levels." *Annals of the New York Academy of Science* 844 (1998): 201–207.

Jeanne Spurlock

Spurlock, J. "Current Trends in Child Psychiatry." *Illinois Medical Journal* 130 (1966): 31–35.

Spurlock, J., and R. S. Cohen. "Should the Poor Get None?" *Journal of the American Academy of Child Psychiatry* 8 (1969): 16–35.

Spurlock, J. "Problems of Identification in Young Black Children—Static or Changing?" *Journal of the National Medical Association* 61 (1969): 504–507.

Spurlock, J. "Social Deprivation in Childhood and Character Formation." *Journal of the American Psychoanalytical Association* 18 (1970): 622–630.

Spurlock, Jeanne. *A Reappraisal of the Role of Black Women: The Eighth Annual Institute*

of the Pennsylvania Hospital Aware Lecture in Memory of Edward A. Strecker, M.D.
Philadelphia: Institute of Philadelphia Hospital, 1971.

Spurlock, J., J. Green, E. Kessler, K. H. Martin, J. V. Wallinga, and L. J. Wise. "The Private Practice of Child Psychiatry by Members of the American Academy of Child Psychiatry." *Journal of the American Academy of Child Psychiatry* 10 (1971): 53–64.

North, A. F., Jr., R. E. Helfer, J. Spurlock, D. Fink, F. C. Green, T. B. Brazelton, and M. W. Piers. "White House Conference on Children: Personal Reflections." *Clinical Pediatrics (Philadelphia)* 10 (1971): 502–512.

Spurlock, J. "New Recognition for Women in APA." *American Journal of Psychiatry* 132 (1975): 647–648.

Spurlock, J., and K. Rembold. "Child Psychiatry Perspectives: Women at Fault: Societal Stereotypes and Clinical Conclusions." *Journal of the American Academy of Child Psychiatry* 17 (1978): 383–386.

Spurlock, J. "Assessment and Therapeutic Intervention of Black Children." *Journal of the American Academy of Child Psychiatry* 24 (1985): 168–174.

Spurlock, J. "Survival Guilt and the Afro-American of Achievement." *Journal of the National Medical Association* 77 (1985): 29–32.

Spurlock, J. "Development of Self-Concept in Afro-American Children." *Hospital and Community Psychiatry* 37 (1986): 66–70.

Coner-Edwards, Alice F., and Jeanne Spurlock, editors and contributors. *Black Families in Crisis: The Middle Class.* New York: Brunner/Mazel, 1988.

Nightingale, E. O., K. Hannibal, H. J. Geiger, L. Hartmann, R. Lawrence, and J. Spurlock. "Apartheid Medicine: Health and Human Rights in South Africa." *Journal of the American Medical Association* 31 (1990): 2097–2102.

Spurlock, Jeanne, editor. *Adolescent Suicide.* Thorofare, N.J.: SLACK, 1990.

Spurlock, Jeanne, and Carolyn Robinowitz, editors and contributors. *Women's Progress: Promises and Problems.* New York: Plenum Press, 1990.

Canino, Ian A., and Jeanne Spurlock. *Culturally Diverse Children and Adolescents: Assessment, Diagnosis, and Treatment.* New York: Guilford Press, 1994.

Spurlock, J. "Multiple Roles of Women and Role Strains." *Health Care for Women International* 16 (1995): 501–508.

Baker, F. M., A. Bondurant, C. Pinderhughes, R. Fuller, S. Kelley, S. P. Kim, E. Triffleman, and J. Spurlock. "Survey of the Cross-Cultural Content of U.S. Psychiatry Residency Training Programs." *Cultural Diversity and Mental Health* 3 (1997): 215–218.

Canino, I. A., and J. Spurlock. "Mental Health Issues of Culturally Diverse Underserved Children." *Journal of the Association of Academic Minority Physicians* 8 (1997): 63–66.

Spurlock, Jeanne, editor. *Black Psychiatrists and American Psychiatry.* Washington, D.C.: American Psychiatric Association, 1999.

Marie Clark Taylor

Taylor, Marie C. "More Dynamic Botany." *The Science Counselor* 19 (1956): 78–79.

Taylor, Marie C., and Victor A. Greulach. "Botany in the National Science Foundation Institutes for High School Teachers." *Plant Science Bulletin* 4 (1958): 6–8.

Taylor, Marie C. "Live Specimens." *The American Biology Teacher* 27 (February 1965): 116–117.

Alberta Banner Turner

Turner, Alberta Banner. "The Effects of Practice on the Perception and Memorization of Digits Presented in Single Exposure."*

Turner, Alberta Banner. "The Psychologist at the Juvenile Diagnostic Center: Past, Present, Future."*

Turner, Alberta Banner. "Incest Cases of Female Adolescents."*

Jane Cooke Wright

Wright, J. C., J. I. Plummer, R. S. Coidan, and L. T. Wright. "The in Vivo and in Vitro Effects of Chemotherapeutic Agents on Human Neoplastic Diseases." *The Harlem Hospital Bulletin* 6 (1953): 58–63.

Wright, J. C., P. Foster, B. Billow, S. S. Gumport, and J. P. Cobb. "The Effect of Triethylene Thiophosphoramide on Fifty Patients with Incurable Neoplastic Diseases." *Cancer* 10 (1957): 239–245.

Wright, J., J. P. Cobb, S. L. Gumport, F. M. Golomb, and D. Safadi. "Investigation of the Relationship between Clinical and Tissue Response to Chemotherapeutic Agents on Human Cancer." *New England Journal of Medicine* 257 (1957): 1207–1211.

Cobb, J. P., F. M. Golomb, S. L. Gumport, D. Safadi, and J. C. Wright. "Chemotherapy of Human Breast Cancer." *Seventh International Cancer Congress*, 1958.*

Cobb, J. P., and J. C. Wright. "Studies on a Craniopharyngioma in Tissue Culture." *Journal of Neuropathology and Experimental Neurology* 18 (1959): 563–568.

Wright, J. C., J. P. Cobb, F. M Golomb, S. L. Gumport, D. Lyall, and D. Safadi. "Chemotherapy of Disseminated Carcinoma of the Breast." *Annals of Surgery* 150 (1959): 221–240.

Cobb, J. P., D. G. Walker, and J. C. Wright. "Observations on the Action of Triethylene Thiophosphoramide within Individual Cells." *Acta of VII International Cancer Congress*, 1960: 567–583.

Golomb, F. M., J. C. Wright, J. P. Cobb, S. L. Gumport, A. Postal, and D. Safadi. "The Chemotherapy of Human Solid Tumors by Perfusion Techniques." *Proceedings of the American Association for Cancer Research* 3 (1960).

Wright, J. C., J. P. Cobb, S. L. Gumport, D. Safadi, D. G. Walker, and F. M. Golomb. "Further Investigation of the Relation between the Clinical and Tissue Culture Response to Chemotherapy Agents on Human Cancer." *Cancer* 15 (1962): 284–293.

Golomb, F. M., J. P. Cobb, D. G. Walker, and J. C. Wright. "In Vitro Selection of Chemotherapeutic Agents for Perfusion Therapy of Human Cancer." *Surgery* 51 (1962): 639–644.

Golomb, F. M., A. C. Solowey, A. Postel, S. L. Gumport, and J. C. Wright. "Induced Remission of Malignant Melanoma with Actinomycin D: Immunologic Implications." *Cancer* 20 (1967): 656–662.

Wright, J. C. "The Role of Chemotherapy in Pelvic Malignancies." *Hospital Topics* 45 (1967): 89–93.

Malkin, R. B., T. Strax, and J. C. Wright. "Cancer in Advancing Age." *Journal of the National Medical Association* 69 (1969): 255–257.

Wright, J. C. "Chemotherapy: Changing Concepts in Therapy." *Minnesota Medicine* 53 (1970): 373–377.

Wright, J. C., and R. E. Madden. "Cancer Chemotherapy in Man." *Review of Surgery* 31 (1974): 217–237.

Wright, J. C., and D. Walker. "A Predictive Test for the Selection of Cancer Chemotherapeutic Agents for the Treatment of Human Cancer." *Journal of Surgical Oncology* 7 (1975): 381–393.

Wright, J. C. "Cancer Chemotherapy: Past, Present, and Future—Part I." *Journal of the National Medical Association* 76 (1984): 773–784.

Wright, J. C. "Cancer Chemotherapy: Past, Present, and Future—Part II." *Journal of the National Medical Association* 76 (1984): 865–876.

Wright, J. C. "Update in Cancer Chemotherapy: General Considerations and Breast Cancer, Part I." *Journal of the National Medical Association* 77 (1985): 617–625.

Wright, J. C. "Update on Cancer Chemotherapy: General Considerations and Breast Cancer, Part II." *Journal of the National Medical Association* 77 (1985): 691–703.

Wright, J. C. "Update in Cancer Chemotherapy, Part III: Lung Cancer, Part 1." *Journal of the National Medical Association* 77 (1985): 815–827.

Wright, J. C. "Update in Cancer Chemotherapy, Part IV: Lung Cancer, Part 2." *Journal of the National Medical Association* 77 (1985): 907–919.

Wright, J. C. "Update in Cancer Chemotherapy: Gastrointestinal Cancer, Cancer of the Pancreas." *Journal of the National Medical Association* 78 (1986): 519–527.

Wright, J. C. "Update in Cancer Chemotherapy: Gastrointestinal Cancer, Cancer of the Stomach and Carcinoid Tumors." *Journal of the National Medical Association* 78 (1986): 623–632.

Wright, J. C. "Update in Cancer Chemotherapy: Gastrointestinal Cancer—Colorectal Cancer, Part 1." *Journal of the National Medical Association* 78 (1986): 295–304.

Wright, J. C. "Update in Cancer Chemotherapy: Gastrointestinal Cancer—Colorectal Cancer, Part 2." *Journal of the National Medical Association* 78 (1986): 395–408.

Wright, J. C. "Update in Cancer Chemotherapy: Head and Neck Cancer, Part 1." *Journal of the National Medical Association* 78 (1986): 955–961.

Wright, J. C. "Update in Cancer Chemotherapy: Head and Neck Cancer, Part 2." *Journal of the National Medical Association* 78 (1986): 1041–1052.

Wright, J. C. "Update in Cancer Chemotherapy: Gastrointestinal Cancer, Cancer of the Small Intestines, Gallbladder, Liver, and Esophagus." *Journal of the National Medical Association* 78 (1986): 753–766.

Wright, J. C. "Update in Cancer Chemotherapy: Genitourinary Tract Cancer, Part 1." *Journal of the National Medical Association* 79 (1987): 1249–1258.

Wright, J. C. "Update in Cancer Chemotherapy: Genitourinary Tract Cancer, Part 2: Wilms' Tumor and Bladder Cancer." *Journal of the National Medical Association* 80 (1988): 169–181.

Wright, J. C. "Update in Cancer Chemotherapy: Genitourinary Tract Cancer, Part 3: Cancer of the Prostate." *Journal of the National Medical Association* 80 (1988): 305–314.

Wright, J. C. "Update in Cancer Chemotherapy: Genitourinary Tract Cancer, Part 4: Testicular Cancer." *Journal of the National Medical Association* 80 (1988): 425–435.

Wright, J. C. "Update in Cancer Chemotherapy: Genitourinary Tract Cancer, Part 5: Ovarian Cancer." *Journal of the National Medical Association* 80 (1988): 565–576.

Wright, J. C. "Update in Cancer Chemotherapy: Genitourinary Tract Cancer, Part 6: Cancer of the Uterus and Vulva." *Journal of the National Medical Association* 80 (1988): 657–667.

Wright, J. C. "Update in Cancer Chemotherapy: Genitourinary Tract Cancer, Part 7: Gestational Trophoblastic Neoplasms." *Journal of the National Medical Association* 80 (1988): 753–761.

Roger Arliner Young

Young, Roger Arliner. "On the Excretory Apparatus in Paramecium." *Science*, 12 September 1924.

Bibliography

Books

Abajian, James de T. *Blacks in Selected Newspapers, Censuses, and Other Sources: An Index to Names and Subjects.* Boston: G. K. Hall, 1977. Supplement, 1985.

Abir-am, Pnina G., and Dorinda Outram. *Uneasy Careers and Intimate Lives: Women in Science 1789–1979.* New Brunswick, N.J.: Rutgers University Press, 1987.

ABMS Compendium of Medical Specialists. 7 vols. Chicago: American Board of Medical Specialists, 1986.

African-American Heritage. Inglewood, Calif.: Dellco Publishing, 1976–.

Allen, Will W., and Daniel Murray. *Banneker: The Afro-American Astronomer.* 1921. Reprint, The Black Heritage Library Collection. New York: Books for Libraries Press, 1971.

Ambrose, Susan A., Kristin L. Dunkle, Barbara B. Lazarus, Indira Nair, and Deborah A. Harkus. *Journeys of Women in Science and Engineering: No Universal Constants.* Philadelphia: Temple University Press, 1997.

American Men of Science. 1st–11th eds. [New York: Bowker], 1906–1968. The 1st through 8th editions published by Science Press; 9th edition by Bowker and Science Press; 10th edition by Jacques Cattell Press; 11th edition by Bowker. See also *American Men and Women of Science.*

American Men and Women of Science: The Physical and Biological Sciences. New York: Bowker, 1971–. Continuation of *American Men of Science*, 12th edition to present.

Anderson, Gloria Long. "May Edward Chinn." In *Notable Black American Women*, edited by Jessie Carney Smith. Vol. 1, pp. 183–185. Detroit: Gale Research, 1992.

Anderson, Matthew. *Presbyterianism: Its Relation to the Negro.* Philadelphia: John McGill, White & Company, 1899.

Aptheker, Bettina. "Quest for Dignity: Black Women in the Professions, 1885–1900." In *Woman's Legacy: Essays on Race, Sex, and Class in American History.* Amherst: University of Massachusetts Press, 1982.

Baker, Henry Edwin. *The Colored Inventor: A Record of Fifty Years.* New York: Crisis Publishing, 1913.

Bedini, Silvio A. *Early American Scientific Instruments and Their Makers.* Washington, D.C.: Smithsonian Institution, 1964.

———. *The Life of Benjamin Banneker.* New York: Scribner, 1972.

———. *Thinkers and Tinkers: Early American Men of Science.* New York: Scribner, 1975.

Bekin, Carol, and Mary Beth Norton. *Women of America: A History.* Boston: Houghton, Mifflin, 1979.

Bison. 1923 edition. Howard University Archival Collection. Moorland-Spingarn Research Center, Howard University, Washington, D.C.

Black Americans Information Directory. Boulder, Colo.: Numbers & Concepts, 1990–.

Black Contributors to Science and Energy Technology. Washington, D.C.: U.S. Department of Energy, Office of Public Affairs, 1979.

Blackwell, J. *The Black Community: Diversity and Unity.* New York: Dodd, Mead, 1975.

Bond, Horace M. *A Study of Factors Involved in the Identification and Encouragement of*

Unusual Academic Talent among Underprivileged Populations. Project No. 5-0859, Contract SAE 8028. Washington, D.C.: U.S. Department of Health, Education, and Welfare, Office of Education, 1967.

Booth, Nicholas. *The Encyclopedia of Space.* New York: Mallard Press, 1990.

Branch, Taylor. *Parting the Waters: America in the King Years.* New York: Simon & Schuster, 1988.

Branson, Herman R. "The Negro Scientist." In *The Negro in Science,* edited by Julius H. Taylor, Clyde R. Dillard, and Nathaniel K. Proctor. Baltimore: Morgan State College Press, 1955.

Brown, Hallie Q., compiler. *Homespun Heroines and Other Women of Distinction.* Xenia, Ohio: Aldine Publishing, 1926.

Bruce, Robert V. *The Launching of Modern American Science: 1846–1876.* Ithaca, N.Y.: Cornell University Press, 1987.

Carwell, Hattie. *Blacks in Science: Astrophysicist to Zoologist.* Hicksville, N.Y.: Exposition Press, 1977.

Cazort, Jean Elder. "Jane C. Wright." In *Notable Black American Women,* edited by Jessie Carney Smith. Vol. 2, pp. 1283–1285. Detroit: Gale Research, 1992.

Changing America: The New Face of Science and Engineering. Final Report of the Federal Task Force on Women, Minorities, and the Handicapped in Science and Technology. Washington, D.C.: U.S. Government Printing Office, 1988.

Chess, Stella. "Very Gifted and Black: Mamie Phipps Clark." In *The Women of Psychology,* edited by Gwendolyn Stevens and Sheldon Gardner. Vol. 2, p. 191. Cambridge, Mass.: Schenkman, 1982.

Clark, Mamie Phipps. Personal account. In *Models of Achievement: Reflections of Eminent Women in Psychology,* edited by Agnes N. O'Connell and Nancy Felipe Russo, pp. 266–278. New York: Columbia University Press, 1983.

Clewell, Beatriz C., and Bernice Anderson. *Women of Color in Mathematics, Science, and Engineering: A Review of the Literature.* Washington, D.C.: Center for Women Policy Studies, 1991.

Cottle, Thomas J. *Black Children, White Dreams.* Boston: Houghton, Mifflin, 1974.

——. "Show Me a Scientist Who's Helped Poor Folks and I'll Kiss Her Hand." In *Science and Society: Past, Present, and Future,* edited by N. H. Steneck. Ann Arbor: University of Michigan Press, 1974.

Dannett, Sylvia G. L. *Profiles of Negro Womanhood.* Vol. 2, *Twentieth Century.* Negro Heritage Library. Yonkers, N.Y.: Educational Heritage, [1964–1966].

Davenport, Charles H., and Morris Steggerda. *Race Crossing in Jamaica.* Washington, D.C.: Carnegie Institution of Washington, 1929.

Davis, Elizabeth L. *Lifting as They Climb: The National Association of Colored Women.* Washington, D.C.: National Association of Colored Women's Clubs, 1933.

Davis, Marianna W., editor. *Contributions of Black Women to America.* 2 vols. Columbia, S.C.: Kenday Press, 1982.

DeKruif, Paul. *The Microbe Hunters.* New York: Harcourt, Brace, 1926.

Dictionary of International Biography. (Cambridge, England: International Biographical Centre, 1963–).

DuBois, W. E. B. *The Souls of Black Folk.* Chicago: A. C. McClug, 1904. Reprint, New York: Fawcett World Library, 1961.

Dunn, Lois J. "Dorothy L. Brown." In *Notable Black American Women,* edited by Jessie Carney Smith. Vol. 1, pp. 114–116. Detroit: Gale Research, 1992.

Dupree, A. Hunter. *Science in the Federal Government: A History of Policies and Activities.* Baltimore: Johns Hopkins University Press, 1986.

Ebony Success Library. 3 vols. Chicago: Johnson Publishing, 1973.

Ellison, Aaron M. *Expedition Briefing: Life among the Mangroves.* Earthwatch, 1988.

Embree, Edwin R. *Investment in People: The Story of the Julius Rosenwald Fund.* New York: Harper, 1949.

Emerson, Alfred Edwards, and Eleanor Fish. Foreword by William Beebe. *Termite City.* New York: Rand McNally, 1937.

Flexner, Abraham. *Medical Education in the United States and Canada.* Bulletin No. 4. New York: The Carnegie Foundation, 1910.

Flexner, Eleanor. *Century of Struggle: The Women's Rights Movement in the United States.* Cambridge: Harvard University Press, Belknap Press, 1959.

Fort, Deborah C., editor. *A Hand Up: Women Mentoring Women in Science.* Washington, D.C.: Association for Women in Science, 1993.

Fosdick, Raymond B. *Adventures in Giving: The Story of the General Education Board.* New York: Harper & Row, 1962.

Franklin, Eleanor Ison. "Cultural Conflicts in Professional Training of Minority Women." In *The Minority Woman in America: Professionalism at What Cost?* San Francisco: University of California Press, 1979.

Franklin, John H., and August Meier, editors. *Black Leaders of the 20th Century.* Urbana: University of Illinois Press, 1982.

Fredrickson, George M. *The Black Image in the White Mind: The Debate on Afro-American Character and Destiny, 1817–1914.* New York: Harper Torch Books, 1971.

Gilkes, Cheryl Townsend. "Religion." In *Notable Black American Women*, edited by Jessie Carney Smith. Vol. 2, pp. 967–972. Detroit: Gale Research, 1992.

Glazer, Penina Magdal, and Miriam Slater. *Unequal Colleagues: The Entrance of Women into the Professions, 1890–1940.* New Brunswick, N.J.: Rutgers University Press, 1987.

Goodman, Mary Ellen. *Race Awareness in Young Children.* Cambridge, Mass.: Addison-Wesley, 1952.

Gornick, Vivian. *Women in Science.* New York: Simon & Schuster, 1983.

Gossett, T. F. *Race: The History of an Idea in America.* New York: Schocken Books, 1965.

Green, John C. *American Science in the Age of Jefferson.* Ames: Iowa State University Press, 1984.

Greene, Harry W. *Holders of Doctorates among American Negroes: An Educational and Social Study of Negroes Who Have Earned Doctoral Degrees in Course, 1876–1943.* Boston: Meador, 1946.

Greene, Lorenzo J., and Carter G. Woodson. *The Negro Wage Earner.* Washington, D.C.: The Association for the Study of Negro Life and History, 1930.

Grinstein, Louise S., and Paul J. Campbell, editors. *Women of Mathematics: A Bibliographic Sourcebook.* New York: Greenwood Press, 1987.

Guthrie, Robert V. *Even the Rat Was White: An Historical View of Psychology.* New York: Harper & Row, 1976.

Guy-Sheftall, Beverly. "Johnnetta Betsch Cole." In *Black Women in America: An Historical Encyclopedia*, edited by Darlene Clark Hine, Elsa Barkley Brown, and Rosalyn Terborg-Penn. Vol. 1, p. 261. Brooklyn, N.Y.: Carlson, 1993; Bloomington: Indiana University Press, 1994.

Haber, Louis. *Black Pioneers of Science and Invention.* New York: Harcourt, Brace & World, 1970.

———. *The Role of the Negro in the Fields of Science*. New York: Harcourt, Brace & World, 1966.

———. *Women Pioneers of Science*. New York: Harcourt Brace Jovanovich, 1979.

Haller, J. S., Jr. *Outcasts from Evolution: Scientific Attitudes of Racial Inferiority, 1859–1900*. Urbana: University of Illinois Press, 1971.

Hammonds, Evelyn. "Roger Arliner Young." In *Black Women in America: An Historical Encyclopedia*, edited by Darlene Clark Hine, Elsa Barkley Brown, and Rosalyn Terborg-Penn. Vol. 2, pp. 1298–1299. New York: Carlson, 1993; Bloomington: Indiana University Press, 1994.

Harding, Sandra, editor. *The "Racial" Economy of Science: Toward a Democratic Future*. Bloomington: Indiana University Press, 1993.

Harris, Barbara J. *Beyond Her Sphere: Women and the Professions in American History*. Westport, Conn.: Greenwood Press, 1978.

Hart, Jamie, and Elsa Barkley Brown, comps. "Black Women in the United States: A Chronology." In *Black Women in America: An Historical Encyclopedia*, edited by Darlene Clark Hine, Elsa Barkley Brown, and Rosalyn Terborg-Penn. Vol. 2, pp. 1309–1332. New York: Carlson, 1993; Bloomington: Indiana University Press, 1994.

Hawkins, Walter L. *African American Biographies: Profiles of 558 Current Men and Women*. Jefferson, N.C.: McFarland, 1992.

Hayden, Robert C. *Seven Black American Scientists*. Reading, Mass.: Addison-Wesley, 1970.

Hayden, Robert C., and Jacqueline Harris. *Nine Black American Doctors*. Reading, Mass.: Addison-Wesley, 1976.

Hill, Susan. *Blacks in Undergraduate Science and Engineering Education*. Washington, D.C.: Division of Science Resources Studies, National Science Foundation, 1992.

Hine, Darlene Clark. *Black Women in the Nursing Profession: A Documentary History*. New York: Garland, 1985.

———. *Black Women in White: Racial Conflict and Cooperation in the Nursing Profession, 1890–1950*. Bloomington: Indiana University Press, 1989.

———, editor. *Black Women's History: Theory and Practice*. Brooklyn, N.Y.: Carlson Publishing, 1990.

———. "Co-Laborers in the Work of the Lord: Nineteenth Century Black Women Physicians." In *Send Us a Lady Physician: Women Doctors in America, 1835–1920*, edited by Ruth J. Abrams, pp. 249–253. New York: W.W. Norton, 1985.

———, editor. *The State of Afro-American History: Past, Present, and Future*. Baton Rouge: Louisiana State University Press, 1986.

———, Elsa Barkley Brown, and Rosalyn Terborg-Penn, editors. *Black Women in America: An Historical Encyclopedia*. 2 vols. Brooklyn, N.Y.: Carlson, 1993; Bloomington: Indiana University Press, 1994.

Ho, James K. *Black Engineers in the United States—A Directory*. Washington, D.C.: Howard University Press, 1974.

Howard, Ruth Winifred. Personal account. In *Models of Achievement: Reflections of Eminent Women in Psychology*, edited by Agnes N. O'Connell and Nancy Felipe Russo, pp. 55–67. New York: Columbia University Press, 1983.

Howard University Directory of Graduates, 1870–1985. White Plains, N.Y.: Bernard C. Harris, 1986.

Hunt, Carolyn L. *The Life of Ellen H. Richards*. Boston: Whitcomb & Barrows, 1912.

Innes, Doris F., and Julianna Wu, editors. *Profiles in Black: Biographical Sketches of 100 Living Black Unsung Heroes*. New York: CORE Publications, 1976.

Ives, Patricia Carter. *Creativity and Inventions: The Genius of Afro-Americans and Women in the United States and Their Patents.* Arlington, Va.: Research Unlimited, 1987.

Ivin, Dona L. "Jewel Plummer Cobb." In *Notable Black American Women,* edited by Jessie Carney Smith. Vol. 1, pp. 195–198. Detroit: Gale Research, 1992.

Jackson, George L. *Black Women Makers of History: A Portrait.* Sacramento, Calif.: Fong and Fong, 1977.

Jackson, Jacquelyne Johnson. *These Rights They Seek.* Washington, D.C.: Public Affairs Press, 1962.

———. *Minorities in Aging.* Belmont, Calif.: Wadsworth Publishing, 1980.

James, Joy, and Ruth Farmer, editors. Foreword by Angela Davis. *Spirit, Space, and Survival: African American Women in (White) Academe.* New York: Routledge Press, 1993.

James, Portia P. *The Real McCoy: African-American Invention and Innovation, 1619–1930.* Washington, D.C.: Smithsonian Institution Press, Anacostia Museum, 1989.

Jay, James M. *Negroes in Science: Natural Science Doctorates, 1876–1969.* Detroit: Belamp, 1971.

———. *Minority Groups among United States Doctorate-Level Scientists, Engineers, and Scholars.* Washington, D.C.: Commission on Human Resources of the National Research Council, National Academy of Sciences, 1974.

Johns, Robert L. "Josephine Silone Yates." In *Notable Black American Women,* edited by Jessie Carney Smith. Vol. 2, pp. 1286–1287. Detroit: Gale Research, 1992.

Kelly, Mary, editor. *Woman's Being, Woman's Place: Female Identity and Vocation in American History.* Boston: G. K. Hall, 1979.

Kennedy, John A. *The Negro in Medicine.* Tuskegee, Ala.: Tuskegee Institute, 1912.

Kenschaft, Patricia Clark. "Blacks and Women in Mathematics." Paper presented at the Science and Technology Seminar, The City University of New York, 3 March 1983. In *Educational Policy Seminar Papers,* pp. 22–23. New York: City University of New York, Office of Special Programs, 1986.

———. "Evelyn Boyd Granville." In *Black Women in America: An Historical Encyclopedia,* edited by Darlene Clark Hine, Elsa Barkley Brown, and Rosalyn Terborg-Penn. Vol. 1, p. 499. New York: Carlson, 1993; Bloomington: Indiana University Press, 1994.

———. "Evelyn Boyd Granville." In *Women of Mathematics: A Bibliographic Sourcebook,* edited by Louise S. Grinstein and Paul J. Campbell, pp. 57–60. New York: Greenwood Press, 1987.

———. "Gloria Conyers Hewitt." In *Black Women in America: An Historical Encyclopedia,* edited by Darlene Clark Hine, Elsa Barkley Brown, and Rosalyn Terborg-Penn. Vol. 1, pp. 557–558. New York: Carlson, 1993; Bloomington: Indiana University Press, 1994.

———. "Marjorie Lee Browne." In *Black Women in America: An Historical Encyclopedia,* edited by Darlene Clark Hine, Elsa Barkley Brown, and Rosalyn Terborg-Penn. Vol. 1, p. 186. New York: Carlson, 1993; Bloomington: Indiana University Press, 1994.

Kessler, James H., J. S. Kidd, Renee A. Kidd, and Katherine A. Morin. *Distinguished African American Scientists of the Twentieth Century.* Phoenix, Ariz.: Oryx Press, 1996.

Kevles, Daniel J. *The Physicists: The History of a Scientific Community in Modern America.* New York: Knopf, 1977. Reprint with new preface, Cambridge: Harvard University Press, 1987.

Klein, Aaron E. *The Hidden Contributors: Black Scientists and Inventors in America.* Garden City, N.Y.: Doubleday, 1971.

LaCapra, Dominick, editor. *The Bounds of Race: Perspectives on Hegemony and Resistance.* Ithaca, N.Y.: Cornell University Press, 1991.

Leaders in American Science. Nashville, Tenn.: Who's Who in American Education, 1954–1969.

Lerner, Gerda, editor. *Black Women in White America: A Documentary History.* New York: Random House, 1973.

Levy, Elizabeth, and Mara Miller, editors. *Doctors for the People: Profiles of Six Who Serve.* New York: Alfred Knopf, 1977.

———. "Dorothy Brown: A Doctor for the People." In *Doctors for the People: Profiles of Six Who Serve.* New York: Knopf, 1977. Reprinted in *Ms,* May 1978, pp. 65–68.

Lightfoot, Sarah Lawrence. *Balm of Gilead: Journey of a Healer.* Reading, Mass.: Addison-Wesley, 1989.

Litwack, Leon, editor. *Black Leaders of the 19th Century.* Urbana: University of Illinois Press, 1988.

Logan, Rayford W., and Michael R. Winston, editors. *Dictionary of American Negro Biography.* New York: W. W. Norton, 1982–.

Lunardini, Christine A. "Dorothy Lavinia Brown." In *Black Women in America: An Historical Encyclopedia,* edited by Darlene Clark Hine, Elsa Barkley Brown, and Rosalyn Terborg-Penn. Vol. 1, p. 175. New York: Carlson, 1993; Bloomington: Indiana University Press, 1994.

———. "Mae C. Jemison." In *Black Women in America: An Historical Encyclopedia,* edited by Darlene Clark Hine, Elsa Barkley Brown, and Rosalyn Terborg-Penn. Vol. 1, pp. 633–635. New York: Carlson, 1993; Bloomington: Indiana University Press, 1994.

Lynk, Miles Vandahurst. *Afro-American School Speaker and Gems of Literature.* Jackson, Tenn.: M. V. Lynk, 1896.

———. *The Black Troopers, Or the Daring Heroism of the Negro Soldiers in the Spanish-American War.* Jackson, Tenn.: M. V. Lynk, 1899. Reprint, New York: AMS Press, [1971].

———. *The Negro Pictorial Review of the Great World War: A Visual Narrative of the Negro's Glorious Part in the World's Greatest War.* Memphis, Tenn.: Twentieth Century Art Company, [1919].

Majors, Monroe A. *Noted Negro Women: Their Triumphs and Activities.* Chicago: Donohue & Henneberry [1893]. Reprint, The Black Heritage Library Collection, Freeport, N.Y.: Books for Libraries Press, 1971.

Malcolm, Shirley. "Increasing the Participation of Black Women in Science and Technology." In *The "Racial" Economy of Science: Toward a Democratic Future,* edited by Sandra Harding. Bloomington: Indiana University Press, 1993.

Manning, Kenneth W. *Black Apollo of Science: The Life of Ernest Everett Just.* New York: Oxford University Press, 1983.

Marrett, Cora Bagley. *Minority Females in High School Mathematics and Science.* Madison: Wisconsin Center for Education Research, University of Wisconsin, 1982.

Mather, Frank Lincoln, editor. *Who's Who of the Colored Race: A General Biographical Dictionary of Men and Women of African Descent.* Chicago, Ill., 1915. Reprint, Detroit: Gale Research, 1976.

Melnick, Vilaya L., and Franklin D. Hamilton, editors. *Minorities in Science: The Challenge for Change in Biomedicine.* New York: Plenum Press, 1977.

Merrill, Marlene Deahl. "Oberlin College." In *Black Women in America: An Historical Encyclopedia,* edited by Darlene Clark Hine, Elsa Barkley Brown, and Rosalyn Terborg-Penn. Vol. 2, pp. 897–899. New York: Carlson, 1993; Bloomington: Indiana University Press, 1994.

"Methods and Values in Science." In *On Being a Scientist.* Washington, D.C.: National Academy Press, 1989.

Mickens, Ronald E. "Shirley Ann Jackson." In *Notable Black American Women*, edited by Jessie Carney Smith. Vol. 1, pp. 565–566. Detroit: Gale Research, 1992.

Minority Groups among United States Doctorate-Level Scientists, Engineers, and Scholars. 1973 Commission on Human Resources of the National Research Council. Washington, D.C.: National Academy of Sciences, 1974.

Morais, Herbert M. *The History of the Negro in Medicine.* 3rd ed. New York: Association for the Study of Negro Life and History/Publishers, 1969.

Mossell, Mrs. Nathan F. *The Work of the Afro-American Woman.* 1894. 2nd ed. Philadelphia: G. S. Ferguson, 1908. Reprint, Freeport, N.Y.: Books for Libraries Press, 1971. Reprint with an introduction by Joanne Braxton, New York: Oxford University Press, 1988.

National Faculty Directory. Detroit: Gale Research, 1970–.

Negro Year Book. Tuskegee Institute, Ala.: Negro Year Book Publishing, 1915–.

Neverdon-Morton, Cynthia. *Afro-American Women of the South and the Advancement of the Race, 1895–1925.* Knoxville: University of Tennessee Press, 1989.

Newby, I. A. *Challenge to the Court: Social Scientists and the Defense of Segregation, 1954–1966.* Baton Rouge: Louisiana State University Press, 1969.

Newell, Virginia K., et al., editors. *Black Mathematicians and Their Works.* Ardmore, Penn.: Dorrance, 1980.

Newspapers and Periodicals By and About Black People. North Carolina Central University School of Library Science, African-American Materials Project. Boston: G. K. Hall, 1978–.

Noble, Jeanne L. *The Negro Woman's College Education.* New York, N.Y.: Teachers College/Columbia University, 1956.

Novick, Peter. *That Noble Dream: The "Objectivity Question" and the American Historical Profession.* Cambridge: Cambridge University Press, 1988, 1990.

Oakes, Jeannie. *Lost Talent: The Underparticipation of Women, Minorities, and Disabled Persons in Science.* Santa Monica, Calif.: Rand, 1990.

O'Brien, Eileen M. "Without More Minorities, Women, Disabled, U.S. Scientific Failure Certain Fed Study Says." In *The "Racial" Economy of Science: Toward a Democratic Future,* edited by Sandra Harding. Bloomington: Indiana University Press, 1993. Reprinted from *Black Issues in Higher Education* 6, no. 20 (December 21, 1989).

O'Connell, Agnes N., and Nancy Felipe Russo, editors. *Models of Achievement: Reflections of Eminent Women in Psychology.* New York: Columbia University Press, 1983.

Organ, Claude H., Jr., and Margaret M. Kosiba, editors. *A Century of Black Surgeons: The USA Experience.* Norman, Okla.: Transcript Press, 1987.

Pearson, Willie, Jr., and H. Kenneth Bechtel. *Blacks, Science, and American Education.* New Brunswick, N.J.: Rutgers University Press, 1989.

Ploski, Harry A., and James Williams, editors. *The Negro Almanac: A Reference Work on the African American.* 5th ed. Detroit: Gale Research, 1989.

Rader, Karen. "Making Mice: C. C. Little, the Jackson Laboratory, and the Standardization of *Mus musculus* for Research." Ph.D. dissertation, Indiana University, March 1995.

Robinson, Wilhelmena S. *Historical Negro Biographies.* New York: Publishers Company, 1968.

Rollins, Charlemae Hill. *They Showed the Way: Forty American Negro Leaders.* New York: Crowell, 1964.

Rossiter, Margaret W. *Women in Science: Struggles and Strategies to 1940.* Baltimore: Johns Hopkins University Press, 1982.

———. *Women Scientists in America: Before Affirmative Action, 1940–1972.* Baltimore: Johns Hopkins University Press, 1995.

Royster, Jacqueline Jones. *Women as Healers: A Noble Tradition.* Atlanta: Spelman Col-
lege Women's Research and Resource Center, 1983.
Rywell, Martin, comp., and Charles H.Wesley, et al., editors. *Afro American Encyclope-
dia.* 10 vols. North Miami, Fla.: Educational Book Publishers, 1974.
Sammons, Vivian Ovelton. *Blacks in Science and Medicine.* New York: Hemisphere,
1990.
Sammons, Vivian Ovelton, and Denise Dempsey, comps. *Blacks in Science and Related
Disciplines.* Washington, D.C.: Science Reference Section, Science and Technology
Division, Library of Congress, 1989.
Scarborough, Elizabeth, and Laurel Furumoto. *Untold Lives: The First Generation of
American Women Psychologists.* New York: Columbia University Press, 1987.
Schiebinger, Londa L. *The Mind Has No Sex? Women in the Origins of Modern Science.*
Cambridge: Harvard University Press, 1989.
"Science and Black People." In *The "Racial" Economy of Science: Toward a Democratic
Future,* edited by Sandra Harding, pp. 456–457. Bloomington: Indiana University
Press, 1993. Reprinted from *The Black Scholar* 5, no. 6 (March 1974), p. i.
Shifrin, Susan. "May Edward Chinn." In *Black Women in America: An Historical En-
cyclopedia,* edited by Darlene Clark Hine, Elsa Barkley Brown, and Rosalyn Terborg-
Penn. Vol. 1, pp. 235–236. New York: Carlson, 1993; Bloomington: Indiana University
Press, 1994.
Simmons, William J. *Men of Mark: Eminent, Progressive, and Rising.* 1887. Reprint, New
York: Arno Press, 1968.
Smith, Jessie Carney, editor. *Epic Lives: One Hundred Black Women Who Made a Dif-
ference.* Detroit: Visible Ink Press, 1993.
———. "Geraldine 'Jerry' Pittman Woods." In *Notable Black American Women,* edited by
Jessie Carney Smith. Vol. 2, pp. 1278–1280. Detroit: Gale Research, 1992.
———. "Mae C. Jemison." In *Notable Black American Women,* edited by Jessie Carney
Smith. Vol. 1, pp. 571–573. Detroit: Gale Research, 1992.
———, editor. *Notable Black American Women.* 2 vols. Detroit: Gale Research, 1992.
Spencer, Gerald A. *A Medical Symphony: A Study of the Contributions of the Negro to
Medical Progress in New York.* New York: Arlain Printing, 1947.
Spradling, Mary Mace, editor. *In Black and White: A Guide to Magazine Articles, News-
paper Articles, and Books Concerning More Than 15,000 Black Individuals and Groups.*
2 vols. Detroit: Gale Research, 1980.
Stanton, W. *The Leopard's Spots: Scientific Attitudes towards Race in America, 1815–
1859.* Chicago: University of Chicago Press, 1960.
Steneck, Nicholas H. *Science and Society: Past, Present, and Future.* Ann Arbor: Univer-
sity of Michigan Press, 1974.
Stepan, Nancy Leys, and Sander L. Gilman. "Appropriating the Idioms of Science: The
Rejection of Scientific Racism." In *The Bounds of Race: Perspectives on Hegemony and
Resistance,* edited by Dominick LaCapra. Ithaca, N.Y.: Cornell University Press, 1991.
Sterne, Emma Gelders. *Blood Brothers: Four Men of Science.* New York: Knopf, 1959.
Stevens, Gwendolyn, and Sheldon Gardner, editors. *The Women of Psychology.* Cam-
bridge, Mass.: Schenkman Publishing, 1982.
Takaki, Ronald T. "Aesculapius Was a White Man: Race and the Cult of True Woman-
hood." In *The "Racial" Economy of Science: Toward a Democratic Future,* edited by
Sandra Harding, pp. 201–209. Bloomington: Indiana University Press, 1993.
Taylor, Julius H., Clyde R. Dillard, and Nathaniel K. Proctor, editors. Introduction by
Herman R. Branson. *The Negro in Science.* Baltimore: Morgan State College Press,
1955.

Toppin, Edgar Allen. *A Biographical History of Blacks in America since 1528.* New York: McKay, 1971.

U.S. Department of Energy. See *Black Contributors to Science and Energy Technology.*

U.S. Department of Health, Education, and Welfare. See Bond, Horace M.

Van Sertima, Ivan, editor. *Blacks in Science: Ancient and Modern.* New Brunswick, N.J.: Transition Books, 1984.

Wallace, Susan Brown. "Jeanne Spurlock." In *Notable Black American Women,* edited by Jessie Carney Smith. Vol. 2, pp. 1072–1074. Detroit: Gale Research, 1992.

———. "Margaret Morgan Lawrence." In *Notable Black American Women,* edited by Jessie Carney Smith. Vol. 1, pp. 658–660. Detroit: Gale Research, 1992.

Warren, Wini Mary Edwina. "Hearts and Minds: Black Women Scientists in the United States, 1900–1960." Ph.D. dissertation, Department of History and Philosophy of Science, Indiana University, 1997. Available on University Microfilm.

Wesley, Charles Harris. *The History of the National Association of Colored Women's Clubs.* Washington, D.C.: National Association of Colored Women's Clubs, 1984.

West Virginia State High School Reunion Booklet, 1994–6. Reunion Booklet Committee: Ethel H. Andrews, Helen S. Reynolds, Ann B. Turpeau. Institute, W.Va.: West Virginia State College Alumni Association, 1996.

Who's Who among African Americans. See *Who's Who among Black Americans.*

Who's Who among Black Americans. Northbrook, Ill.: Who's Who among Black Americans Publishing, 1975–1988; Detroit: Gale Research, 1990–.

Who's Who in Aviation and Aerospace. Boston: National Aeronautical Institute, 1983–.

Who's Who in Colored America: A Biographical Dictionary of Notable Living Persons of Negro Descent in America. New York: Who's Who in Colored America, 1927–.

Who's Who in Engineering. New York: American Association of Engineering Societies, 1977–.

Who's Who in Science and Technology. Chicago: Marquis Who's Who in America, 1984–.

Who's Who in the World. 7th ed. Chicago: Marquis Who's Who, 1984–1985.

Who's Who of American Women. Chicago: Marquis Who's Who, 1958–.

Woodson, Carter G., and Charles H. Wesley. *The Negro in Our History.* 10th ed. Washington, D.C.: Associated Publishers, [1962].

Yates, Josephine Silone. "1904 Address before the National Association of Colored Women's Clubs Convention, St. Louis, Missouri." In *The History of the National Association of Colored Women's Clubs: A Legacy of Service,* by Charles Harris Wesley. Washington, D.C.: National Association of Colored Women's Clubs, 1984.

Young, Herman A., and Barbara H. Young. *Scientists in the Black Perspective.* Louisville, Ky.: Lincoln Foundation, 1974.

Articles

Alexander, Leslie L. "Early Medical Heroes: Susan Smith McKinney Steward, M.D., 1847–1918: First Afro-American Woman Physician in New York State." *Journal of the National Medical Association* 67 (March 1975): 173–175.

"Alumnae News," *Spelman Messenger,* February 1972, p. 32.

"Alumnae News: Spelman Alumnae Elect New President." *Spelman Messenger,* August 1972, pp. 25–26.

"The American Negro in College, 1932–1933." *The Crisis,* August 1933, p. 181.

"The American Negro in College, 1939–1940." *The Crisis,* August 1940, pp. 233–237.

"The American Negro in College." *The Crisis,* August 1942.

"The American Negro in College, 1946." *The Crisis,* August 1946, p. 243.

"The American Negro in College." *The Crisis,* January 1953.

"The American Negro in College, 1953." *The Crisis*, March 1955, p. 184.

Anderson, Karen Tucker. "Last Hired, First Fired: Black Women Workers during World War II." *Black Woman and U.S. History* 5 (June 1982): 17–32.

"Atomic Mathematician." *Ebony*, May 1954, p. 5.

"Bachelor Mother: Unmarried Meharry Surgeon Adopts a Child." *Ebony*, September 1958, pp. 92–96.

Bennett, Lerone, Jr. "The Making of Black America, Part 11: Money, Merchants, Markets: The Quest for Economic Security." *Ebony*, February 1974, pp. 66–70, 72, 74–78.

Bigglestone, William. "Oberlin College and the Negro Student, 1865–1940." *Journal of Negro History* 56 (July 1971): 198–219.

Bims, Hamilton. "Charles Drew's 'Other' Medical Revolution." *Ebony*, February 1974, pp. 88–92.

"Biographical Sketch." *Africa Feature*, IPS/Africa, June 1963. (The Flemmie P. Kittrell Collection. Moorland-Spingarn Research Center, Special Collections. Founders' Library, Howard University, Washington, D.C.)

Black, Charlie J. "Upward Mobility." *New Observer*, January 1991.

Bozeman, Sylvia T. "Black Women Mathematicians: In Short Supply." *Sage* 6 (fall 1989): 18–23.

Brown, Dorothy L. "History and Evolution of Abortion Laws in the United States." *Southern Medicine* 61 (August 1973): 11–14.

Brown, Sara W. "Colored Women Physicians." *Southern Workman* 52 (1923): 586.

Brush, Stephen G. "Women in Science and Engineering." *American Scientist* 79 (September/October 1991): 404–419.

Chapin, Joanna. "Interview with Dr. Margaret Morgan Lawrence." *Association for Psychoanalytic Medicine Bulletin* 28 (March 1989): 116–122.

Cobb, Jewel Plummer. "A Life in Science: Research and Service." *Sage* 6 (fall 1989): 39–43.

Cohen, Robert. "A Positive Reaction: Rutgers Physicist Sails through Hearing as Nuclear Nominee." *The (Newark, N.J.) Star Ledger*, 17 February 1995, p. 11.

"Current Lack of Minority Doctoral Candidates Will Lead to Future Minority Faculty Shortage." *Black Issues in Higher Education* 4, no. 23 (1988).

Dejoie, Mike. "Doctor Anna Cherrie Epps, Ph.D.: Medical Educator and Miracle Worker." *The New Orleans Tribune*, June 2, 1996, pp. 14–15.

"Dr. Jane Wright Honored." *The Crisis*, May 1965, p. 328.

"Doctor to Long Island Ducks: N.Y. Woman Scientist Fights Disease That Causes $250,000 Damage Yearly." *Ebony*, September 1964, pp. 76–78, 80, 82.

Eckman, Fern Marja. Article on Jane Cooke Wright, *New York Post*, 22 April 1957.

Eldridge, Adda. "The Need for a Sound Professional Preparation for Colored Nurses." *Proceedings of the Annual Congress on Medical Education, Medical Licensure, and Hospitals* 94 (1930): 168–171.

Ennis, Thomas W. "Dr. May Edward Chinn, 84, Long a Harlem Physician." *New York Times*, 3 December 1980.

Evans, Gaynelle. "The 'Crown Jewel' of California State University, Fullerton." *Black Issues in Higher Education* (September 1988): 5.

"Faculty and Staff Notes." *Spelman Messenger*, February 1971, p. 24.

"Faculty Items." *Atlanta University Bulletin*, No. 95, July 1956, p. 19. (Atlanta University Archives, Archives and Special Collections Division, Atlanta, Ga.)

"Faculty News." *Spelman Messenger*, November 1966, p. 38.

Falconer, Etta Z. "A Story of Success: The Sciences at Spelman College." *Sage* 6 (fall 1989): 36–38.

Flanagan, Brenda. "Essence Women." *Essence*, August 1980, pp. 37–38.

"Flemmie Kittrell." Obituary. *Washington Afro-American*, 11 October 1980.

"Flemmie Kittrell." Obituary. *Washington Post*, 5 October 1980.

"From Here and There during the Month: Young Negro Scientist Engaged in Cancer Research." *Interracial Review* 24 (June 1951): 110.

"The Future-Makers: They Clone Cells and Smash Atoms in Search of Tomorrow's World Today." *Ebony*, August 1985, pp. 62–64, 66.

George, Yolanda Scott. "The Status of Black Women in Science." *The Black Collegian*, May/June 1979, p. 70.

Gibbons, Ann. "Pathbreakers: Gaining Standing—By Standing Out." *Science* 260 (16 April 1993): 393.

Gindick, Tia. "Able Advocate of Higher Education: Dr. Geraldine Woods Has Words for 'Qualifying Minorities.'" *Los Angeles Times*, 18 July 1980, pp. 1, 12.

Glitz, William. "The Minority Biomedical Support Program." *The Black Collegian*, October–November 1981, pp. 38–40.

Goethals, Henry. "Research and Researchers Change with Times." *Quest (National Museum of Natural History, The Smithsonian)* 4 (spring 1995): 10–11.

Goodman, Mary Ellen. "Evidence Concerning the Genesis of Interracial Attitudes." *American Anthropologist* 48 (1946): 624–630.

Granville, Evelyn Boyd. "My Life as a Mathematician." *Sage* 6 (fall 1989): 44–46.

Harvey, Basil C. H. "Provision for Training Colored Medical Students." *Journal of the American Medical Association*, 94 (May 1930): 1415.

"Her Heart Is Strong." *New York Times*, 6 January 1980.

"Her Paper Helped to Track Astronaut: Math Expert Who Aided Spaceman Is 'Thrilled.'" *New York Amsterdam News*, 24 February 1962, pp. 1, 20.

Hewitt, Gloria C. "1992 Survey of Black Graduate Students in Mathematics." A Report for the National Association of Mathematicians, 1993.

Hicks, James L. "Negroes in Key Roles in U.S. Race for Space: Four Tan Yanks on Firing Team." *New York Amsterdam News* (City Edition), 8 February 1958, pp. 1, 15.

Hine, Darlene Clark. "From Hospital to College: Black Nurse Leaders and the Rise of Collegiate Nursing Schools." *Journal of Negro Education* 51 (summer 1982): 223–227.

"Homecoming for Jane Wright." *Ebony*, May 1968, pp. 72–74, 76–77.

Hunter-Gault, Charlayne. "Black Women M.D.s." *New York Times*, 16 November 1977.

"Husband-Wife Team's Project Draws International Interest." *Afro-American*, 10 September 1960, p. 10.

Jay, James M. "Michigan Negro Science Doctorates." *Michigan Challenge*, June 1968.

Johnson, Hope. Article on Jane Cooke Wright. *New York World Telegram and Sun*, 16 December 1959.

Johnson, Maria C. "Upward with Worldly Lessons." *Greensboro News and Record*, 28 January 1991, p. B5.

Kenschaft, Patricia Clark. "Black Women in Mathematics." *Association for Women in Mathematics Newsletter* 8 (September 1978): 5–7.

———. "Black Women in Mathematics in the United States." *American Mathematical Monthly* 88 (October 1981): 599–600.

———. "Black Women in Mathematics in the United States." *Journal of African Civilizations* 4 (April 1982): 63–83.

———. "Marjorie Lee Browne: In Memoriam." *Association for Women in Mathematics Newsletter* 10 (September/October 1980): 8–11.

King, Reatha Clark. "Becoming a Scientist: An Important Career Decision." *Sage* 6 (fall 1989): 47–50.

Kittrell, Flemmie P. "University of Baroda Establishes Home Economics in Higher Education." *Journal of Home Economics* 44 (February 1952): 97–100.

"Ku Klux Klan: In Herrin." *Time*, 2 February 1925.

Lewis, Shawn D. "A Veteran at Meharry Medical College." *Ebony*, August 1977, p. 116.

———. "She Lives with Wind Tunnels." *Ebony*, August 1977, p. 116.

"Long Island Negroes Built Guts of Missile." *New York Amsterdam News* (City Edition), 8 February 1958.

Malcolm, Shirley. "Increasing the Participation of Black Women in Science and Technology." *Sage* 6 (fall 1989): 15–17.

———. "Reclaiming Our Past." *Journal of Negro Education* 59 (summer 1990): 246–259.

"The Mammalian Physiology in Action Program." *Spelman Messenger*, May 1971, p. 37.

Manning, Kenneth R. "Roger Arliner Young, Scientist." *Sage* 6 (fall 1989): 3–7.

Marshall, Marilyn. "Child of the '60s Set to Become First Black Woman in Space." *Ebony*, August 1989, pp. 50, 52, 54–55.

"Mary Logan Reddick." Obituary. *Atlanta Daily World*, 11 October 1996.

"Mary Logan Reddick." Obituary. *Jet*, 3 November 1966, p. 17.

Mayes, Vivienne Malone. "Black and Female." *Association for Women in Mathematics Newsletter* 5 (1975): 4–6.

———. "Lee Lorch at Fisk: A Tribute." *American Mathematical Monthly* 83 (November 1976): 708–711.

McHenry, Susan. "Spelman College Gets Its First Sister President." *Ms.*, October 1987, pp. 58–61, 98–99.

"Medical Profession in Harlem Gets New Addition." *New York Amsterdam News*, 16 June 1926.

"Meet the New Members of the Faculty." *Atlanta University Bulletin*, no. 84 (December 1953), p. 15. (Atlanta University Archives, Archives and Special Collections Division, Atlanta, Ga.)

"Meharry Gets Woman Chief of Surgery." *Afro-American*, 25 July 1953.

Miller, Ronnie. "Spotlight: Dorothy Brown: Strong Will Overcomes Obstacles to Career." *Nashville Banner*, 28 July 1986, pp. B3–B7. (Fisk University Library, Special Collections.)

"Monument Recognizing Black Aviators Is Unveiled." *Jet*, September 1990, p. 34.

Moragne, Lenora. "Black Women Marketing Executives." *Black Business Digest*, February 1973.

Morris, Steven. "How Blacks View Mankind's 'Giant Step.'" *Ebony*, September 1970, pp. 33–36, 38, 40, 42.

"Negroes Aid in Space Research." *Chicago Defender*, 10 May 1958.

"Negroes Who Help Conquer Space: Over 1,000 Negroes Are in the Satellite, Missile Field." *Ebony*, May 1958, pp. 19–22, 24, 26.

New York Age, 1 July 1944. (Atlanta University Archives, Archives and Special Collections Division, Atlanta, Ga.)

"News of the College." *The Morehouse Alumnus*, July 1944, p. 17. (Atlanta University Archives, Archives and Special Collections Division.)

"1988 Essence Awards." *Essence*, October 1988, pp. 59–60.

"Nuclear Physicist at Fermi Lab: Young Ph.D. Holds Her Own in White Male–Dominated World." *Ebony*, November 1974, pp. 114–116, 118, 120, 122.

"Oil Firm Paleontologist." *Ebony*, March 1979, p. 4.

Okon, May. Article on Jane Cooke Wright. *New York Sunday News*, 19 November 1967.

"Outstanding Women Doctors: They Make Their Mark in Medicine." *Ebony*, May 1964, pp. 68–69, 72–74, 76.

"Padmore Top Minority Educator." *Tallahassee Democrat*, 6 March 1996, p. 3B.

Patrick, Jennie R. "Trials, Tribulations, Triumphs." *Sage* 6 (fall 1989): 51–53.

Patterson, Rosalyn Mitchell. "Black Women in the Biological Sciences." *Sage* 6 (fall 1989): 8–14.

Pearce, Ponchitta. "Science Pacemaker." *Ebony*, April 1967, pp. 52–54, 58, 60–62.

People. *Jet*, 5 February 1959, p. 25.

People. *Jet*, 11 February 1960, p. 18.

People. *Jet*, 29 July 1976, p. 14.

Preece, Harold. "The Team Who Challenges Cancer." *Sepia*, January 1961, pp. 54–57.

Rather, Joan. "Programs Provide a History Lesson—Blacks and Women in Science." *Science* 23 (April 1983): 186–187.

"Research Scientist Has Beauty, Brains." *Jet*, 3 April 1958, p. 21.

Riddle, Estelle Massey. "The Training and Placement of Negro Nurses." *Journal of Negro Education* 4 (1935): 42–48.

"Rocket Fuel Researcher." *Ebony*, September 1960, p. 6.

"'Salute to Women's Fete: Governor Is Speaker." *New York Amsterdam News*, 20 October 1962, p. 12.

Saville, Kirk. "What Was the First Computer Model?—A Woman." *Newport News Daily Press*, 14 December 1990, pp. A–3, A–16.

"Scientific Couple Finds Success in Albuquerque: Chemists Kenneth and Katheryn Lawson Say Desert Town Is Ideal Community for Them." *Ebony*, June 1965, pp. 67–70, 72, 73.

"Shaper of Young Minds." *Ebony*, August 1982, pp. 97–98, 100.

"The South: Making of a Crisis in Arkansas." *Time*, 16 September 1957, pp. 23–25.

Southerland, Daniel. "Equation for Success: A Life of Struggle Takes D.C.'s Shirley Jackson to NRC's Helm." *The Washington Post*, 4 May 1995, pp. B10–13.

"Space Age Research Chemist." *Ebony*, April 1961, p. 6.

"Space Computing Mathematician." *Ebony*, August 1960, p. 7.

"Space Is Her Destination." *Ebony*, October 1987, pp. 93–98.

Speaking of People. *Ebony*, October 1960, p. 7.

"The Status of Blacks in Graduate and Professional Schools: A Report by the National Advisory Committee on Black Higher Education and Black Colleges and Universities." *The Black Collegian*, October/November 1981, pp. 86–87.

Stepan, Nancy Leys. "Race and Gender." *Isis* 77 (1986): 261–277.

Stewart, Isabel M. "Next Step in the Education of Nurses." *National News Bulletin* 13 (December 1939).

"A Surgeon Goes to Legislature." *Washington Post*, 4 June 1967.

"Two Atlanta Educators Win Fellowships to Study Abroad." *Atlanta Daily World*, April 2, 1952. (Atlanta University Archives, Archives and Special Collections Division, Atlanta, Ga.)

"U.S. Science Wars against an Unknown Enemy: Cancer." *Life*, 1 March 1973, pp. 11–17.

"Vison Lab Director." *Ebony*, September 1970, p. 7.

Washburn, S. L. "The Study of Race." *American Anthropologist* 65 (June 1963): 521–531.

"Water Plant Tester." *Ebony*, March 1956, p. 5.

Waters, Enoc P., Jr. "They Helped Track Glenn in Orbit." *Afro-American*, 3 March 1962, pp. 1–2.

Williams, Lillian S. "Making a Way Out of No Way: Black Women's Clubs and Philanthropy 1900–1940." Paper delivered at the Conference on Philanthropy in the African-American Experience, Rockefeller Archive Center, Sleepy Hollow, N.Y., 24–26 September 1992.

"Young Negro Scientist Engaged in Cancer Research." *Interracial Review* 24 (July 1951): 110.

Zeyher, Allen. "A Nuclear News Interview: Jackson Takes the Reins." *Nuclear News* 38 (November 1995): 26–29.

Periodical Information

Aim Magazine. Chicago: Aim Publishing.

The Black Collegian. New Orleans: Black Collegiate Services, 1969–.

Chicago Defender. Chicago: R. S. Abbott, 1955–1973. Black newspaper.

The Crisis. Baltimore: Crisis Publishing, 1910–. Official organ of the National Association for the Advancement of Colored People.

Ebony. Chicago: Johnson Publishing, 1945–.

Jet. Chicago: Johnson Publishing, 1951–.

Journal of Black Studies. Newbury Park, Calif.: Sage Publications, 1970–.

The Journal of Negro History. New York: United Publishing, 1969–. Issued by the Association for the Study of Negro Life and History (ASNLF).

Negro History Bulletin. Washington, D.C. Issued by the Association for the Study of Negro Life and History (ASNLF).

New York Amsterdam News. New York: Amsterdam News, 1919–.

Spelman Messenger. Atlanta, Ga.: Spelman College. Atlanta University Archives.

Multimedia

"Exceptional Black Scientists." Poster, number 16. CIBA-GEIGY Corporation, February 1985.

Run to Live, 16 mm, 22 min. Nashville, Tenn.: United Methodist Film Service, 1981. Revised edition of *Run to Live,* 16 mm, 30 min. Nashville, Tenn.: United Negro College Fund/United Methodist Board of Higher Education, 1976.

Science: Woman's Work. Video produced by the National Science Foundation with Image Associates (NSF05). Available through Media Design Associates (VHS, 28 min., 1990), P.O. Box 3189, Boulder, Colorado 80307; and Current Affairs Multimedia (VHS, 27 min., 1990), St. Petersburg, Fla.

Archives

Atlanta University Archives, Archives and Special Collections Division, Robert Woodruff Library, Atlanta, Ga. Collection contains clipping file of the National Organization for the Professional Advancement of Black Chemists and Chemical Engineers; Atlanta University archives and publications; Morris Brown College publications; Morehouse College publications; and Spelman College publications.

Cornell University Library, Archives & Special Collections, Ithaca, New York. Collection contains the Ethel B. Waring Collection; a portion of the Flemmie P. Kittrell and Ethel B. Waring correspondence is housed here.

Dow Chemical Company Archives, 205 Post St., Midland, Mich. Collection contains personnel department material on Bettye W. Greene (I am indebted to Doug Draper and archivist Kathy Thomas of the Post Street Archives for their help).

Fisk University Library, Special Collections, Nashville, Tenn. Collection contains the
    Meharry Medical College archives, and the Dorothy L. Brown Collection, including
    her autobiographical manuscript, "Thus Would I Live."
Medical College of Pennsylvania Library, Philadelphia, Penn. Collection contains an
    oral history project on women in medicine, which includes a transcript of Joyce Antler's
    interview with Jeanne Spurlock, 2 June 1978.
Moorland-Spingarn Research Center, Special Collections, Founders' Library, Howard
    University, Washington, D.C. Collection contains Howard University archives and
    publications, including the *Howard University Bulletin, Personnel Register*; the Ernest
    Everett Just Collection (a limited access collection); and the Flemmie P. Kittrell Col-
    lection (uncatalogued). The Kittrell Collection includes papers pertaining to Cecile
    H. Edwards and Marianna Beck Sewell.
Schlesinger Library, Radcliffe College, 10 Garden Street, Cambridge, Mass. Contains
    the Women's Oral History Project in the Gerda Lerner Collection (complete file).
Schomberg Center for Research in Black Culture, a division of the New York Public
    Library, 515 Malcolm X Blvd., New York. Contains extensive archival holdings, verti-
    cal files, audiotapes, prints, and photographs; also includes the Kaiser Index to Black
    Resources, with material from 1948 to 1986; and the May Edward Chinn Collection,
    including an unpublished autobiography. (I am indebted to my student, Faruk Curtis,
    SUNY, College at Old Westbury, for assistance in accessing the Schomberg materials.)
Smithsonian Institution Archives, Washington, D.C. Contains the Alfred E. Emerson
    Collection.
The Women's Oral History Project, The Gerda Lerner Collection, Schlesinger Library,
    Radcliffe College, 10 Garden Street, Cambridge, Mass. Collection contains a clipping
    file, as well as Merze Tate's transcript of his interview with Flemmie P. Kittrell, 29–30
    August 1977.

Index

WINI WARREN teaches in the American Studies program at the
State University of New York, College at Old Westbury.